TECHNICAL WRITING

SITUATIONS AND STRATEGIES

MICHAEL H. MARKEL *Drexel University*

ST. MARTIN'S PRESS *New York*

TECHNICAL WRITING

SITUATIONS AND STRATEGIES

Library of Congress Catalog Card Number: 83-61617

Manufactured in the United States of America.
876
fed

For Information, write St. Martin's Press, Inc.,
175 Fifth Avenue, New York, New York 10010

Interior design: Martin Lubin/Betty Binns Graphics

ISBN: 0-312-78777-4

ACKNOWLEDGMENTS

Diagrams from Reader's Digest Complete Do-it-Yourself Manual, 1973, pp. 221, 228; Copyright © 1973 The Reader's Digest Association, Inc. Reprinted by permission of the publisher.

"How to use Pollution Abstracts" reprinted with the permission of the publisher from Pollution Abstracts, V. 14, N. 3, page iv. (May 1983). © Cambridge Scientific Abstracts 1983. All rights reserved.

"What Price Safety? The 'Zero-Risk' Debate" in Dun's Review, September 1979. Reprinted with the special permission of Dun's Business Month (formerly Dun's Review), September 1979, Copyright 1979, Dun & Bradstreet Publications Corporation.

Citation examples reprinted with permission from the January/February 1983 issue of the CUMULATIVE INDEX TO NURSING & ALLIED HEALTH LITERATURE. © Cumulative Index to Nursing & Allied Health Literature Corporation, Glendale Adventist Medical Center, Glendale, CA, 1983.

Flow chart on solar energy from SOLAR ENERGY by Tim Michels. Copyright © 1979 by Van Nostrand Reinhold Company. Reprinted by permission of the publisher.

Extract from "Change Your Own Motor Oil, It's Easy," reprinted with the permission of Kendall Refining Company, Division of Witco Chemical Corporation, Bradford, Pa. 16701.

Acknowledgments continue on page 531, which constitutes an extension of the copyright page.

To My Parents

PREFACE

Technical Writing: Situations and Strategies is intended to provide a clear and comprehensive teaching tool for technical writing instructors while at the same time preparing students to approach the working world's various writing situations. The idea behind the book is that the best way to learn to write is to write and rewrite. Accordingly, the text contains not only numerous samples of actual technical writing but also dozens of writing and revising exercises that enable students to apply what they have learned in realistic technical writing situations.

The first two chapters explain the concept on which the rest of the book is based: that the difficult part of technical writing is not learning to solve the complicated, little problems (such as how to decide when to write out numbers) but to handle successfully the simple, overriding problem involved in every kind of factual communication. This problem could be called "How to Write About Your Subject, Not About Yourself." Or it might be called "How to Write to Your Readers, Not to Yourself." But by any name, the common problem in technical writing is that most writers simply fail to consider *why* and *for whom* they are writing. They write as if they were still in school and their readers were their instructors. The results of this misunderstanding are all too predictable: long, complicated documents that few people have the time, desire, or knowledge to read.

If *Technical Writing* has a single goal, it is to impress upon students the need to analyze the writing situation—the audience and purpose of the communication—and how this situation affects the structure, development, and style of the document. Only with such attention to audience and purpose will students produce clear, simple, and readable documents.

The words, sentences, and paragraphs of technical writing are also crucial components. Therefore, they are treated early in the text, in a chapter on technical writing style. A discussion of how to find and use information is also included in Part One, as is a chapter on outlines and rough drafts.

The techniques of technical writing are the subject of Part Two. The five chapters in this part discuss not only the rhetorical techniques used in technical writing—definition, analysis, mechanism description, process description, and instructions—but the graphic aids that are an integral part of effective technical presentations.

Part Three concentrates on the technical report, beginning with a chapter on format and moving on to types of reports. The treatment of the memo in this book is substantially different from that in most other technical writing texts. Here it is discussed not as a brief in-house news item, but rather as a short internal report. The chapters on proposals, progress reports, and completion reports have a dual purpose: first, to describe how these documents function in the world of business and industry, and second, to provide a logical and realistic framework for student writing in the course. Many technical writing instructors have found that these three types of reports, considered in sequence, help their students understand the basic future-present-past movement of most on-the-job writing:

- the proposal: what I want to investigate, and why
- the progress report: how the investigation is proceeding
- the completion report: what the investigation revealed, and what the findings mean

Part Four covers other technical writing applications: technical articles, oral presentations, correspondence, and the job-application procedure.

All of the examples in this book—from single sentences to complete reports—are real. Some were written by my students at Clarkson College in Potsdam, New York. Others were written by Drexel University students, either while they were in school in

Philadelphia or while they were off campus working as co-op students. The bulk of the writing, however, was done by engineers, scientists, and managers with whom I have worked as a consultant over the last ten years. Because most of the information in the writing is proprietary, I have changed brand names, company names, and other identifying nouns. I would like to thank the dozens of people—students and professionals alike—who have graciously allowed me to print their writing.

My thanks go, too, to Thomas Broadbent, Nancy Perry, Charles Thurlow, and Marilyn Moller of St. Martin's Press, who provided many invaluable ideas and offered extraordinary assistance throughout this project. For their perceptive reviews and helpful suggestions, I am grateful to the following technical writing instructors: Michael Connaughton, St. Cloud State University; Kevin Dungey, University of Maryland; William Evans, Kansas State University; Marc Glasser, Morehead State University; Dean G. Hall, Kansas State University; Kathryn Hume, Pennsylvania State University; Patrick Hundley, University of Arkansas at Little Rock; Louis Janda, Northeast Wisconsin Technical Institute; Wayne A. Losano, University of Florida; Donald Skarzenski, Framingham State College; Kathryn P. Stege, Eastern Washington University; and Wanda D. Thilsted, Oklahoma State University. Mrs. Lee Endicott and Ms. Maga Wogman deserve thanks for typing the manuscript. Dr. Martha Montgomery and Dean Thomas Canavan of Drexel University kindly arranged for clerical support.

My greatest debt, however, is to my wife, Rita, who over the course of many months tried to make me say what I mean. To her should go praise for the good things in this book. For not always following her advice, I take responsibility.

Michael H. Markel

CONTENTS

TECHNICAL WRITING

SITUATIONS AND STRATEGIES

PART ONE

THE TECHNICAL WRITING PROCESS

CHAPTER ONE

THE ROLE OF TECHNICAL WRITING IN BUSINESS AND INDUSTRY

CHARACTERISTICS OF EFFECTIVE TECHNICAL WRITING

- CLARITY
- ACCURACY
- ACCESSIBILITY
- CONCISENESS
- CORRECTNESS

THE STRATEGY OF TECHNICAL WRITING

Technical writing can be defined as writing that conveys specific information to a specific audience for a specific purpose. Much of what you read every day is technical writing—textbooks (including this one), the owner's manual for your car, cookbooks. The words and graphic aids of technical writing are meant to be practical: that is, to communicate a body of factual information that will help an audience understand a subject or carry out a task. For example, an introductory biology text helps students understand the fundamentals of plant and animal biology and carry out basic experiments. An automobile owner's manual describes how to operate and maintain that particular car.

THE ROLE OF TECHNICAL WRITING IN BUSINESS AND INDUSTRY

The working world depends on written communication. Virtually every action taken in a modern organization is documented in writing (whether on paper or in a computer's memory). The average company has dozens of different forms to be filled out for routine activities, ranging from purchasing office supplies to taking inventory. If you need some filing cabinets at a total cost of $300, for instance, you can purchase them after an office manager signs a simple form. But expenditures over a specified amount—such as $2,000 or $10,000—require a brief report showing that the purchase is necessary and that the supplier you want to buy from is offering a better deal (in terms of price, quality, service, etc.) than the competition. Similarly, most technicians record their day-to-day activities on relatively simple forms. But when these same technicians travel to a client's plant to teach a class on a new procedure, they have to write up a detailed memo or report when they return. That report describes the purpose of the trip, analyzes any problems that arose, and offers any suggestions that would make future trips more successful.

When a major project is being contemplated within an organization, a proposal must be written to persuade management that the project would be not only feasible but also in the best interests of the organization. Once the project is approved and under way, progress reports must be submitted periodically to inform the supervisors and managers about the current status of the project and about any unexpected developments. If, for instance, the project is not going to be completed on schedule or is going to cost more than anticipated, the upper-level managers need to be able to reassess the project in the broader context of the organization's goals. Often, projects are radically altered on the basis of the information contained in progress reports. When the project is completed, a completion report must be written to document the work and enable the organization to implement any recommendations.

In addition to all of this in-house writing, every organization must communicate with other organizations and the public. The letter is the basic format for this purpose. Inquiry letters, sales letters, goodwill letters, and claim and adjustment letters are just some of the types of daily correspondence. And if a company performs contract work for other companies, then proposals, progress reports, and completion reports are called for again. The world of business and industry is a world of writing.

If you are taking a technical writing course now, the chances are that it was created in response to the needs of the working world. You have undoubtedly seen articles in newspapers and magazines about the importance of writing and other communication skills in a professional's career. The first step in *getting* a professional-level position, in fact, is to write an application letter and a résumé. If there are other candidates as well qualified as you—and in the job market today there almost certainly are—your writing skills will help determine whether an organization decides to interview you. At the interview, your oral communication skills will be examined along with your other qualifications. Once you start work, you will write memos, letters, and short reports, and you might be asked to contribute to larger projects, such as proposals. During your first few months on the job, your supervisors will be looking not only at your technical abilities, but also at your ability to communicate information. The facts of corporate life today are simple. If you can communicate well, you are very valuable to your organization; if you cannot communicate well, you are much less valuable. The tremendous growth in continuing-education courses in technical and business writing reflects this situation. Organizations are paying to send their professionals back to the classroom to improve their writing skills. New employees who can write and speak well bring invaluable skills to their positions.

CHARACTERISTICS OF EFFECTIVE TECHNICAL WRITING

Technical writing is meant to get a job done. Everything else is secondary. If the writing style is interesting, so much the better. But keep in mind the five basic characteristics of technical writing:

1. clarity
2. accuracy
3. accessibility
4. conciseness
5. correctness

CLARITY The primary characteristic of effective technical writing is clarity. That is, the written document must have one meaning, and only one meaning, for the reader. In addition, the reader must be able to understand that meaning easily.

Unclear technical writing is expensive. The cost of a typical business letter today, for instance, counting the time of the writer and the typist and the cost of equipment, stationery, and postage, is more than eight dollars. Every time an unclear letter is sent out, the reader has to write or phone the writer and ask for a clarification. The letter then has to be rewritten and sent again to confirm the correct information. On a larger scale, clarity in technical writing is essential because of the cooperative nature of most projects in business and industry. One employee is assigned to investigate one aspect of the problem, while other employees work on other aspects of it. The vital communication link among the various employees is usually the report. If one link in the communication chain is weak, the entire project will be delayed or, even worse, destroyed. Such a breakdown could represent a huge financial loss for the organization.

More important, unclear technical writing can be dangerous. Poorly written warnings on bottles of medication are a common example, as are unclear instructions on how to operate machinery safely. A carelessly drafted building code tempts contractors to save money by using inferior materials or techniques. The events that led to the accident at the Three Mile Island nuclear plant might have been prevented had the plant's operators received clear notification about similar events that had occurred previously at another facility.

ACCURACY

All of the problems that can result from unclear writing can also be caused by inaccurate writing.

Accuracy is a simple concept in one sense: you must record your facts carefully. If you mean to write *4,000*, don't write *40,000*. If you mean to refer to Figure 3–1, don't refer to Figure 3–2. The slightest error in accuracy will at least confuse and annoy your readers. Inaccurate instructions, naturally, can be dangerous.

In another sense, however, accuracy is a more difficult concept. Technical writing must be every bit as objective and free of bias as a well-conducted scientific experiment. If the readers suspect that you are slanting the information—by overstating the significance of a particular fact or by omitting reference to an important point—they have every right to doubt the validity of the entire document. Technical writing must be effective by virtue of its clarity and organization, but it must also be reasonable, fair, honest, and comprehensive.

ACCESSIBILITY

The word *accessibility* refers to the ease with which the readers can locate the information they are seeking. One of the major differences between technical writing and other kinds of nonfiction writing is that most technical documents are made up of small, independent sections. Some readers are interested in only one or several of the sections; other readers might want to read many or most of the sections. Because relatively few people will pick up the document and start reading from the first page all the way through, the writer's job is to make the various parts of the document accessible. That is, the readers should not be forced to flip through the pages to find the appropriate section.

To increase the accessibility of the writing, use headings and lists (see Chapter 5). For reports, include a detailed table of contents (see Chapter 11).

CONCISENESS

Technical writing is meant to be useful. The longer a document is, the more difficult it is to use, for the obvious reason that it takes more of the reader's time. In a sense, the characteristic of conciseness works against the characteristic of clarity. If a technical explanation is to be absolutely clear, you must describe every aspect of the subject in great detail. The solution to this conflict is to balance the claims of clarity and conciseness. The document must be just long enough to be clear—given the audience, purpose, and subject—but not a word longer. You can shorten most writing by 10–20% simply by eliminating unnecessary phrases, choosing short words rather than long ones, and using economical grammatical forms.

The battle for concise writing, however, often is more a matter of psychology than it is of grammar. A utility company, for instance, recently experienced a fairly serious management problem caused by its employees' writing. Each branch manager of the company was required to submit a semiannual status report, informing the corporate managers about any technical problems that had occurred during the previous six months. The length of the report was clearly specified as three pages, maximum.

As it turned out, the corporate headquarters was paralyzed for almost a month after the status reports came in. They averaged 17 pages. The branch managers had read the "three pages, maximum" directive; they just didn't believe it. They wanted to impress their supervisors. So each put a good many weeks into creating reports that would cover everything. No detail was too small to be included. In their attempt to produce the ultimate status report, they forgot the simple fact that somebody had to read it. From the reader's point of view, few things are more frustrating than having to read 17 pages that say, in effect, that no problems occurred. The corporate managers had to spend hours chopping the 17-page reports down to three pages so that they could be filed for future reference. Justifiably enough, these managers felt they were doing the branch managers' jobs.

The utility company hired a writing consultant to make one point to the branch managers: not only is it acceptable to write three-page status reports, but more than three pages is less than good. Once this point was communicated, the rest of the time was spent discussing ways to get all the necessary information into three pages.

CORRECTNESS

Good technical writing is correct: it observes the conventions of grammar, punctuation, and usage, as well as any appropriate format standards.

Many of the "rules" of correctness are clearly important. If you mean to write, "The three inspectors—Bill, Harry, and I—attended the session," but you use commas instead of dashes, your readers might think six people (not three) attended. If you write, "While feeding on the algae, the researchers captured the fish," some of your readers might have a little trouble following you—at least for a few moments.

Most of the rules, however, make a difference primarily in that readers will form an impression of your writing based on how it looks and sounds. If your report is sloppy or contains a number of grammar errors, your readers will start to doubt the accuracy of your information, or will at least lose their concentration. You will still be communicating, but the message won't be the one you had intended. As a result, the document will not achieve its purpose.

The rest of this book describes ways to help you say what you want to say.

CHAPTER TWO

ASSESSING THE WRITING SITUATION

Chapter 1 defines technical writing as writing that conveys specific information to a specific audience for a specific purpose. In other words, the content and form of any technical document is determined by the situation that calls for that document. This situation includes the reasons the writer has for composing the document and the reasons its recipient has for reading.

THE TWO COMPONENTS OF THE WRITING SITUATION

The writing situation comprises the audience and the purpose of a document. These elements are hardly mysterious: apart from idle chit-chat, most everyday communication is the product of the same two-part environment. When you write, of course, you must be more precise than you are in conversation: your audience cannot stop you to ask for clarification of a complex concept, nor can you rely on body language and intonation to convey your purpose.

Typical everyday examples of the two-part writing situation abound. For example, if you have recently moved to a new apartment and wish to invite a group of friends over, you might duplicate a set of directions explaining how to get there from some well-known landmark. In this case, your writing situation would be clearly identifiable:

AUDIENCE

the group of friends

PURPOSE

to direct them to your apartment

Given your comprehension of this writing situation, you would include in your directions the information that each member of the group might need in order to get to your apartment and arrange this information in a useful sequence.

When a classified advertisement describes a job opening for prospective applicants, the writing situation of the advertiser can be broken down similarly:

AUDIENCE

prospective applicants

PURPOSE

to describe the job opening

When a notice is posted on a departmental bulletin board advising students interested in enrolling in a popular course to sign up on a waiting list, the department's writing situation also conforms to the two-part model:

AUDIENCE

students interested in enrolling in the course

PURPOSE

to advise signing up on the waiting list

Once you have established the two basic elements of your writing situation, you must analyze each before deciding upon the content that your document should include and the form that it should take. Effective communication satisfies the demands of both situation elements. Although you might assume that purpose would be your primary consideration, an examination of audience is in fact more helpful as a first step: the audience element tends to be the more elusive and complicated element of the two, and an exploration of its characteristics often influences the course you take in refining your initial and general conception of purpose.

The separation of audience and purpose is to some extent artificial—you cannot realistically banish one from your mind as you contemplate the other—but it is nonetheless useful. The more thoroughly you can isolate first audience and then purpose, the better the grasp that you will have on the writing situation and the more exactly you will be able to tailor your document to the situation.

AUDIENCE

Identifying and analyzing the audience can be difficult. You have to put yourself in the position of people you might not know reading something you haven't yet started to write. Most writers want to concentrate first on content—the nuts and bolts of what they have to say. Their knowledge and opinions are understandably what they feel most comfortable with. They want to write something first and shape it later.

Resisting this temptation is especially important. Having written mostly for teachers, students do not automatically think much about the audience. Students in most cases have some idea of what their teachers want to read. Further, teachers establish guidelines and expectations for their assignments—and usually know what students are trying to say. The typical teacher is a known quantity. In business and industry, however, you will often have to write to different audiences, some of whom you will know almost nothing about and some of whom will have very limited technical knowledge of your area of expertise.

What are the basic types of technical writing audiences? How do they differ from one another? How do these differences affect how you write to them? These questions have no hard-and-fast answers, of course, because every reader is unique. Still, some useful generalizations can be drawn. Three basic types of audience can be identified:

1. the technical person
2. the manager
3. the general reader

THE TECHNICAL PERSON

The term *technical person* is used here to cover a fairly broad range of reader, from the expert who carries out original research and writes articles for technical journals to the technician who operates equipment. Roughly in the middle of this range is the technically trained professional—the engineer, the biologist, the accountant—who analyzes and solves problems as they arise.

Technical people devise and implement computer programs and fix the computer when it isn't working properly. They carry out audits to determine if proper accounting procedures are being followed. They analyze water samples to determine if the ecology of a stream is threatened. All of these people can be grouped in the same category on the basis of a shared characteristic—a technical understanding of a subject area and an interest in that area. There are differences, of course. The expert understands the theory to a much greater degree than either the middle-level professional or the technician. The middle-level professional is well equipped to carry out experiments to measure productivity or quality on a production line but might not be able to conduct basic research. The technician is the person who identifies a problem in a piece of equipment, improvises a temporary solution until the defective part can be replaced, and installs the replacement part when it arrives.

When you write to a technical person, keep in mind his or her needs. The expert feels quite at home with the technical vocabulary and the formulas. You can get to the details of the technical subject right away, without sketching in the background; the expert already knows it. The middle-level technical person, such as the engineer, is likely to need a brief orientation to the subject; unlike the expert, the engineer isn't always familiar with theoretical background. The technician, however, needs schematic diagrams, parts lists, and step-by-step instructions to apply to a concrete task.

Following is a paragraph, from an article in a professional journal (Hindin 1981: 160), that illustrates the needs and interests of the middle-level professional audience. The subject of the article is a laboratory analysis of volcanic ash from Mount St. Helens.

> Although the chemical composition of the ash was similar to the average composition of the earth's crust, the physical characteristics of the ash were site specific. Hooper et al. (1980), in a report on the composition of the Mount St. Helens ash fall in the Moscow-Pullman area, found that the ash had a density of 2.14 kg/m^2 (dried at 100° C). The ash was bimodal, consisting of an initial dark ash followed by a pale ash. The mass of the pale ash was two to four times that of the dark ash. The two types of ash had similar particle sizes, ranging between 10 and 100 μm, although small-sized particles (0.1–1.0 μm) were found in the air weeks after the May 18 eruption.

This paragraph exhibits several characteristics of technical writing addressed to the middle-level professional. First, it assumes that the readers are familiar with the standard laboratory procedures used in the discipline; the author does not feel it necessary to describe any of his research methods. Second, it contains a lot of technical data (on ash density and particle size). And third, it refers to another technical source, which the readers are likely to want to consult.

THE MANAGER The manager is harder to define than the technical person, for the word *manager* describes what a person does more than what a person is or knows. A manager's job is to make sure the operations of the organization are smooth and efficient. The manager of the procurement department at a manufacturing plant is responsible for seeing that raw materials are purchased and delivered on time so that production will not be interrupted. The manager of the sales department of that same organization is responsible for seeing that salespeople are out in the field, creating interest in the products and following up leads. In other words, managers coordinate and supervise the day-to-day activities of the organization.

Management is a popular subject in the business curriculum, and many managers today have a background in general business, psychology, and sociology. Often, however, managers are trained in the technical areas their organizations work in. An experienced chemical engineer, for instance, might manage the engineering division of a consulting company. Although he has a solid background in chemical engineering, he earned his managerial position because of his broad knowledge of engineering and his ability to deal with colleagues effectively. In his daily work he might have little opportunity to use his specialized skills in chemical engineering.

Although generalizing about the average manager's background is difficult, identifying the manager's needs is an easier task. Managers are primarily interested in the "bottom line." They have to get a job done on schedule; they don't have time to study and admire a theory the way an expert does. Rather, managers have to juggle constraints—money, knowledge, and organizational priorities—and make logical and reasonable decisions quickly.

When you write to a manager, try to determine his or her technical background and then choose an appropriate vocabulary and sentence length. Regardless of the individual's background, however, focus on the practical information the manager will need to carry out his or her job. For example, an engineer, describing to the sales manager a new product line that the research-and-development department has created, would want to provide some theoretical background on the product so that the sales representatives can communicate effectively with potential clients. For the most part, however, the description would concentrate on the product's capabilities and its advantages over the competition.

Following is a paragraph addressed to a managerial audience. The subject is the effect of the Mount St. Helens ash fall on the state of Washington's sewage treatment plants (Kish 1980: 2310).

> The majority of treatment plant problems in Washington revolve around ash deposits in plant processes, line plugging, pump and other equipment failures, hydraulic overloading, biomass washout, and digester upsets. Cleanup costs are another matter. The federal government, through the Federal Emergency Management Agency, is assisting by providing funds to cover 75% of the eligible costs. Washington and/or the local community must pick up the remaining 25% of the cleanup costs. Damage survey reports have not been completed at the time of this writing, but interviews with various treatment plant officials indicate that plant restoration costs could total at least $300,000.

This paragraph is characteristic of technical writing addressed to managers. It begins with a brief listing of the technical problems affecting the treatment plants. Notice that the writer doesn't concentrate on the specifics of the problems. Rather, he focuses on the managerial implications of the ash fall: how much the repairs will cost, and who will pay for them.

THE GENERAL READER

Occasionally you will have to address the general reader, sometimes called the layman (or layperson). In such cases, you must avoid technical language and concepts and translate jargon—however acceptable it might be in your specialized field—into standard English idiom. A nuclear scientist reading about economics is a general reader, as is a homemaker reading about new drugs used to treat arthritis.

The layman reads out of curiosity or self-interest. The average article in the magazine supplement of the Sunday paper—on attempts to increase the populations of endangered species in zoos, for example—will attract the general reader's attention if it seems interesting and well written. The general reader may also seek specific information in an unfamiliar field of expertise: someone interested in buying a house might read articles on new methods of alternative financing.

In writing to a general audience, use a simple vocabulary and relatively short sentences when you are discussing areas in which the layman might be confused. Use analogies and examples to clarify your discussion. Sketch in any special background—historical or ethical, for example—so that your reader can follow your discussion easily. Concentrate on the implications for the general reader. For example, in discussing a new substance that removes graffiti from buildings, focus on its effectiveness and cost, not on its chemical composition.

Following is an example of a paragraph addressed to the general reader (Evans 1980: 388).

> The ash from this and the three previous fallouts is a major concern. Its silicate component resembles ground glass and, when airborne, tiny particles can literally bring machinery to a grinding halt. Car engines have been immobilized, and timber harvesting of felled trees has become a Herculean task because the ash on the logs dulls a chainsaw in moments. All agricultural machinery is vulnerable to engine breakdown, as are the turbine blades at hydroelectric plants that receive waterborne ash. Profits for agriculturalists will come from the volcano, but they must wait with Job's patience for the weather to break down ash in the soil and so release fertilising minerals.

This paragraph, from an English journal, is characteristic of technical writing aimed at a general but educated audience. Rather than concentrating on a description of the ash itself, the writer examines the effects of the ash fall on the daily life of the affected area. The major industries—such as logging and agriculture—are mentioned, as is the ordinary automobile. The writer also appeals to the tastes of the general reader by cleverly exploiting the literal sense of the phrase "grinding halt." A classical and a biblical reference—"Herculean task" and "Job's patience"—help clarify the writer's ideas by drawing on a common stock of established concepts. A further benefit is gained by the reference to Job at the end of the paragraph: the author subtly makes the point that the ultimate effects of the ash fall will be long-lived but that they can also be endured.

ACCOMMODATING THE MULTIPLE AUDIENCE

These profiles of the three basic kinds of audiences—technical persons, managers, and general readers—assume that you are writing to a single person or to a group of people who share the same basic educational background and job responsibilities. Often, however, you will be writing to a multiple audience made up of both managers and middle-level technical people.

Twenty years ago, the need to address a multiple audience placed no special demands on the writer, because most managers were technical people who had risen from the technical staff. Today, however, the knowledge explosion and the increasing complexity of business operations have combined to cause a fairly wide gap between the knowledge of the technical staff and that of most managers. Whereas two decades ago the manager of a paint manufacturing company almost certainly would have known all about paint, today that manager might have worked most recently for a brokerage house or a bakery.

Another factor that complicates the situation is an innocent-looking piece of hardware—the photocopy machine. In the days of carbon paper, a secretary could make only three or four legible copies of a typed page. Thus the number of readers—and the likelihood of having different kinds of readers—was relatively limited. But today, running off a few dozen copies is so easy and inexpensive that more people (and more kinds of people) see almost everything that is written on company stationery. In many companies the manager of an engineering division will routinely receive copies of *everything* written by its one hundred engineers. And, with the increasingly standard electronic office, writing has become dispersed even more widely.

Because of this increase in the number and types of readers, you have to assess the full range of your audience carefully before you write. If you think you might have a multiple audience, structure the document accordingly. For memos and reports, include a preliminary section addressed to the manager and a detailed section addressed to the technical reader. (See Chapter 11 for a further discussion.) Clearly distinguish the two sections so that the managers know exactly where their discussion begins and ends. Use a vocabulary and a sentence length and structure appropriate to your different readers.

UNDERSTANDING THE PERSONAL ELEMENT

Identifying your readers' backgrounds helps you understand who they are and why they are reading your document. Don't forget, however, that your readers are unique individuals. If you know your readers, consider their character traits—their biases and eccentricities as well as their particular interests—which will probably tell you more about them as readers than their backgrounds do. When you don't know your readers, remember as you try to "put yourself in their shoes" that biases exist and that you want to avoid at all costs offending delicate sensibilities.

Personal preferences range from the important to the trivial. Some readers appreciate an introductory summary of the basic problem and the results of the project. Others want the summary in a letter of transmittal (see Chapter 11). Some readers are annoyed by the use of the pronoun *I* in a technical document; others appreciate it as a means of avoiding the passive voice. Some readers are distracted by the use of computer jargon, such as *feedback* and *input*; others find it a useful shorthand.

Common sense suggests that you should accommodate as many of the readers' preferences as possible. Sometimes, of course, such general accommodation is impossible, either because several of the preferences are contradictory or because of some special demands of the subject. In any case, never forget that technical writing is intended to get a job done; it is not a personal statement. Whenever possible, try to avoid alienating or distracting your readers.

PURPOSE

Once you have identified and scrutinized your audience, reexamine your general purpose in writing. Ask yourself this simple question: "What do I want this document to accomplish?" When your readers have finished reading what you have written, what do you want them to *know* or to *do*? Think of your writing not as an attempt to say something about the subject but as a way to help others understand it or act on it.

To come to a clearer definition of your purpose, think in terms of verbs. Try to isolate a single verb that represents what you are trying to do and keep it in mind throughout the writing process. Almost without exception, you will find that to some extent your purpose can be expressed in a simple English verb. (Of course, in some cases a technical document has several purposes, and therefore you might want to choose several verbs.) Here are a few examples of verbs used to define "understanding" purposes and "action" purposes:

"UNDERSTANDING" VERBS	"ACTION" VERBS
to explain	to request
to describe	to authorize
to review	to propose
to forecast	to recommend

Following are a few examples of how "understanding" and "action" verbs can be used in clarifying the purpose of the document. (The verbs are italicized.)

1. This report *describes* the research project to determine the effectiveness of the new waste-treatment filter.

2. This report *reviews* the progress in the first six months of the heat-dissipation study.

3. This letter *authorizes* the purchase of six new word processors for the Jenkintown facility.

4. This memo *proposes* that we study new ways to distribute specification revisions to our sales staff.

Make sure, as you define your purpose in writing, that you understand your audience's needs. Most technical writing is used either to document a project—so that the organization will have a record of what it did and why—or to communicate some information necessary for carrying out a current project. If the document is intended to initiate or facilitate a technical task, the structure must help the audience to understand why and how to begin. If the document is intended as a reference, other structures—such as "problem-methods-solutions" (see Chapter 4)—might be most appropriate. The important point to remember is that your purpose in writing should answer your audience's needs. Accordingly, a thorough understanding of audience is necessary before purpose can be defined precisely.

AUDIENCE, PURPOSE, AND STRATEGY

With a clear idea in mind of what you want to accomplish and what your audience needs, you can begin to determine the scope, or boundaries, of the document. If, for example, you have some information on the history of computers, but your audience needs an instruction manual that explains how to operate a particular system, you might well decide that the historical information is not relevant.

More importantly, a good sense of audience and purpose will remind you that some kinds of information *must* be included. For example, if you are proposing that your company purchase some computer equipment, you know that you must define the need for the equipment, survey the available equipment, and make the case that buying it is less expensive than not buying it. The following example shows how an understanding of the audience and purpose would determine the content and form of such a proposal.

You would start by sketching in some basic facts about your reader:

1. He is a middle-level manager who cannot authorize large capital expenditures but can recommend—or not recommend—your proposal to his boss.

2. He is not an expert on the kind of equipment you want to buy, but he isn't hostile to the idea of new high-technology equipment.

3. He knows that your department is productive and that you have never requested unnecessary or inappropriate purchases of equipment. On the other hand, the company did make an unwise purchase just three months ago, and nobody has forgotten it.

4. He would prefer simply to attach a covering memo to your proposal (if he approves of it) and send it on to his boss; he does not want to rewrite it in his own words.

These facts tell you that you have to accommodate the needs of both *your* reader and *his* reader. You have to be direct, straightforward, and objective. You must avoid excessive technical details and vocabulary. You should focus clearly on the practical advantages the new equipment would provide the company. You should include a brief summary for the convenience of your two readers.

Next, you define your purpose in writing. In this case, the purpose is clear: to convince your boss that your proposal is reasonable, so that he will endorse it and pass it on to his supervisor, who in turn will authorize the purchase of the equipment.

What do your audience and purpose tell you about what to include in the document? Because your readers will be particularly careful about large capital expenditures, you must clearly show that the type of equipment you want is necessary and that you have recommended the most effective and efficient model. Your proposal will have to answer several questions that will be going through your readers' minds:

1. What system is being used now?

2. What is wrong with that system, or how would the new equipment improve our operations?

3. On what basis should we evaluate the different kinds of available equipment?

4. What is the best piece of equipment for the job?

5. How much will it cost to purchase (or lease), maintain, and operate the equipment?

6. Is the cost worth it? At what point will the equipment pay for itself?

7. What benefits and problems have other purchasers of the equipment experienced?

8. How long would it take to have the equipment in place and working?

9. How would we go about getting it?

A careful definition of the writing situation is the first step in planning any technical writing assignment. If you know whom you are writing to, and why, you can decide on your strategy—what to include and how to shape it. The result will be a document that works.

WRITING SITUATION CHECKLIST

Following is a checklist of questions you should ask yourself when you profile the writing situation.

1. To whom are you writing?
   - **a.** Is your audience one person, several people, a large group, or several groups with various needs? _____
   - **b.** What do you know about your reader's (readers') job responsibilities? _____
   - **c.** What do you know about your reader's (readers') personal preferences that will help you plan the writing? _____
   - **d.** What will your readers be doing with the document? Why will they be reading it? _____
2. What is your purpose in writing?
   - **a.** What is the document intended to accomplish? _____
   - **b.** Is your purpose consistent with your audience's needs? _____
3. How does your understanding of your audience and of your purpose determine the scope of your document? How do the audience and purpose determine other elements of your document, such as structure, organization, tone, and vocabulary? _____

EXERCISES

1. Choose two articles on the same subject, one from a general-audience periodical, such as *Reader's Digest* or *Newsweek*, and one from a more technical journal, such as *Scientific American* or *Forbes*. Write a 500-word essay comparing and contrasting the two articles from the point of view of the authors' assessment of the writing situation: the audience and the purpose. As you plan the essay, keep in mind the following questions:

   a. What is the background of each article's audience likely to be? Does either article require that the reader have specialized knowledge?
   b. What is the author's purpose in each article? In other words, what is each article intended to accomplish?
   c. How do the differences in audience and purpose affect the following elements in the two articles?

(1) scope
(2) sentence length and structure
(3) vocabulary
(4) number and type of graphic aids
(5) references within the articles and at the end

2. Choose a 200-word passage from a technical article addressed to an expert audience. Rewrite the passage so that it is clear and interesting to the general reader.

3. Audience is the primary consideration in many types of non-technical writing. Choose a magazine advertisement for an economy car, such as a Honda, and one for a luxury car, such as a Mercedes. In a 500-word essay, contrast the audiences for the two ads in terms of age, sex, economic means, general way of life, etc. In contrasting the two audiences, consider the explicit information in the ads—the writing—as well as the implicit information—hidden persuaders such as background scenery, color, lighting, angles, and the situation portrayed by any people photographed.

REFERENCES

Evans, M. 1980. The volcano that won't lie down. *New Scientist* (26 June): 388.

Hindin, H. 1981. Treatment of Mt. St. Helens volcanic ash suspensions by plain sedimentation, coagulation and flocculation. *American Water Works Association Journal* 73 (March): 160.

Kish, T. 1980. Mt. St. Helens' effects on Washington's treatment plants. *Journal Water Pollution Control Federation* 52 (September): 2310.

CHAPTER THREE

FINDING AND USING INFORMATION

We live in what is called the information age. The mark of a true professional in every technical field is the ability to find and use effectively the massive amounts of information available. This chapter could not hope to provide a comprehensive look at how to conduct research: there are literally thousands of different kinds of research tools and techniques ranging from familiar reference books to computerized mathematical models. Rather, this discussion will introduce the basic methods of finding and using printed information. The techniques used in questionnaires and interviews will also be discussed.

CHOOSING A TOPIC

Whereas very few professionals have the opportunity to choose their topics, you as a student are likely, at least on occasion, to have this freedom. Like most other freedoms, this one is a mixed blessing. An instructor's request that you "come up with an appropriate topic" is for many people frustratingly vague; they would rather be told what to write about—how the Soviet Union's Sputnik changed the American policy on space exploration, for instance—even though the topic assigned might not interest them.

As long as you don't spend weeks agonizing over the decision, the freedom to choose your own topic and approach is a real advantage: if you are interested in your topic, you'll be more likely to want to read and write about it. Therefore, you will do a better job.

Start by forgetting topics such as the legal drinking age, the draft, and abortion. These topics are unrelated to most practical writing situations. Ask yourself what you are *really* interested in: perhaps something you are studying at school, some aspect of your part-time job, something you do during your free time. Browse through three or four recent issues of *Time* or *Newsweek*, and you will find dozens of articles that suggest interesting, practical topics involving technical information. The January 17, 1983, issue of *Time*, for example, contained the following articles:

SUBJECT OF THE ARTICLE	SUGGESTED RESEARCH TOPICS
the eruption of the Kilauea volcano in Hawaii	early-warning techniques for volcanoes
	regulations for making buildings volcano-proof
	the Mount St. Helens eruptions and their impact on the environment
	flood-prevention and flood-control techniques
the explosion at gasoline storage tanks in Newark, New Jersey	the safety element in the transportation and storage of petroleum products
	new techniques of oil drilling
waste from a pizza factory backing up a sewage-treatment plant in Wellston, Ohio	techniques of sewage treatment and disposal
	industrial pollution in the water, soil, and air
the return of the sports car	new techniques for increasing the fuel economy in automobiles
	new techniques for increasing the power of auto engines
	new techniques for improving auto safety
an attempt to steal information from the Federal Reserve Board's computers	efforts to combat computer theft
a nuclear-powered Soviet spy satellite plunging back to Earth	civilian and military uses of space
	techniques of verifying arms agreements
the progress of the first artificial-heart patient	new advances in medical technology

Many of these topics are not focused enough for a research report of 10 to 15 pages, but they are good starting points. The topic of civilian and military uses of space, for instance, could be narrowed to a study of the kinds of industrial processes that could be carried out effectively in a weightless environment, such as on future space missions. The point is that good topics are all around. If you are interested in music, you might write about what happened to eight-track tapes or quadrophonic recording, or the use of new materials in the manufacture of instruments, or the Japanese dominance in audio components. If you are interested in health care, you might write about hospices, or birthing rooms, or the nursing shortage, or the regulation of so-called quack medicine.

Once you have chosen a tentative topic for your research report, the next step is to see if you have the resources necessary to do the job well. Basically, there are three resources you must consider:

1. information on the subject
2. your own knowledge of the subject
3. time

Most of this chapter discusses how to find information, both in the library and through personal interviews and questionnaires. Actually, the job of finding information involves two separate tasks: determining if the information exists, and trying to obtain it. Everyone at one time or another makes the mistake of assuming that if a book is listed in the card catalog, it can be found on the shelf. But fully 10% of a library's holdings might be missing: miscatalogued, misplaced, stolen, or checked out indefinitely by a faculty member.

After you have looked at some of the information on your topic, you will be able to determine if you already know enough to understand it. As a student you will have to decide how much background reading you are willing to do just to understand the technical information.

Time is the third crucial resource. You might not have the two or three weeks needed to get a book through interlibrary loan, or the month required to send out a questionnaire. You might not have the time for extensive background reading. And even if you have all of your information in hand, it might not be possible for you to write a good report in time. Before you start to do any serious research, skim through your sources to determine if you will have enough time: changing topics is much easier *before* you've put in a lot of hours.

Some writers like to have an absolutely precise topic in mind before they begin to search the literature. A few examples of precise topics follow:

1. the effect that energy-efficiency tax credits have had on the home-improvement industry
2. the effect of salt-water encroachment on the aquifer in southern New Jersey
3. the market, during the coming decade, for videodisc instructional systems designed to be used in elementary schools

The advantage of having a precise topic is that you can begin your research quickly, because you know what you're looking for.

Other writers prefer to begin their research with only a general topic in mind. For example, you might know that you want to write about computers but be unsure about what aspect of computers to focus on. The basic research techniques described in this chapter will enable you to discover—quickly and easily—what aspects of computers will make the most promising topics. Perhaps you didn't know about the large amount of research being carried out to help companies reduce the theft of computerized data. That topic might appeal to you. Or you might want to research new techniques used in computer graphics. The advantage of being flexible about your topic is obvious: in fifteen minutes you can discover dozens of possible topics from which to choose.

FINDING THE INFORMATION

The best place to find information about most topics is the library. Learning to use the library is essential. However, some topics require that you interview a person or persons or that you send questionnaires.

USING THE LIBRARY

You should become familiar with the different kinds of libraries. The local public library in a small city is a good source for basic reference works, such as general encyclopedias, but it is unlikely to have more than a few specialized reference works. Most college libraries have substantially larger reference collections and receive the major professional journals. Large universities, of course, have comprehensive library collections. Many large universities have specialized libraries that complement selected graduate programs, such as those in zoology or architecture. Large cities often have special scientific or business libraries that you can use as well.

REFERENCE LIBRARIANS The most important information sources at any library are the reference librarians. Although as a college student you are expected to know how to use the library, reference librarians are there to help you solve special problems. They are invariably willing to suggest new ways to search for what you need—specialized directories, bibliographies, or collections that you didn't know existed. Perhaps most important, they will tell you if the library *doesn't* have the information you need and suggest other libraries to try. Reference librarians can save you a lot of time, effort, and frustration. Don't be afraid to ask them questions when you run into problems.

CARD CATALOGS With few exceptions, every general library maintains a main card catalog, which lists almost all of the library's books, microforms, films, phonograph records, tapes, and other materials. Some libraries list periodicals (newspapers and journals) in the main catalog; others keep separate "serials" catalogs. The library's nonfiction books and pamphlets, as well as many non-literary resources, are usually entered on three types of cards (author, title, and subject) in the main catalog. Periodicals, whether in the main catalog or in the serials catalog, are catalogued by title: to find an article you must know the name and date of the journal in which it appears (see the following discussion of periodicals indexes).

Accordingly, if you know the title or author of the nonserial item for which you are looking, you can determine easily whether the library has it. Your search is more difficult if you have no specific work in mind and are simply looking for a discussion of a topic by consulting the subject cards. (Some libraries separate their subject cards from their author and title cards.) In such cases you must determine the likeliest subject headings under which relevant publications might be classified. If you have trouble finding appropriate materials, a reference librarian should be able to suggest other subject headings to look under. The broader the subject, the greater the number of cards there will be that you will have to go through. However, broad subjects are often subdivided: *biology*, for example, might be subdivided into *cytology*, *histology*, *anatomy*, *physiology*, and *embryology*, following the range of general *biology* entries. Also, at the end of a range of subject cards, a "see also" card often suggests other subject headings under which to look.

All three types of catalog cards provide the same basic information about an item, as Figure 3–1 shows. In addition to the basic bibliographic information—author, title, place of publication, publisher, number of pages, International Standard Book Number (ISBN), and cataloguing codes—each card also lists the call number (the number that indicates where on the shelves the item is located) in the upper left-hand corner. Call numbers are based on one of two major classification systems. The older, the Dewey Decimal system, uses numbers (such as 519.402462) to designate subject areas. The newer system, the Library of Congress classification, uses combinations of letters and numbers (such as TA330.H68). Some libraries have converted completely from Dewey Decimal to Library of Congress; others are still in the process. In Dewey Decimal, for example, 519 is the category for engineering mathematics; in Library of Congress, engineering mathematics is TA. Also posted in every library is a map that will direct you to the area where your books are shelved.

REFERENCE BOOKS Some of the books in the card catalog will have call numbers preceded by the abbreviation *Ref.* These books are part of the reference collection, a separate grouping of books that normally may not be checked out of the library.

In the reference collection are the general dictionaries and encyclopedias, biographic dictionaries (*International Who's Who*), almanacs (*Facts on File*), atlases (*Rand McNally Commercial Atlas and Marketing Guide*), and dozens of other general research tools. In addition, the reference collection contains subject encyclopedias (*Encyclopedia of Banking and Finance*), dictionaries (*Psychiatric Dictionary*), and handbooks (*Biology Data Book*). These specialized reference books are especially useful when you begin a writing project, for they can provide an overview of the subject and often list the major works in the field.

How do you know if there is a dictionary of the terms used in a given field, such as nutrition? The answer, as you might have guessed, can be found in the reference collection. It would be impossible to list in this chapter even the major reference books in science, engineering, and business, but you should be familiar with the reference books that list the available reference books. Among these guides-to-the-guides are the following:

Downs, R. B., and C. D. Keller. 1975. *How to do library research.* 2nd ed. Urbana, Ill.: University of Illinois Press.

Guide to reference books. 1976. 9th ed., ed. E. P. Sheehy. Chicago: Amercian Library Association.

Guide to reference material. 1973. 3rd ed. 3 vols., ed. A. J. Walford. London: The Library Association.

AUTHOR CARD

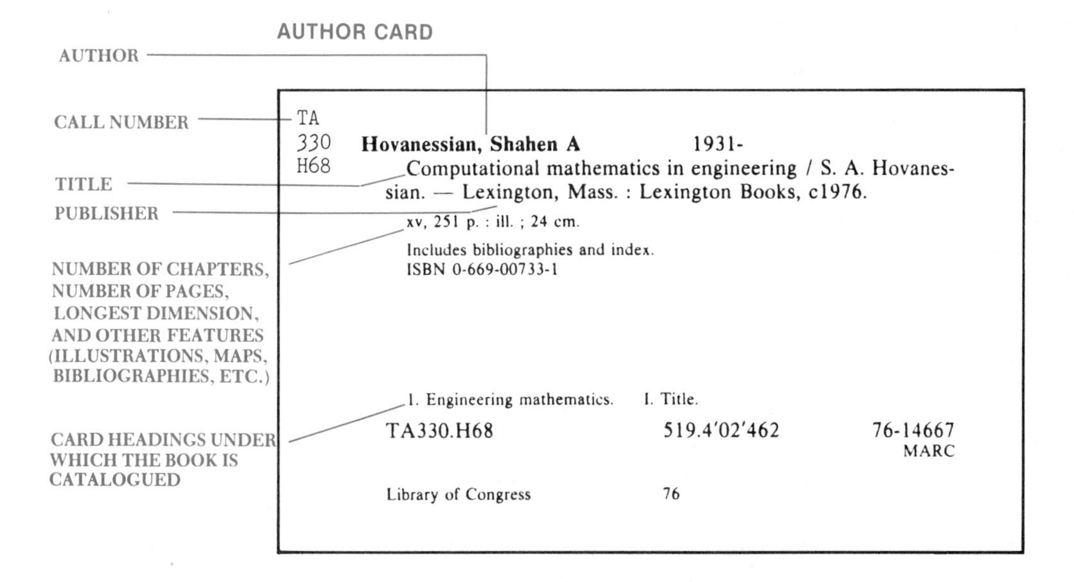

TITLE CARD

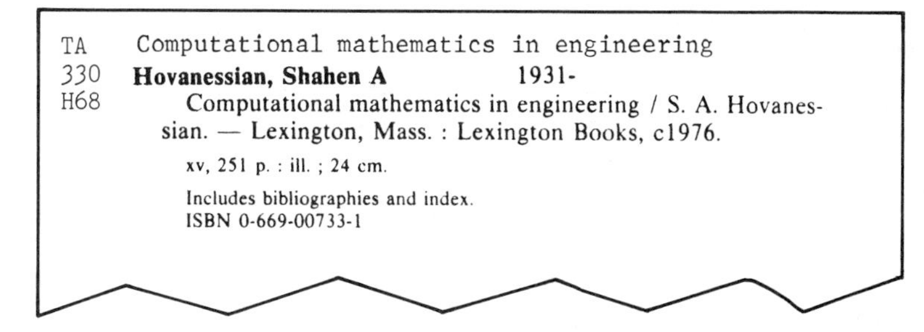

TA Computational mathematics in engineering
330 Hovanessian, Shahen A 1931-
H68 Computational mathematics in engineering / S. A. Hovanessian. — Lexington, Mass. : Lexington Books, c1976.
xv, 251 p. : ill. ; 24 cm.
Includes bibliographies and index.
ISBN 0-669-00733-1

SUBJECT CARD

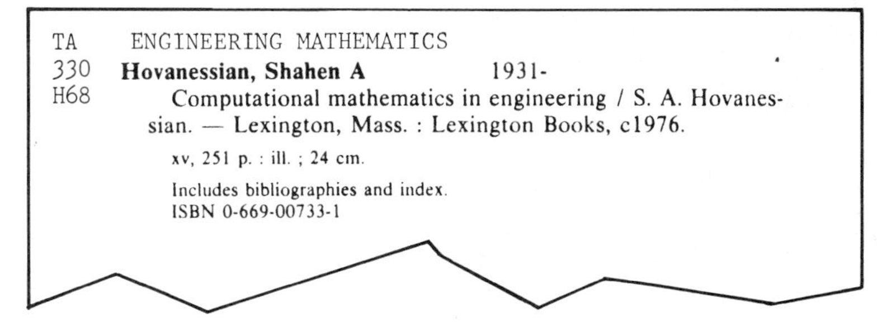

TA ENGINEERING MATHEMATICS
330 Hovanessian, Shahen A 1931-
H68 Computational mathematics in engineering / S. A. Hovanessian. — Lexington, Mass. : Lexington Books, c1976.
xv, 251 p. : ill. ; 24 cm.
Includes bibliographies and index.
ISBN 0-669-00733-1

FIGURE 3–1

Author, Title, and Subject Cards

Look under *Reference* in the subject catalog to see which guides the library has. The most comprehensive is Sheehy's *Guide to Reference Books* (also see its 1980 *Supplement*), an indispensable resource that lists bibliographies, indexes, abstracting journals, dictionaries, directories, handbooks, encyclopedias, and many other sources. The items are classified according to specialty (for example, *organic chemistry*) and annotated. One of the most useful features of Sheehy's book is that it directs you to other guides (such as Henry M. Woodburn's *Using the Chemical Literature: A Practical Guide*) geared to your own specialty. Read the prefatory materials in Sheehy; you can save yourself many frustrating hours.

PERIODICALS INDEXES Periodicals are the best source of information for most research projects, because they offer recent discussions of subjects whose coverage is often otherwise limited. The hardest aspect of using periodicals is identifying and locating the dozens of pertinent articles that are published each month. Although there may be only a half-dozen major journals that concentrate on your field, a useful article might appear in one of a hundred other publications. A periodical index, which is simply a listing of articles classified according to title, subject, and author, can help you determine which journals you want to locate.

Figure 3–2, an explanatory page from the *Cumulative Index to Nursing and Allied Health Literature*, demonstrates how most indexes work.

Some periodical indexes are more useful than others. A number of indexes—such as *Engineering Index* and *Business Periodicals Index*—are very comprehensive. However, if you are going to rely on a narrower, more specialized index when compiling a preliminary bibliography for your report, you should determine if it is accurate and comprehensive. The prefatory material—the publisher's statement and the list of journals indexed—will supply answers to the following questions:

1. Is the index compiled by a reputable organization? Many of the better indexes, such as *The Readers' Guide to Periodical Literature*, are published by the H. W. Wilson Company. Almost all of the reputable indexes are sponsored by well-known professional societies or associations.

2. What is the scope of the index? An index is not very useful unless it includes all of the pertinent journals. Scan the list of journals that the index picks up. Also, determine if the index includes materials other than articles, such as annual reports or proceedings of annual conferences. Does the index include articles in foreign languages? In foreign journals? Ask the reference librarian if you have questions.

Citation Examples

In the Subject Section Citation, the bibliographic data is identified as follows:

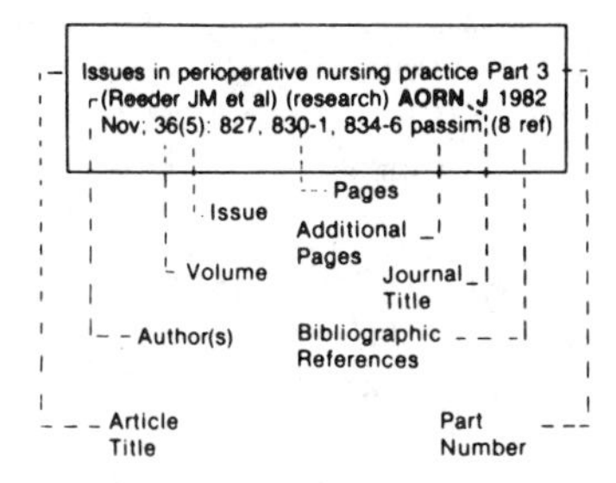

And may include descriptors:

Smoking stress and nurses (Hawkins L et al) (research, statistics) **NURS MIRROR** 1982 Oct 13; 155(15): 18-22 (27 ref)

(See transition guide for complete list of descriptors)

When an article title is not definitive, CINAHL indexers add a descriptive subtitle, as in the case below:

"... an early outpatient hospital/clinic based exercise program for cardiac patients".

Get em reconditioned ...an early outpatient hospital/clinic based exercise program for cardiac patients. (Porter GH et al) **CVP** 1982 Oct-Nov; 10(6): 15-16, 18-20, 22-24 (5 ref)

Frequently, subject headings have subdivisions:

INFECTION

INFECTION — PREVENTION AND CONTROL

Georgraphic subheadings are used when the article content includes significant geographic data or the historical importance requires geographic identification:

HOSPITALS — AUSTRALIA

"See" references direct the user from terms **not** used to terms used (annual only):

PERSONNEL DISCIPLINE
see EMPLOYEE DISCIPLINE

"See also" references direct the user from a general term to a more specific term (annual only).

KIDNEY FAILURE, ACUTE
see also HEMODIALYSIS

The Author Section lists a mximum of three authors:

JOHNSON DH, HAINSWORTH JD, GRECO FA: Pancoast's syndrome and small cell lung cancer (care study, pictorial) **CHEST** 1982 Nov; 82(5): 602-6 (22 ref)

with "See" references listed alphabetically for the second and third authors:

HAINSWORTH JD **See**: Johnson DH

Biographies and Obituaries are included in the Author Section. The name of the biographee or deceased appears alphabetically in lower case letters in parentheses.

(Heimlich H) (biography): Henry Heimlich, M.D: the man behind the maneuver... by Grossman J. **HEALTH** 1982 Dec; 14(12): 46-7, 56

FIGURE 3–2

Explanatory Page from the Cumulative Index to Nursing and Allied Health Literature

3. Are the listings clear? Does the index define the abbreviations, and do the listings provide enough bibliographic information to enable you to locate the article?
4. How current is the index? How recent are the articles listed, and how frequently is the index published? Don't restrict your search to the bound annual compilations; most good indexes are updated monthly or quarterly.
5. How does the index arrange its listings? Most of the better indexes are heavily cross-indexed: that is, their listings are arranged by subject, author, and (where appropriate) formula and patent.

The better subject indexes include *Applied Science and Technology Index* (New York: H. W. Wilson Company, 1913 to date), *Science Citation Index* (Philadelphia: Institute for Scientific Information, 1961 to date), and *Index Medicus* (Washington, D.C.: National Library of Medicine, 1960 to date). The *New York Times* and *Wall Street Journal* also publish indexes of their own articles.

ABSTRACT JOURNALS Abstract journals not only contain bibliographic information for articles listed, but also provide abstracts—brief summaries of the articles' important results and conclusions. (See Chapter 11 for a discussion of abstracts.) The advantage of having the abstract is that in most cases it will enable you to decide whether to search out the full article. The title of an article, alone, is often a misleading indicator of its contents.

Whereas some abstract journals, such as *Chemical Abstracts*, cover a broad field, many are specialized rather than general. *Adverse Reaction Titles*, for instance, covers research on the subject of adverse reactions to drugs. *Applied Mechanics Review*, too, is relatively narrow in scope. A major drawback of abstract journals is that they take longer to compile than indexes (about a year, as opposed to about two months); therefore, the articles listed are not so recent. In many fields, however, this disadvantage is more than offset by the usefulness of the abstracts.

In evaluating the quality of an abstract journal, apply the same criteria that you apply to indexes, but add one other: How clear and informative is the abstract? To be useful, abstracts have to be brief (usually no more than 250 words), and summarizing a long and complex article in so few words requires great skill. Poorly written abstracts can be very confusing. Most abstract journals reserve the right to revise or rewrite the abstracts provided by authors. Read a few of the abstracts to see if they make sense.

Figure 3–3 shows a guide included in *Pollution Abstracts* that provides an excellent explanation of how to use that journal. (No-

HOW TO USE
pollution abstracts®

TO RESEARCH A SPECIFIC SUBJECT: Determine which terms best describe the subject you are researching such as scrubbers, wastewater outfalls, pyrolysis, etc. Look up the terms in the Subject Index at the back of the issue. This is a rotating subject index that lists alphabetically all controlled terms assigned to the abstract; each entry refers you by accession number to the corresponding abstract. If you cannot find a listing for a certain term, try a synonym or related term. A complete listing of the controlled terms used in indexing *Pollution Abstracts* appears in the Annual Cumulative Index.

EXAMPLE: You are seeking literature about sulfur dioxide emissions from Mt. St. Helens. You look up **Sulfur dioxide** in the subject index and find the following entry:

Sulfur dioxide, Emission measurements, Sulfur compounds, Air pollution measurements, Explosions, Washington State, Air sampling, **Mt St Helens** 81-00180

Citation 81-00180 would also be listed under the controlled terms **Emission measurements, Sulfur compounds, Air pollution measurements, Explosions, Washington State, Air sampling, Ash, Distribution,** and **Particle size.** Free terms such as Mt. St. Helens further describe the material in the abstract but are not listed alphabetically in the index.

You then turn to the abstract numbered 81-00180:

Text language
Summary language

Citation number **81-00180**

Original title **Preliminary airborne observations of the Mt. St. Helens eruptions.** [En;en]

Author L. F. Radke (Univ. of Washington, Atmospheric Sciences Dept., Aerosol Research Group, Seattle, WA 98195), P. V. Hobbs, M. W. Eltgroth, D. A. Hegg.

Publication Air Pollution Control Association. Journal, Vol. 30, No. 8, Aug. 1980, pp. 904-905.

Abstract Ongoing airborne measurements began on Mar. 28, 1980, using the University of Washington's twin engine research aircraft which is equipped for comprehensive measurements of size, concentration, and size segregated elemental and chemical characteristics of airborne particles of $> \approx 0.01$ μm-4.5 mm, and of the nature and concentration of a large number of trace gases. Use of automated bag samplers allows large volumes to be sampled without extended periods of ash plume exposure. Emissions of $\approx 2{,}000$ t/d of SO_2 were observed on June 18. A significant fraction of the smaller particles now appears to be condensed soluble materials with quantities of SO_4^{-2}, Br^-, F^-, and Cl^- in contrast to the main eruptions of May 18 and May 19 where the soluble fraction was small and ash pH near neutral. Size distributions from the eruptions of Mar. 28, May 28, and May 25 are contrasted with a representative ambient air sample. The most prominent feature of the eruption cloud size distributions is the several orders of magnitude increase in concentration of highly respirable micron size S-rich ash.

Language abbreviations
text summaries

Af	af	Afrikaans	Ee	ee	Estonian	Ja	ja	Japanese	Ro	ro	Romanian
Ar	ar	Arabic	En	en	English	Ko	ko	Korean	Ru	ru	Russian
Be	be	Belorussian	Es	es	Spanish	Li	li	Lithuanian	Sh	sh	Serbo-croat
Bg	bg	Bulgarian	Fi	fi	Finnish	Lv	lv	Latvian	Sk	sk	Slovak
Ch	ch	Chinese	Fr	fr	French	Ma	ma	Macedonian	Sn	sn	Slovenian
Cs	cs	Czech	Gr	gr	Greek	Nl	nl	Dutch	Sv	sv	Swedish
Da	da	Danish	He	he	Hebrew	No	no	Norwegian	Tr	tr	Turkish
De	de	German	Hu	hu	Hungarian	Pl	pl	Polish	Uk	uk	Ukrainian
			It	it	Italian	Pt	pt	Portuguese			

FIGURE 3-3

Guide to an Abstract Journal

tice, too, how well written the abstract is.) Most abstract journals are organized in this way.

GOVERNMENT PUBLICATIONS The United States government publishes about twenty thousand documents annually. In researching any field of science, engineering, or business, you are likely to find that a government agency has produced a relevant brochure, report, or book.

Government publications usually are not listed in the indexes and abstract journals. The *Monthly Catalog of United States Government Publications*, put out by the U.S. superintendent of documents, provides extensive access to these materials. The *Monthly Catalog* is indexed by author, title, and subject. If you are doing any research that requires information published by the government prior to 1970, you should know about the *Cumulative Subject Index to the Monthly Catalog, 1900–1971* (Washington, D.C.: Carrollton Press, 1973–76). This fifteen-volume index eliminates the need to search through the early *Monthly Catalogs.*

Government publications are usually catalogued and shelved separately from the other publications. They are classified according to the Superintendent of Documents system, not the Dewey Decimal or the Library of Congress system. See the reference librarian or the government documents specialist for information about finding government publications in your library.

State government publications are indexed in the *Monthly Checklist of State Publications* (1910 to date), published by the Library of Congress.

If you would like more information on government publications, consult the following two guides:

Morehead, J. 1978. *Introduction to United States public documents.* 2nd ed. Littleton, Col.: Libraries Unlimited.

Schmeckebier, L. F., and R. B. Eastin. 1969. *Government publications and their use.* Washington, D.C.: The Brookings Institution.

GUIDES TO BUSINESS AND INDUSTRY Many kinds of technical research require access to information about regional, national, and international business and industry. If, for example, you are researching a new product for a report, you may need to contact its manufacturer and distributors. For information such as to whom to write to discover the range of the product's applications, you could consult one of the following two directories:

Thomas' Register of American Manufacturers. New York: Thomas. 1905 to date.

Poor's Register of Corporations, Directors, and Executives. New York: Standard and Poor's. 1928 to date.

These annual directories provide information on products and services and on corporate officers and executives. In addition to these national directories, many local and regional directories are published by state governments, local chambers of commerce, and business organizations. These resources are particularly useful for students and professionals who wish to communicate with potential vendors or clients in their own area.

Also valuable as resources are investment services—publications that provide balance sheets, earnings data, and market prices for corporations. Two major investment services exist:

Moody's Investor's Service. New York: Moody's.

Standard and Poor's Corporation Records. New York: Standard and Poor's.

Two government periodicals contain valuable information about national business trends:

The Federal Reserve Bulletin. Washington, D.C.: U.S. Board of Governors of the Federal Reserve System. A monthly publication that outlines banking and monetary statistics.

Survey of Current Business. Washington, D.C.: U.S. Office of Business Economics. Details general business indicators.

For more information on how to use business reference materials, consult the following guides:

Daniells, L. M. 1971. *Business Reference Sources.* Boston: Baker Library, Harvard University.

Coman, E. T. 1964. *Sources of Business Information.* rev. ed. Berkeley: Univ. of California Press.

COMPUTERIZED INFORMATION RETRIEVAL Most college and university libraries—and even some public libraries—have facilities for computerized information retrieval. Technological improvements in computer science, along with continuing growth in the amount of scholarly literature produced, ensure that in the foreseeable future computerized information retrieval will be the primary means of gaining access to information.

According to the 1980 *Supplement* to Sheehy's *Guide to Reference Books*, some seventy million references were contained in the more than five hundred data bases available in 1979. A data base is a machine-readable file of information from which a computer searches and retrieves specific data. Libraries today lease access to one or more data bases, with which they communicate by computer terminal.

Most computerized literature searches require the assistance of a trained librarian, but the concept is relatively simple. You decide what key words (or phrases) to use in your search, and what limitations you wish to place on it. For example, you might wish to limit your search to articles, excluding other types of literature, or you might wish to retrieve only those articles written over a limited period. When you enter a key word, the screen will show you how many items in the data base are filed according to that key word. You can ask the computer to combine several key words and show the number of items that include all of the key words. For example, the key word "solar energy" might elicit 735 entries. The key word "home heating" might elicit 1,438. Combined, the two key words might elicit 459 entries—that is, 459 items dealing with solar energy *and* home heating.

By trying out various combinations of key words, you can come up with an effective search strategy. You can then ask the computer to print the bibliography—or in some cases the actual abstracts—at the terminal or at the computer site.

Computerized information retrieval offers several advantages over manual searching:

1. It is faster. The search can take as little as a few minutes.
2. It is more comprehensive. The computer doesn't overlook any items: it follows your instructions to the letter.
3. It is more flexible. You can devise your strategy at the terminal, modifying it effortlessly.
4. It is more up-to-date. Data bases are produced and updated more quickly than printed indexes or abstract services.
5. It is more accurate. The printout of items contains no mechanical errors or indecipherable handwriting.

However, computerized information retrieval also has its disadvantages:

1. It is expensive. An average search can cost $10 to $20 in computer time. Students usually have to assume at least part of this expense.

2. It can produce a lot of irrelevant information. Unless you tell it not to, the computer will print out items you haven't anticipated—such as speeches or films. In addition, even a well-planned search strategy can yield items that are not useful. For instance, an article filed under the key words "Japan" and "whale hunting" might relate to whale hunting by the Japanese—as you had hoped—but deal almost exclusively with Japanese consumption of whale products.

Logistical problems also detract from the usefulness of computerized literature searches: your library might not have access to the data base you need, and in most cases you cannot perform the search without assistance.

Despite these problems, computerized literature searches are now standard in a number of fields, especially in science. Excellent data bases—such as the *Geological Reference File*, which encompasses six different printed indexes related to the field—make computerized literature searching simple and effective.

For more information on the problems and promise of the computerized literature search, see William Katz's *Introduction to Reference Work*, vol. II, *Reference Services and Reference Processes* (New York: McGraw-Hill, 1978).

PERSONAL INTERVIEWS

Personal interviews are a good source of information on subjects too new to have been discussed in the professional literature or inappropriate for widespread publication (such as local political questions). Most students are inexperienced at interviewing and hence reluctant to conduct interviews. Interviewing, like any other communications technique, is a skill that requires practice. The following discussion is meant to make interviewing straightforward and productive.

CHOOSING AN INTERVIEWEE Start by defining on paper what you want to find out. Only then can you begin to search for a person who can provide the necessary information.

The ideal interviewee can be defined as an expert who is willing to talk. Many times the nature of your subject will dictate whom you should ask. If you are writing about research being conducted at your university, for instance, the logical choice would be a person involved in the project. Sometimes, however, you might be interested in a topic about which a number of people could speak knowledgeably, such as the reliability of a particular kind of office equipment or the reasons behind the growth in the number of adult students. Use directories, such as local industrial guides, to locate the names and addresses of potential interviewees.

Once you have located an expert, find out if he or she is willing to be interviewed. On the phone or in a letter, indicate what you want to ask about; the person might not be able to help you but might be willing to refer you to someone who can. And be sure to tell the potential interviewee why you have decided to ask him or her: a well-chosen compliment will be more effective than an admission that the person you really wanted to interview is out of town. Don't forget to mention what you plan to do with the information: write a report, give a talk, etc. Then, if the person is willing to be interviewed, set up an appointment at his or her convenience.

PREPARING FOR THE INTERVIEW Never give the impression that the reason you are conducting the interview is to avoid library research. If you ask questions that are already answered in the professional literature, the interviewee might become annoyed and uncooperative. Make sure you are thoroughly prepared for the interview: research the subject carefully in the library.

Write out your questions in advance, even if you think you know them by heart. Frame your questions so that the interviewee won't simply answer yes or no. Instead of "Are adult students embarrassed about being in class with students much younger than themselves?" ask, "How do the adult students react to being in class with students much younger than themselves?" You don't want to give the impression that all you want is a simple confirmation of what you already know.

CONDUCTING THE INTERVIEW Arrive on time for your appointment. Thank the interviewee for taking the time to talk with you. Repeat the subject and purpose of the interview and what you plan to do with the information.

If you wish to tape-record the interview, ask permission ahead of time; taping makes some people uncomfortable. Have paper and pens ready. Even if you are taping the interview, you will want to take brief notes as you go along.

Start by asking your first prepared question. Listen carefully to the interviewee's answer. Be ready to ask a follow-up question or request a clarification. Have your other prepared questions ready, but be willing to deviate from them. In a good interview, the interviewee probably will lead you in directions you had not anticipated. Gently return the interviewee to the point if he or she begins straying unproductively, but don't interrupt rudely or show annoyance.

After all of your questions have been answered (or you have run out of time), thank the interviewee again. If a second meeting would be useful—and you think the person would be willing to talk with you further—this would be an appropriate time to ask.

AFTER THE INTERVIEW After you have left, take the time to write up the important information while the interview is still fresh in your mind. (This step is, of course, unnecessary if you have recorded the interview.)

It's a good idea to write a brief thank-you note and send it off within a day or two of the interview. Show the interviewee that you appreciate the courtesy extended to you and that the information you learned will be of great value to you. Confirm any previous offers you made, such as to send a copy of the report.

LETTERS OF INQUIRY

A letter of inquiry is often a useful alternative to a personal interview. If you are lucky, the person who responds to your inquiry letter will provide detailed and helpful answers. Keep in mind, however, that the person might not understand what information you are seeking or might not want to take the trouble to help you. Also, you can't ask follow-up questions in a letter, as you can in an interview. Although the strategy of the inquiry letter is essentially that of a personal interview—persuading the reader to cooperate and phrasing the questions carefully—inquiry letters in general are less successful, because, unlike interviewees, the readers have not already agreed to provide information.

For a full discussion of inquiry letters, see Chapter 18.

QUESTIONNAIRES

A questionnaire enables you to solicit information from a large group of people. Although they provide a useful and practical alternative to interviewing dozens of people spread out over a large geographical area, questionnaires rarely yield completely satisfactory results. For one thing, some of the questions, no matter how carefully constructed, are not going to work: the respondents will misinterpret them or supply answers that are not useful for your purposes. In addition, you probably will not receive nearly as many responses as you had hoped. The response rate will almost never exceed 50%; in most cases, it will be closer to 10 or 20%. And you can never be sure how representative the responses will be; in general, people who feel strongly about an issue are much more likely to respond than are those who do not. For this reason, be careful in drawing conclusions based on a small number of responses to a questionnaire.

When you send a questionnaire, you are asking the recipient to do you a favor. If the questionnaire requires only two or three minutes to complete, you are of course more likely to receive a response than if it requires an hour to fill out. Your goal, then, should be to construct questions that will elicit the information you need as simply and efficiently as possible.

CREATING EFFECTIVE QUESTIONS Effective questions are unbiased and clearly phrased. Avoid "charged" language and slanted questions. Don't ask, "Do you think the greedy oil companies should be allowed to rip off the American consumer?" Instead, ask, "Are you in favor of the windfall profits tax that would be imposed on the oil companies?" Make sure the questions are worded as specifically as possible, so that the reader understands exactly what information you are seeking. If you ask, "Do you favor improving the safety of automobiles?" only an eccentric would answer, "No." However, if you ask, "Do you favor requiring automobile manufacturers to equip new cars with air bags, at a cost to the consumer of $300 per car?" you are more likely to get a useful response.

As you make up the questions, keep in mind that there are several formats from which to choose:

MULTIPLE CHOICE

Would you consider joining a company-sponsored sports team?

Yes ____ No ____

How do you get to work? (Check as many as apply.)

my own car ____

car/van pool ____

bus ____

train ____

walk ____

other ____ (please specify)

The flextime program has been a success in its first year.

____	____	____	____
Agree strongly	Agree more than disagree	Disagree more than agree	Disagree strongly

RANKING

Please rank the following work schedules in order of preference. Put a "1" next to the schedule you would most like to have, a "2" next to your second choice, etc.

8:00-4:30 ____

8:30-5:00 ____

9:00-5:30 ____

flexible ____

SHORT ANSWER

What do you feel are the major advantages of the new parts-requisitioning policy?

1. ________________________________

2. ________________________________

3. ________________________________

SHORT ESSAY

The new parts-requisitioning policy has been in effect for a year. How well do you think it is working?

____

____

____

____

____

Remember that you will receive fewer responses if you ask for essay answers; moreover, essays, unlike multiple-choice answers, cannot be quantified.

After you have created the questions, write a letter or memo to accompany the questionnaire. A letter to someone outside of your organization is basically an inquiry letter (sometimes with the questions themselves on a separate sheet); therefore, it must clearly indicate who you are, why you are writing, what you plan to do with the information, and when you will need it. (See Chapter 18 for a discussion of inquiry letters.) For people within your

FIGURE 3–4

Questionnaire

September 6, 1983

To: All employees

From: William Bonoff, Vice-President of Operations

Subject: Evaluation of the Lunches Unlimited food service

As you may know, every two years we evaluate the quality and cost of the food service that caters our lunchroom. We would like you to help by sharing your opinions about the food service. Your anonymous responses will help us in our evaluation. Please drop the completed questionnaires in the marked boxes near the main entrance to the lunchroom.

1. Approximately how many days per week do you eat lunch in the lunchroom?

   0____ 1____ 2____ 3____ 4____ 5____

2. At approximately what time do you eat in the lunchroom?

   11:30-12:30____ 12:00-1:00____ 12:30-1:30____

   varies____

3. Do you have trouble finding a clean table?

   often____ sometimes____ rarely____

4. Are the Lunches Unlimited personnel polite and helpful?

   always____ usually____ sometimes____ rarely____

5. Please comment on the quality of the different kinds of food you have had in the lunchroom.

organization, a memo accompanying a questionnaire should answer the same questions.

Figure 3–4 shows a sample questionnaire.

SENDING THE QUESTIONNAIRE Drafting the questions is only part of the task. The next step is to administer the questionnaire. Determining whom to send it to can be simple or difficult. If you want to know what the residents of a particular street think about a proposed construction project, your task is easy. But if you want to

a. Hot meals (daily specials)

excellent____ good____ satisfactory____ poor____

b. Hot dogs and hamburgers

excellent____ good____ satisfactory____ poor____

c. Sandwiches

excellent____ good____ satisfactory____ poor____

d. Salads

excellent____ good____ satisfactory____ poor____

e. Desserts

excellent____ good____ satisfactory____ poor____

6. What foods would you like to see served that are not served now?

____

____

____

7. What beverages would you like to see served that are not served now?

____

____

____

know what mechanical engineering students in colleges across the country think about their curricula, you will need some background in sampling techniques in order to isolate a representative sample.

Before you send *any* questionnaire, show it and the accompanying letter or memo to a few people whose backgrounds are similar to those of your real readers. In this way you can sample your questionnaire's effectiveness before "going public."

Be sure also to include a self-addressed, stamped envelope with questionnaires sent to people outside your organization.

FIGURE 3–4
(Continued)

8. Please comment on the prices of the foods and beverages served.
   a. Hot meals (daily specials)
      too high____ fair____ a bargain____
   b. Hot dogs and hamburgers
      too high____ fair____ a bargain____
   c. Sandwiches
      too high____ fair____ a bargain____
   d. Desserts
      too high____ fair____ a bargain____
   e. Beverages
      too high____ fair____ a bargain____
9. Would you be willing to spend more money for a better-quality lunch, if you thought the price was reasonable?
   yes, often____ sometimes____ not likely____
10. Please provide whatever comments you think will help us evaluate the catering service.
    ____
    ____
    ____
    ____
    ____

Thank you for your cooperation.

USING THE INFORMATION

Once you have gathered your books and articles, conducted your interviews, and received responses from your questionnaires, you should start planning a strategy for using the information. The first step in working with any group of materials is to evaluate each individual source.

EVALUATING THE SOURCES

Authority is sometimes difficult to determine. Your interviewees are likely to be authoritative (you would not have deliberately chosen suspect sources in the first place), but you must still judge the quality of the information they have given. Your questionnaires, once you have discarded eccentric or clearly extreme responses, should contain authoritative material. With books and articles, however, this question is not so clear-cut. Check to see if the author and the publisher or journal are respected.

Many books and journals include biographical sketches. Does the author appear to have solid credentials in the subject he or she is writing about? Look for academic credentials, other books or articles written, membership in professional associations, awards, and so forth. If no biographical information is provided, consult a "Who's Who" of the field.

Evaluate the publisher, too. A book should be published by a reputable trade, academic, or scholarly house. A journal should be sponsored by a university or professional association. Read the list of editorial board members; they should be well-known names in the field. If you have any doubts about the authority of a book or journal, ask the reference librarian or a professor in the appropriate field to comment on the reputation.

Finally, check the date of publication. In high-technology fields, in particular, a five-year-old article or book is likely to be of little value (except, of course, for historical studies).

SKIMMING

Inexperienced writers often make the mistake of trying to read every potential source. The result, all too commonly, is that they get halfway through one of their several books when they realize they have to start writing immediately in order to submit their reports on time. Knowing how to skim is an invaluable skill.

To skim a book, read the prefatory materials—the preface and introduction. This will give you a basic idea of the writer's approach and methods. The acknowledgements section will tell you about any assistance the author received from other experts in the field, or about his or her use of important primary research or other resources. Read the table of contents carefully to get an idea of the scope and organization of the book. A glance at the notes at the end of the chapters or the end of the book will help you understand the nature of the author's research. Check the index for clues about the coverage that the information you need receives.

To skim an article, focus on the abstract. Check the notes and references as you would for a book. Read the headings throughout the article.

For both books and articles, sample a few paragraphs from different portions of the text to gauge the quality and relevance of the information.

Skimming will not assure you that a book or article *is* going to be useful. A careful reading of a work that looks useful might prove disappointing. However, skimming can tell you if the work is *not* going to be useful: because it doesn't cover your subject, for example, or because its treatment is too superficial or too advanced. Eliminating the sources that you don't need will give you more time to spend on the ones that you do.

TAKING NOTES

Note taking is the first step in the actual writing of the report. Your notes will provide the vital link between what your sources have said and what you are going to say. You will refer to them over and over. For this reason, it is smart to take notes logically and systematically.

Adopting a few mechanical conventions can help you to establish a system. Before you start to take notes, buy two packs of note cards: one 4 inches by 6 inches or 5 inches by 8 inches, the other 3 inches by 5 inches. (There is nothing sacred about ready-made note cards; you can make up your own out of scrap paper.) The major advantage of using cards is that they are easy to rearrange later, when you want to start outlining the report.

On the smaller cards, record the bibliographic information for each source from which you take notes. For a book, record the following information:

author
title
publisher
place of publication
year of publication
call number

Figure 3–5 shows a sample bibliography card for a book.

For an article, record the following information:

author
title of the article
periodical
volume
number
date
pages on which the article is included
call number

Figure 3–6 shows a sample bibliography card for an article.

FIGURE 3–5

Bibliography Card for a Book

KA

31.6 Honeywell, Alfred R.

.H306 The Meaning of Saturn II

N.Y.: The Intersteller Society,

1983

On the larger cards, record the notes. To simplify matters, write on one side of a card only and limit each card to a narrow subject or discrete concept so that you can easily reorder the information to suit the needs of your document.

Although many writers have devised their own systems of note taking, most notation involves two different kinds of activities: paraphrasing and quoting.

PARAPHRASING A paraphrase is a restatement, in your own words, of someone else's words. "In your own words" is crucial: if you simply copy someone else's words—even a mere two or three in a row—you must use quotation marks.

FIGURE 3–6

Bibliography Card for an Article

HZ

102.3

Hastings, W.

"The Skylab debate"

The Modern Inquirer 19, 2 (Fall, 1983)

106-113

What kind of material should be paraphrased? Any information that you think *might* be useful in writing the report: background data, descriptions of mechanisms or processes, test results, and so forth.

To paraphrase accurately, you have to study the original and understand it thoroughly. Once you are sure you follow the writer's train of thought, rewrite the relevant portions of the original. If you find it easiest to rewrite in complete sentences, fine. If you want to use fragments or merely list information, make sure you haven't compressed the material so much that you'll have trouble understanding it later. Remember, you might not be looking at the card again for a few weeks. Put a title on the card so that you'll be able to identify its subject at a glance. The title should include the general subject the writer describes—such as "Open-sea pollution-control devices"—and the author's attitude or approach to that subject—such as "Criticism of open-sea pollution-control devices." To facilitate the later documentation process, also include the author's last name, a short title of the article or book, and the page number of the original.

Figure 3–7 shows a paraphrased note card based on the following discussion of rattlesnake bites (Larson 1970: 170).

> The rattlesnake will, of course, also use its bite for defense, and it is here that man—and fear—enter the picture. A rattlesnake bite is serious, and occasionally fatal. Laurence Klauber, author of a monumental two-volume work on rattlesnakes published in 1956, came to the conclusion that each year approximately 1000 rattlesnake bites occur in the United

FIGURE 3–7

Paraphrased Notes

<u>Statistics</u> <u>on</u> <u>rattlesnake</u> <u>bites</u> <u>in</u> <u>U.S.</u>

in U.S., ≈ 1,000 bites/year

fatalities: 30/year

mortality rate: 0.02/100,000 population annually

Larson, <u>Deserts</u>, p. 170, citing Laurence Klauber's research (1956)

States. Of these, thirty cases are fatal. The resulting mortality rate from snakebite is about 0.02 per 100,000 population per year. In other words, the odds are excellent that you will never be bitten, and even if you are, the odds still greatly favor your recovery. Klauber lists the sixty-five species and subspecies of rattlesnakes which are found in North and South America. These vary widely in size and other characteristics. The severity of a rattlesnake bite to a human is dependent upon many factors: the species, the size, the virulence of the venom, the amount of venom received, the site of the bite, the physical condition and age of the person bitten, the care received, and other factors. Most interestingly, it has been found that a rattlesnake does not always inject venom when it strikes! This, however, you cannot depend upon.

The title of the notes in Figure 3-7—"Statistics on rattlesnake bites in U.S."—is important, because it shows the logic behind the writer's choice of what to paraphrase. He wanted only the statistics on bites. He was *not* interested in other details of Klauber's research, so he did not record that information. There is no one way to paraphrase: you have to decide what to paraphrase—and how to do it—based on your analysis of the audience and the purpose of your report.

QUOTING On occasion you will want to quote a source, either to preserve the author's particularly well expressed or emphatic phrasing or to lend authority to your discussion. In general, do not quote passages more than two or three sentences long, or your report will look like a mere compilation. Your job is to integrate an author's words into your own work, not merely to introduce a series of quotations.

The simplest form of quotation is that of an author's exact statement:

```
As Jones states, "Solar energy won't make much of a difference in this
century."
```

To add an explanatory word or phrase to a quotation, use brackets:

```
As Nelson states, "It [the oil glut] will disappear before we under-
stand it."
```

Use ellipses (three spaced dots) to show that you are omitting part of an author's statement:

ORIGINAL STATEMENT:

```
"The generator, which we purchased in May, has turned out to be one of
our wisest investments."
```

ELLIPTICAL QUOTATION:

```
"The generator . . . has turned out to be one of our wisest invest-
ments."
```

For more details on the mechanics of quoting, see the entries under "Quotation Marks," "Brackets," and "Ellipses" in Appendix A.

DOCUMENTATION

Documentation is the explicit identification of the sources of the ideas and quotations used in your report.

For you as the writer, complete and accurate documentation is primarily a professional obligation—a matter of ethics. But it also serves to substantiate the report. Effective documentation helps to place the report within the general context of continuing research and to define it as a responsible contribution to knowledge in the field. Failure to document a source—whether intentionally or unintentionally—is plagiarism. At most universities and colleges, plagiarism means automatic failure of the course and, in some instances, suspension or dismissal. In many companies, it is grounds for immediate dismissal.

For your readers, complete and accurate documentation is an invaluable tool. It enables them to find the source you have relied on, should they want to read more about a particular subject.

What kind of material should be documented? Any quotation from a printed source or an interview, even if it is only a few words, should be documented. In addition, a paraphrased idea, concept, or opinion gathered from your reading should be documented. There is one exception to this rule: if the idea or concept is so well known that it has become, in effect, general knowledge, it need not be documented. Examples of knowledge that is within the public domain would include Einstein's theory of relativity and the Laffer curve. If you are unsure whether an item is within the public domain, document it anyway, just to be safe.

Many organizations have their own preferences for documentation style; others use published style guides, such as the *U.S. Government Printing Office Style Manual*, the American Chemical Society's *Handbook for Authors*, or *The Chicago Manual of Style*. You should find out what your organization's style is and abide by it.

Three basic systems of documentation have become standard:

1. author/date citations and bibliography
2. footnotes and endnotes (with or without bibliography)
3. numbered citations and numbered bibliography

Variations on these systems abound.

AUTHOR/DATE CITATIONS AND BIBLIOGRAPHY In the author/date style, a parenthetical notation that includes the name of the source's author and the date of its publication immediately follows the quoted or paraphrased material:

```
This phenomenon was identified as early as forty years ago (Wilkinson
1943).
```

Sometimes, particularly if the reference is to a specific fact or idea, the page (or pages) from the source are also listed:

```
(Wilkinson 1943: 36-37)
```

If two or more sources by the same author in the same year are listed in the bibliography, the notation may include an abbreviated title, to prevent confusion:

```
(Wilkinson, "Cornea Research," 1943: 36-37)
```

Or the citation for the first source written that year can be identified with a lowercase letter:

```
(Wilkinson 1943a: 36-37)
```

The second source would be identified similarly:

```
(Wilkinson 1943b: 19-21)
```

The simplicity and flexibility of the author/date system make it highly attractive. Of course, because the citations are minimal and because their form is dictated more by common sense than by a style sheet, a conventional bibliography that contains complete publication information must be used in conjunction with them, following the text. Your obligation as a writer when you are using this system is to leave your reader no doubt about which of your many sources you are citing in any particular instance. *The Chicago Manual of Style*, 13th edition (Chicago: University of Chicago Press, 1982), advocates using the author/date system throughout the natural and social sciences and recommends the following bibliographic conventions for use with author/date citations.

The order of basic information included in this style of bibliography is as follows:

FOR BOOKS

1. author (last name followed by initials)
2. date of publication
3. title
4. place of publication
5. publisher

FOR ARTICLES

1. author
2. date of publication
3. article title
4. journal or anthology title
5. volume (of a journal) or place of publication and publisher (of an anthology)
6. inclusive pages

The individual entries are arranged alphabetically by author (and then by date if two or more works by the same author are listed) within the bibliography. Anonymous works are integrated into the alphabetical listing by title. Where several works by the same author are included, they are arranged by date under the author's name. In such cases, a long dash (10 hyphens on a typewriter) takes the place of the author's name in all entries after the first. The first line of a bibliographic entry is flushed left with the margin; each succeeding line is indented (five spaces on a typewriter).

Chapman, D. L. 1981. The closed frontier: Why Detroit can't make cars
    that people will buy. Motorist's Metronome 12(June):17-26.

----------. 1983. *The driver's guide to evaluating compact automobiles*. Athens, Ga.: Consumer Press.

Courting credibility: Detroit and its mpg figures. 1977. *Countercultural Car & Driver* (July): 19.

Following are the standard bibliographic forms used with various types of literature.

A BOOK

Cunningham, W. 1980. *Crisis at Three Mile Island*. New York: Madison.

The author's name (surname first) is given with initials only. The date of publication comes next, followed by the title of the book, underlined (italicized in print). In the natural and social sciences, the style is generally to capitalize only the first letter of the title, the subtitle (if there is one), and all proper nouns. The last items are the location and name of the publisher.

For a book by two or more authors, all of the authors are named.

Cunningham, W., and A. Breyer . . .

Only the name of the first author is inverted.

A BOOK ISSUED BY AN ORGANIZATION

Department of Energy. 1979. *The energy situation in the eighties*. Washington, D.C.: U.S. Government Printing Office, Technical Report 11346-53.

A BOOK COMPILED BY AN EDITOR OR ISSUED UNDER AN EDITOR'S NAME

Morgan, K. E., ed. 1980. *Readings in alternative energies*. Boston: Smith-Howell.

AN EDITION OTHER THAN THE FIRST

Schonberg, N. 1982. *Solid state physics*. 3rd ed. London: Paragon.

AN ARTICLE INCLUDED IN A BOOK

May, B., and J. Deacon. 1981. Amplification systems. In *Third Annual Conference of the American Electronics Association*, ed. A. Kooper, 101-114. Miami: Valley Press.

A JOURNAL ARTICLE

Hastings, W. 1983. The Skylab debate. The Modern Inquirer 13: 311-18.

The title of the journal is underlined (italicized in print). The first letters of the first and last words and all nouns, adjectives, verbs (except the infinitive *to*), adverbs, and subordinate conjunctions are capitalized. After the journal title comes the volume and page numbers of the article.

AN ANONYMOUS JOURNAL ARTICLE

The state of the art in microcomputers. 1983. Newscene 56: 406-421.

Anonymous journal articles are arranged alphabetically by title. If the title begins with a grammatical article such as *the* or *a*, alphabetize it under the first word following the article (in this case, *state*).

A NEWSPAPER ARTICLE

Eberstadt, A. 1983. Why not a Rabbit, why not a Fox? Morristown Mirror and Telegraph. Sunday 31 July: Business and Finance Section.

A PERSONAL INTERVIEW

Riccio, Dr. Louis, Professor of Operations Research, Tulane University. Interview with author. New York City, 13 July 1980.

The name and professional title of the interviewee should be cited first, then the fact that the source was an interview, and finally the place and date of the interview.

A QUESTIONNAIRE CONDUCTED BY SOMEONE OTHER THAN THE AUTHOR

Recycled Resources Corp. Data derived from questionnaire administered to 33 foremen in the Bethlehem plant, November 3-4, 1982.

An individual, an outside firm, or a department might also serve as the questionnaire's "author" for bibliographic purposes. If you are citing data derived from your own questionnaire, a bibliographic entry is unnecessary: you include the questionnaire and appropriate background information in an appendix. When citing your own questionnaire data in the text of your report, refer your readers to the appendix: "Foremen at the Bethlehem plant overwhelmingly favored staggered shifts that would keep the plant in operation sixteen hours per day (Appendix D)."

Figure 3–8 illustrates a coventional bibliography for use with the author/date citation system.

FOOTNOTES AND ENDNOTES (WITH OR WITHOUT BIBLIOGRAPHY) The documentation style with which most readers are familiar uses footnotes or endnotes. A raised number in the text alerts the reader that the material immediately preceding it is derived from a source; the same number precedes the note, either at the foot of the page (a footnote) or at the end of the article, chapter, or paper (an endnote). The number that precedes the note appears either at the left-hand margin, followed by a period, or indented as a paragraph and raised as in the text. (*The Chicago Manual of Style* recommends the former, the *MLA Handbook*—published by the Modern Language Association—recommends the latter: this discussion will follow *Chicago*; *MLA* varies slightly in several respects.)

This system of documentation is a good deal more complicated than the author/date system because the notes duplicate much of the information that appears in the bibliography (for this reason, most journals that use the system omit the bibliography altogether). The system is used most frequently in the humanities, where authors often comment on a particular source after identifying it in a note, making observations that are not germane to the text proper, and also intersperse explanatory notes and citations. This type of note is increasingly rare in the natural and social sciences, where authors most often simply need to document an authoritative source for a fact or a figure.

FIGURE 3–8

Bibliography for Use in Conjunction with the Author/Date Citation System

BIBLIOGRAPHY

Daly, P. H. 1975. Selecting and designing a group insurance plan. Personnel Journal 54:322-23.

Flanders, A. 1973. Measured daywork and collective bargaining. British Journal of Industrial Relations 3:368-92.

Goodman, R. K., J. H. Wakely, and R. H. Ruh. 1972. What employees think of the Scanlon Plan. Personnel 3:22-29.

Trencher, P. 1977. Recent trends in labor-management relations. New York: Madison.

Zwicker, D. Professor of Industrial Relations, Hewlett College. Interview with author. Philadelphia, 19 March 1979.

In spite of its declining popularity among technical writers, traditional footnotes are still occasionally used, and you should be familiar with them and with the bibliographic style that is used in conjunction with them. Following are typical note entries that can be used in the numbered notes system.

A BOOK

1. William Cunningham, Crisis at Three Mile Island (New York: Madison, 1980), 14-16.

The author's name is spelled out in the normal order. The title of the book is underlined (italicized in print). Note that all words in the title are capitalized except articles (*a*, *an*, *the*), coordinating conjunctions (*and*, *or*, *for*, etc.), and prepositions under five letters in length. The facts of publication (city of publication, name of publisher, and year of publication) are enclosed within parentheses. The numbers following the facts of publication refer to the pages cited in the book.

For a book by two or three authors, include all the authors' names:

2. William Cunningham and Nancy Bauer . . .

3. William Cunningham, Nancy Bauer, and Edwin Housman . . .

For a book by four or more authors, use *et al.*

4. William Cunningham, et al.

A BOOK ISSUED BY AN ORGANIZATION

5. Department of Energy, The Energy Situation in the Eighties (Washington, D.C.: U.S. Government Printing Office, 1979), Technical Report 11346-53;15.

A BOOK COMPILED BY AN EDITOR OR ISSUED UNDER AN EDITOR'S NAME

6. Kevin E. Morgan, ed., Readings in Alternative Energies (Boston: Smith-Howell, 1980), 219.

AN EDITION OTHER THAN THE FIRST

7. Nathan Schonberg, Solid State Physics, 3rd ed. (London: Paragon, 1982), 196.

AN ARTICLE INCLUDED IN A BOOK

8. Brian May and John Deacon, "Amplification Systems," in Third Annual Conference of the American Electronics Association, ed. Al Kooper (Miami: Valley, 1981), 101-114.

The page number(s) at the end of the note refers to the citation itself, not to the inclusive pages of the article.

A JOURNAL ARTICLE

9. Warren Hastings, "The Skylab Debate," The Modern Inquirer 13 (March 1983), 311-18.

The title of the journal is underlined (italicized in print). After the journal title comes the volume and (in parentheses) the date of publication of the article.

AN ANONYMOUS JOURNAL ARTICLE

10. "The State of the Art in Microcomputers," Newscene 56 (3 July 1983), 406-421.

Anonymous journal articles are arranged alphabetically by title. If the title begins with a grammatical article, such as *the* or *a*, alphabetize it under the first word following the article (in this case, *state*).

A NEWSPAPER ARTICLE

11. Alan Eberstadt, ''Why Not a Rabbit, Why Not a Fox?'' Morristown Mirror and Telegraph, 31 July 1983, Business and Finance Section.

A PERSONAL INTERVIEW

12. Dr. Louis Riccio, Professor of Operations Research, Tulane University, interview with author. New York City, 13 July 1980.

A QUESTIONNAIRE CONDUCTED BY SOMEONE OTHER THAN THE AUTHOR

13. Recycled Resources Corp. Data derived from questionnaire administered to 33 foremen in the Bethlehem plant, 3-4 November 1982.

An individual, an outside firm, or a department might also serve as a questionnaire's "author." If you are citing data derived from your own questionnaire, a note is unnecessary: you simply refer your reader to the appendix that includes the questionnaire.

SHORTENED FORMS

Following the first full reference to a work, abbreviated notes are permissible. Just be sure that there is no room for confusion: if you have cited more than one work by an author, you must include not only the author's name but also some indication of the title to which you are referring; if among your sources are two authors with the same last name, you must also include first names.

14. Harrison, 26.

15. Harkness, Environment, 318.

16. Edward Smith, 73.

17. Harkness, "Water Pollution," 741.

18. Annabel Smith, 488.

Figure 3–9 shows a typical endnotes page in a research report.

Following are typical bibliography entries that can be used with endnotes or footnotes. Although the forms differ in detail, the alphabetizing and indentation principles are the same as those for bibliographies used with author/date citations.

A BOOK

Cunningham, William. Crisis at Three Mile Island. New York: Madison, 1980.

Note that periods separate the major items and that no page numbers are included. Also note that for a bibliography entry, all of the authors' names are listed; *et al.* is not used.

A BOOK ISSUED BY AN ORGANIZATION

Department of Energy. The Energy Situation in the Eighties. Washington, D.C.: U.S. Government Printing Office, 1979. Technical Report 11346-53.

A BOOK COMPILED BY AN EDITOR OR ISSUED UNDER AN EDITOR'S NAME

Morgan, Kevin E., ed. Readings in Alternative Energies. Boston: Smith-Howell, 1980.

FIGURE 3–9
Typical Endnotes

NOTES

1. Peter H. Daly, "Selecting and Designing a Group Insurance Plan," Personnel Journal 54 (Fall 1975), 324.
2. Allan Flanders, "Measured Daywork and Collective Bargaining," British Journal of Industrial Relations 3 (May 1973), 371.
3. Eleanor Sanchez, "The 'Taft-Hartley' Act and the Warren Court," Dominion Political Science Review 1 (1978), 21.
4. Sanchez, 23.
5. Dr. Daniel Zwicker, Professor of Industrial Relations, Hewlett College, interview with author in Philadelphia, 19 March 1979.
6. Robin K. Goodman, Jeffrey H. Wakely, and Robert H. Ruh, "What Employees Think of the Scanlon Plan," Personnel 3 (January 1972), 26-28.
7. Pauline Trencher, Recent Trends in Labor-Management Relations (New York: Madison, 1977), 211.

AN EDITION OTHER THAN THE FIRST

Schonberg, Nathan. Solid State Physics. 3rd ed. London: Paragon, 1982.

AN ARTICLE INCLUDED IN A BOOK

May, Brian, and John Deacon. "Amplification Systems." In Third Annual Conference of the American Electronics Association. Edited by Al Kooper. Miami: Valley, 1981.

A JOURNAL ARTICLE

Hastings, Warren. "The Skylab Debate." The Modern Inquirer 13 (March 1983): 311-18.

Inclusive page numbers of the article are listed.

AN ANONYMOUS JOURNAL ARTICLE

"The State of the Art in Microcomputers." Newscene 56 (3 July 1983): 406-421.

A NEWSPAPER ARTICLE

Eberstadt, Alan. "Why Not a Rabbit, Why Not a Fox?" Morristown Mirror and Telegraph, 31 July 1983: Business and Finance Section.

A PERSONAL INTERVIEW

Riccio, Dr. Louis, Professor of Operations Research, Tulane University, interview with author in New York City, 13 July 1980.

A QUESTIONNAIRE

Recycled Resources Corp. Data derived from questionnaire administered to 33 foremen in the Bethlehem plant, 3-4 November 1982.

Figure 3–10 shows a bibliography page from a report using endnotes or footnotes.

NUMBERED CITATIONS AND NUMBERED BIBLIOGRAPHY You will occasionally encounter a variation on the author/date notation system, the numbered citation and numbered bibliography system. In this system, the items in the bibliography are arranged either alphabetically or in order of the first appearance of each source in the text and then assigned a sequential number. The citation is the number of the source, enclosed in brackets or parentheses:

According to Hodge [3], "There is always the danger that the soil will shift." However, no shifts have been noted.

As in author/date citations, page numbers may be added:

According to Hodge[3:26], "There is always . . ."

In this example, *3* means that Hodge is the third item in the numbered bibliography; *26* is the page number.

Except that they are numbered, the individual bibliography entries in this system are identical to those used with author/date.

FIGURE 3-10

Bibliography for a Report Using Endnotes or Footnotes

BIBLIOGRAPHY

Daly, Peter H. "Selecting and Designing a Group Insurance Plan." Personnel Journal 54 (Fall 1975): 322-23.

Flanders, Allan. "Measured Daywork and Collective Bargaining." British Journal of Industrial Relations 3 (May 1973): 368-92.

Goodman, Robin K., Jeffrey H. Wakely, and Robert H. Ruh. "What Employees Think of the Scanlon Plan." Personnel 3 (January 1972): 22-29.

Trencher, Pauline. Recent Trends in Labor-Management Relations. New York: Madison, 1977.

Zwicker, Dr. Daniel, Professor of Industrial Relations, Hewlett College. Interview with author in Philadelphia, 19 March 1979.

EXERCISES

1. Choose a topic on which to write a report for this course. Make sure the topic is sufficiently focused that you will be able to cover it in some detail.

a. Using Sheehy's *Guide to Reference Books*, plan a strategy for researching this topic.

(1) Which guides, handbooks, dictionaries, and encyclopedias contain the background information you should read first?

(2) Which basic reference books discuss your topic?

(3) Which major indexes and abstract journals cover your topic?

b. Write down the call numbers of the three indexes and abstract journals most relevant to your topic.

c. Compare two abstract journals in your field on the basis of sponsoring organization, scope, clarity of listings, and timeliness. Also determine who writes the abstracts contained in each.

d. Make up a preliminary bibliography of two books and five articles that relate to your topic.

e. Using the bibliography from Exercise 1d, write down the call numbers of the books and journals (those that your library receives).
f. Find one of the works listed and write a brief assessment of its value.
g. Using a local industrial guide, make up a list of five persons who might have first-hand knowledge of your topic.
h. Make up two sets of questions: one for an interview with one of the persons on your list from Exercise 1g, and one for a questionnaire to be sent to a large group of people.
i. Photocopy the first two pages from an article listed in your bibliography from Exercise 1d. Paraphrase any three paragraphs, each on a separate note card. Also note at least two quotations, each on a separate card: one should be a complete sentence, and one an excerpt from a sentence.

2. Write sample bibliography entries for the following items. Use the style recommended in conjunction with author/date citations.

a. A book called *Cell Biology*, written by David E. Phillips and published in 1982 by Horizon Press, which is located in Dallas, Texas.
b. An article called "Choosing a Micro." The article is included in *New Directions in Office Management*, a book edited by Ruth Walker. The article is on pp. 19–27. The book was published last year in New York by Management Associates, Inc.
c. An interview with John Abrams, who is the vice-president for research for Allied Industries. The interview was carried out in Lexington, Kentucky, on June 3, 1983.

REFERENCE

Larson, P. 1970. *Deserts of America.* Englewood Cliffs, N.J.: Prentice-Hall.

CHAPTER FOUR

OUTLINES AND ROUGH DRAFTS

In a recent television commercial for an automotive oil filter, a garage mechanic, after finishing a major repair on a customer's car, notes that the repair would not have been necessary had the customer only changed the oil and filter a few months earlier. The mechanic shakes his head sadly and, holding a shiny new filter, says to the audience, "You can pay me now—or you can pay me later."

Taking the time to work out a careful outline and write a rough draft is like changing your oil and filter when you're supposed to—it's cheaper in the long run. Many writers, however, want to start writing the "final draft" right away. They want to get it over with. And writing an outline is one of those miserable little chores, anyway. Most people have bad memories of having to hand in outlines to English teachers or history teachers who always seemed to be more interested in "correcting" the format—the Roman numerals and the capital letters and the indentation—than in suggesting ways to help write the papers: sometimes there was not even a paper to write, just the outline.

For these reasons, even people who always change their oil at the right time sometimes jump right into a writing assignment without sketching the skimpiest outline. As a result, they often find themselves staring at a blank page and wasting valuable time.

Some minutes later, the word *Introduction* appears on the top of the page. After another half-hour, they have created an awkward and underdeveloped paragraph. For the rest of the morning, they fiddle with the paragraph, adding a little here, changing a little there, cleaning up the grammar and punctuation, and consulting the thesaurus. After lunch they take a look at what they've written, happy that they've at least made a good start. In 20 seconds, they realize that it's all wrong and toss it out. Total cost: about $50 or $100 of company time, and a slight case of panic.

Because of experiences like this, most practiced writers start with outlines.

PUBLIC AND PRIVATE OUTLINES

An outline is basically a plan—a preliminary verbal sketch for a more formal presentation. It can be a few words written on a note card, or a six-page, step-by-step design for a complicated report. It can be a highly personal bit of hieroglyphics that means absolutely nothing to anyone except its author, or a clear and comprehensive document that someone else could use to write out a report.

In some cases, you might be asked to create a "public" outline. For example, your supervisor might ask you to help her write a report on your department's activities during the past year. Because you are more familiar with the department's research work, she asks you to outline that section of the report and pass it on to her. What exactly is she looking for? You don't really know. But one fact, at least, is clear: you are being asked to write a document that someone else will use, and so you must find out as much as you can about the user's needs. Because the outline will reflect on your professionalism, you will probably want to produce a detailed, typed, formal outline. In other cases, the outline might be a "private" one that no one except you will ever see. If you are the only person who is going to see it, you can be casual about the conventions.

Few people are able to create a perfect outline—either public or private. With private outlines, you will find that as you write your first draft, you have forgotten something or that something in the outline isn't relevant and has to be dropped. This doesn't mean that the whole outline is flawed; it simply means that you never know what you really want to say until you try to put it on paper. Writing forces you to examine your ideas.

BRAINSTORMING

Even when you are writing a private outline, keep your writing situation in mind. The situation defines what information goes into the outline, but even experienced writers can run into trouble determining the content that best satisfies the requirements of audience and purpose.

If it were easy to know what to include, it would be easy to structure the information into a coherent outline. Unfortunately, it isn't so easy. Just as many people make the mistake of trying to write a final draft before a first draft, many people write *Introduction* at the top of the page and hope the rest of the outline will unfold, in order. The human mind is not so structured or orderly.

Start by taking advantage of the way the mind works. Gather the note cards you wrote when you did your preliminary research. Don't bother trying to arrange them in an apparently sensible order. Just read through them. Take a piece of paper and jot down a brief phrase to help you remember any point that you think should be in the document. Don't write out sentences; that wastes time. At this point, just write the shortest phrase that will mean something to you later (often the title of a note card will suffice):

greater efficiency

advantages of Wankel engine

bad EPA ratings

principle of operation

no spark plugs

As you look at your note cards, you will probably think of new ideas that your research did not include. Add them to the list. Do not worry at this point about not having read about them or not having taken notes about them. You can do the additional research later.

This technique of jotting down phrases in whatever order they come to mind is called brainstorming. As the word suggests, when you brainstorm you are not trying to impose order. Because of the random order of the cards, you will probably skip around from one topic to another; that is what you're supposed to do. And do not worry if some of the phrases you jot down involve details of larger ideas embraced by other phrases. You can straighten them out later.

Brainstorming has one purpose: to free your mind, within the limits of your situation, so that you can think of as many items as possible that relate to your topic. When you start to construct the outline, you will probably find that some of the items you listed do not in fact belong in the final document. Just toss them out. The advantage of brainstorming is that it is the most efficient way to catalogue those things that *might* be important to the document. A more structured way of generating ideas, ironically, would miss more of them.

Brainstorming is actually a way of classifying everything you know into two categories: (1) information that probably belongs in the document and (2) information that does not belong. Once you have isolated the material that probably belongs, you can start to write the outline.

Outlining is the act of establishing increasingly precise categories from the range of your brainstorming. If you are writing a general discussion of the rotary engine, for example, you will quickly see that some items are logically related to the principle of operation of the engine, some to its history, some to its future implications for the automobile industry, and so on. And within each of these categories you will discover other relationships. Once you have created a group of, for example, six or eight categories, you have to put them in a sequence that is consistent with your analysis of the writing situation. Then you have a good working outline. You will probably need to revise it as you start to write your rough draft, but at least you will have a plan from which to work.

THE PRINCIPLES OF OUTLINING

The concept of outlining is to distribute items—pieces of information—into increasingly precise categories and then to put them in a sequence. To do this effectively, you should keep in mind three general principles:

1. The choice and arrangement of items must be guided by your analysis of the writing situation.
2. The items included in the outline must be logically related to one another.
3. The categories must be arranged in a logical order.

ANALYZING THE WRITING SITUATION

Chapter 2 discussed the writing situation: the audience and purpose of the document. Although it is important to keep these factors in mind throughout the writing process, you must be absolutely sure at this stage that you have a clear understanding of the writing situation. You are committing yourself to a plan of attack—a direction, a scope, a sequence, a particular tone and level of difficulty, and so on. You must know what you are doing. Otherwise you will eventually have to rewrite substantially—or throw out—long sections of your first draft.

One way to be sure you have a clear understanding of the writing situation is to define it in a complete sentence. Following are several examples:

1. I want to persuade my supervisor that an in-house coronary-fitness program would be an effective way to decrease the incidence of heart disease among our employees.
2. I want to explain to management how a catalog describing the different generators we sell would help our sales staff.
3. I want to describe to the technical staff a new method for removing miscible organic pollutants from streams and rivers.

The central characteristic of brainstorming, as mentioned earlier, is freedom to explore your mind for ideas about your topic without stopping to consider how they fit together logically. When you begin to turn the brainstorming list into an outline, however, you have to become selective. Some of the items on your list will pertain to your topic *but would not be relevant to what you want to do with the topic.* In other words, they will not suit your purpose and audience.

For example, assume that you are researching the ecological damage caused by dredging of riverways, canals, and the like. You will be writing an article for specialists in the ecological effects of dredging. Your brainstroming list includes the following items:

1. earliest canal digging--Sumerians, 4,000 B.C.
2. amount of dredged material dumped
3. impairment of photosynthesis
4. DDT, PCB in dredged material
5. burial of bottom-dwelling communities
6. plain suction pipeline dredges
7. heavy metals in dredged materials
8. hopper dredges
9. pathogens in dredged materials
10. localized effects on fish

11. dustpan hydraulic dredges
12. food-chain magnification
13. sidecasting dredges
14. disruption of feeding, mating patterns

Everything here is related to the subject, but not necessarily to your writing situation. If you are concentrating on the kinds of ecological damage that dredging can cause, you will want to focus on items 2, 3, 4, 5, 7, 9, 10, 12, and 14. If you are concentrating on new technologies created to minimize the ecological damage, you will want to focus on items 6, 8, 11, and 13. Item 1 will probably be discarded in either case.

An analysis of the writing situation, therefore, enables you to answer two crucial questions: (1) Does this item belong in the outline? and (2) How important is this item in relation to the other items in the outline?

RELATING ITEMS TO ONE ANOTHER

Relating items to one another logically involves coordination and subordination.

Coordination is the process of identifying two or more items as roughly equivalent in importance. For instance, in a report on in-house coronary-fitness programs, the discussion of how the programs work might be roughly equivalent to the discussion of how much the programs cost to implement and maintain. Because they are coordinated, these two discussions would have the same "importance" in the outline—that is, they would have the same level of heading:

III. How Do Coronary-Fitness Programs Work?
IV. How Much Do Coronary-Fitness Programs Cost?

In this case, the coordinated items share uppercase Roman numerals. (Keep in mind, as you create your outline, that most readers can easily comprehend and remember up to six or seven coordinated items; any more than that number makes comprehension much more difficult.)

Subordination is the process of identifying two or more items as different in importance. For example, a discussion of start-up costs for a coronary-fitness program is subordinate to (less important than) a discussion of the total costs of the program. A subordinate item would therefore have a less "important" role in the outline:

```
IV. How Much Do Coronary-Fitness Programs Cost?
    A. Start-up costs
    B. Maintenance Costs
```

In this case, the first subordinate item ("Start-up costs") is less prominent than the major item of which it is a part ("How Much Do Coronary-Fitness Programs Cost?").

Two basic problems can occur when items are related to each other: faulty coordination and faulty subordination.

Faulty coordination involves equating items that are not parallel. Look, for example, at the following listing:

```
baseball equipment
  gloves
  balls
  wooden bats
  metal bats
```

Although there are metal bats and wooden bats, this listing is not parallel, because the other items under the heading "baseball equipment" are not differentiated according to the materials they are made from. The listing should read:

```
baseball equipment
  gloves
  balls
  bats
```

A related problem of coordination involves overlapping units:

```
lawn-care companies
  Smith's Lawn Care
  Evergreen Nurseries, Inc.
  Lawn Grow, Inc.
Barron's Lawn Service
```

"Barron's Lawn Service" should be subordinated to "lawn-care companies." However, if you wish to devote more attention to Barron's Lawn Service than to the other companies, an acceptable solution would be to revise the listing as follows:

```
three lawn-care companies
  Smith's Lawn Care
  Evergreen Nurseries, Inc.
  Lawn Grow, Inc.
Barron's Lawn Service
```

Adding the word "three" before "lawn-care companies" prevents the overlapping. Another solution would be to revise the listing as follows:

```
Barron's Lawn Service
other lawn-care companies
  Smith's Lawn Care
  Evergreen Nurseries, Inc.
  Lawn Grow, Inc.
```

Adding the word *other* and reversing the order of the two units eliminates the problem of overlapping.

Faulty subordination occurs when an item is made a subunit of a unit to which it does not belong:

```
power sources for lawnmowers
  manual
  gasoline
  electric
  riding mowers
```

In this excerpt from an outline, "riding mowers" is out of place because it is not one of the power sources of a lawnmower. Whether it belongs in the outline at all is another question, but it certainly doesn't belong here.

A second kind of faulty subordination occurs when only one subunit is listed under a unit:

```
types of sound reproduction systems
  records
    phonograph records
  tapes
    cassette
    eight-track
    open reel
```

It is illogical to list "phonograph records" if there are no other kinds of records. To solve the problem of faulty subordination, incorporate the single subunit into the unit:

```
types of sound reproduction systems
  phonograph records
  tapes
    cassette
    eight-track
    open reel
```

ARRANGING THE CATEGORIES IN A LOGICAL SEQUENCE Once you have selected the categories of your outline, you must arrange them in a sensible order. There is no one single way to sequence the items of an outline. Every document is different. As you think about a logical order, keep in mind the writing situation.

Most documents can be structured according to one of the following five patterns, or some combination of them:

1. chronological
2. spatial
3. general to specific
4. more important to less important
5. problem-methods-solution

CHRONOLOGICAL PATTERN The chronological (time) pattern works effectively when you want to describe a process. If your readers have to follow your discussion in order to perform a task or to understand how something happens, chronology is a natural pattern to use.

The following outline is structured chronologically.

```
Writing Situation: to explain to a general audience the geologic his-
                   tory of St. John's, United States Virgin Islands

  I.  Introduction
 II.  Stage One: Submarine Volcanism
      A.  Logical Evidence: The Pressure Theory
      B.  Empirical Evidence: Evidence from Ram Head Drillings
```

III. Stage Two: Above-water Volcanism
    A. Logical Evidence
        1. The Role of the Lowered Water Level
        2. The Role of Continental Drift
    B. Empirical Evidence: Modern Geological Data
IV. Stage Three: Depositing of Organic Marine Sediments
    A. The Role of the Ice Age
        1. Water Level
        2. Water Temperature
    B. The Role of "Drowned" Reefs
V. Conclusion

Notice that this outline (and the others that follow) is prefaced by a statement of the writing situation. By defining the writing situation at the top of the page, you force yourself to focus on it as you make up the outline. You thus decrease the chance of wandering away from your purpose and audience.

Notice also in this outline that the procedure itself is preceded by an introduction that orients the readers and states the purpose of the information that follows.

SPATIAL PATTERN In the spatial pattern, items are organized according to their physical relationships to one another. The spatial pattern is effective in structuring a description of a physical object.

A spatially organized outline follows.

Writing Situation: a proposal to the Mooreville F.D. for a heating system for the firehouse

I. System for Offices, Dining Area, and Kitchen
    A. Description of Floor-mounted Fan Heater
    B. Principal Advantages
        1. Individual Thermostats
        2. Quick Recovery Time
II. System for Bathrooms
    A. Description of Baseboard Heaters
    B. Principal Advantages
        1. Individual Thermostats
        2. Small Size

```
III. System for Garage
     A. Description of Ceiling-mounted Fan Heater
     B. Principal Advantages
        1. High Power
        2. High Efficiency
```

The basic organizing principle behind this outline is the physical layout of the firehouse. This principle makes sense because the readers—in this case, the firefighters—are familiar with the firehouse. Other spatial patterns could be developed here (such as top-to-bottom: ceiling fans, baseboards, floor fans), but since the readers are unfamiliar with heating systems, they would have a harder time following the discussion.

GENERAL-TO-SPECIFIC PATTERN The movement from general information to more specific information is a pattern that is used often, especially for sales literature and reports intended for a multiple audience. The general information enables the manager to understand the basics of the discussion without having to read all the details. The technical reader, too, appreciates the general information, which serves as an introductory overview.

Following is an outline for a sales brochure promoting a plastic coating used on institutional and industrial floors.

```
Writing Situation: audience--operations managers and maintenance
                   chiefs
                   purpose--to persuade them to request visit by
                   sales rep.

  I. The Story of EDS Plastic Coating
     A. What Is EDS Plastic Coating?
     B. Why Is EDS the Industry Leader?
        1. Strength and Durability
        2. Long-lasting Results
        3. Ease of Application
        4. Low Cost
     C. How to Find Out More about EDS Plastic Coating
 II. How EDS Works
     A. How EDS Bonds on Different Floorings
        1. Wood
```

```
        2.  Tile
            a.  Flat
            b.  Embossed
        3.  Linoleum
            a.  Flat
            b.  Embossed
    B.  How EDS Dries on the Floor
III. How to Apply EDS Plastic Coating
    A.  Preparing the Surface
    B.  Preparing the EDS Mixture
    C.  Spreading EDS Plastic Coating
        1.  Wetting the Applicators
            a.  Spreading Pads
            b.  Paint Brushes
        2.  Determining Where to Begin
            a.  Tiled Flooring
            b.  Continuous Flooring
        3.  Coordinating Several Workers
        4.  Stopping for the Day
    D.  Cleaning Up
        1.  Spreading Pads
        2.  Paint Brushes
```

Parts I and II of this outline are addressed to the person who would decide whether to seek more information about this product. The general information in Part I answers the three basic questions such a reader might have:

1. What is the product?
2. What are its important advantages?
3. How can I find out more about it?

Part II explains how the product works. Part III of the outline, which contains more specific information about how to use the product, is intended for the reader who actually has to supervise the use of the product. The pattern within this part is chronological, which is the obvious choice for a description of a process.

MORE-IMPORTANT-TO-LESS-IMPORTANT PATTERN Another basic pattern is the movement from more-important to less-important information. This pattern is effective even in describing events or processes that would seem to call for a chronological pattern. For example, suppose you were ready to write a report to a client after having performed an eight-step maintenance procedure on a piece of electronic equipment. A chronological pattern—focusing on what you did—would answer the following question: "What did I do, and what did I find?" A more-important-to-less-important pattern—focusing first on the problem areas and then on the no-problem areas—would answer the following question: "What were the most important findings of the procedure?" Most readers would probably be more interested, first, in knowing *what* you found than in *how* you found it.

The outline that follows is an example of the more-important-to-less-important pattern.

Writing Situation: to convey to the client the results of the tests run on the 14-115 Engine

I. Introduction
   A. Customer's Statement of Engine Irregularities
   B. Explanation of Test Procedures
II. Problem Areas Revealed by Tests
   A. Turbine Bearing Support
      1. Oversized Seal Ring Housing Defective
      2. Compressor Shaft Rusted
   B. Intermediate Case Assembly
      1. LP Compressor Packed with Carbon
      2. Oversized LP Compressor Rear Seal Ring Defective
III. Components That Tested Satisfactorily
   A. LP Turbine
   B. HP Turbine
   C. HP Compressor

In most technical writing, bad news is more important than good news, for the simple reason that breakdowns or problems have to be fixed. Therefore, the writer of this outline has described first the components that are not working properly.

PROBLEM-METHODS-SOLUTION PATTERN A basic pattern for outlining a complete project is to begin with the problem, then discuss the methods you followed, and then finish with the results, conclusions, and recommendations.

The following outline was written by an electrical engineer working for a company that makes portable space heaters.

Writing Situation: to recommend to the technical staff the best way to increase the efficiency of the temperature controls in our heaters

I. The Need for Greater Efficiency in Our Heaters
   A. Ethical Aspects
   B. Financial Aspects
II. The Current Method of Temperature Control: The Thermostat
   A. Principle of Operation
   B. Advantages
   C. Disadvantages
III. Alternative Methods of Temperature Control
   A. Rheostat
      1. Principle of Operation
      2. Advantages
      3. Disadvantages
   B. Zero-Voltage Control
      1. Principle of Operation
      2. Advantages
      3. Disadvantages
IV. Recommendation: Zero-Voltage Control
   A. Projected Developments in Semiconductor Technology
   B. Availability of Components
   C. Preliminary Design of Zero-Voltage Control System
   D. Schedule for Test Analysis
V. References

Notice that in this outline the writer has placed the discussion of the recommended alternative—zero-voltage control—after the discussions of the two other alternatives. The recommended alternative does not *have* to come last, but most readers will expect to see it in that position. This sequence *appears* to be the most logical—as if the writer had discarded the unsatisfactory alternatives until he finally hit upon the best one. In fact, we cannot know in what sequence he studied the different alternatives, or even whether he studied only one at a time. But the sequence of the outline gives the discussion a sense of forward momentum.

OUTLINE FORMAT

All of the outlines included so far in this chapter are topic outlines: that is, they consist only of phrases that suggest the topic to be discussed. You should know, however, about another kind of outline, the sentence outline, in which all of the items are complete sentences. Were the outline for the study of ways to increase the efficiency of temperature-control components in heaters a sentence outline, here is how Part IIIA would appear:

A. Temperature could be controlled by a rheostat.
   1. A rheostat operates according to the principle of variable resistance: adjustments to the rheostat would allow more or less current to flow in the heater by decreasing or increasing the amount of resistance in the line.
   2. The principal advantages of rheostats are that they are simple and reliable.
   3. The principal disadvantage of rheostats is their size: a rheostat capable of controlling the current in our 1500-watt heater would have to be as big as the heater itself.

Few writers use sentence outlines, because writing out all the sentences is time-consuming. In addition, most of the sentences will need to be revised later anyway. The main advantage of the sentence outline is that it keeps you honest: if you would be content to list a topic heading—such as "Advantages of the Proposed System"—without knowing what you're going to say, a sentence outline would force you to find out the necessary information. However, most people will happily cut corners and use a topic outline unless another person has specifically requested a sentence outline.

The other aspect of outlines that most writers aren't overly concerned about is the mechanical system for signifying the various levels of headings. Many writers, especially in private outlines, simply indent to set off subunits:

first-level heading
   second-level heading
      third-level heading

In the traditional notation system for formal outlines, numbers and letters go in front of the headings:

I.
  A.
    1.
    2.
  B.
    1.
      a.
      b.
      c.
        (1)
        (2)
          (a)
          (b)
    2.
  C. etc.

A popular variation is the decimal style, which is used by the military and often by scientists:

1.0
  1.1
    1.1.1
    1.1.2
  1.2
2.0 etc.

Whichever system helps you stay organized—or is required by your organization—is the one you ought to use.

THE STRATEGY OF THE ROUGH DRAFT

A rough draft is a preliminary version of the final document. Some rough drafts are rougher than others. An experienced writer might be able to write a draft, fix a comma here and change a word there, and have a perfectly serviceable document. On the other hand, every writer remembers the stubborn report that refused to make any sense at all after several drafts.

Many writers devise their own technique for revising, but for writing the first draft, the key is to write as fast as you can without lapsing into total gibberish. Some writers actually force themselves to write for a specified period, such as an hour, without stopping. Writing the rough draft is closer in spirit to brainstorming than to outlining. When you make up the outline, you are concentrating hard, trying to figure out precise relationships. When you write the draft, you are just trying to turn your outline into paragraphs as fast as you can.

Your goal is to get beyond "writer's block," the mental paralysis that sometimes sets in when you look down at a blank piece of paper. Write *something*: it will be easy to revise later. Once you get rolling, you will be able to see if your outline works. If you really don't know what to say about an item, you will probably want to stop and think it out. Perhaps you will revise your outline at this point. But the virtue of not stopping to worry about your writing is that you don't lose your momentum. Some writers are so insistent on keeping the rough draft flowing that they refuse to stop even when they can't figure out an item on their outlines; they just pick up a new piece of paper and start with the next item that *does* make sense.

A FEW WORDS ABOUT REVISION

Some writers can simply read their drafts and instantly recognize all the problems in them; others devise comprehensive checklists so they will not forget to ask themselves important questions.

Yet all writers agree on one thing: you should not revise your draft right after you have finished writing it. You can identify immediately some of the smaller writing problems—such as errors in spelling or grammar—but you cannot accurately assess the quality of what you have said: whether the information is clear, comprehensive, and coherent.

To be able to revise your draft effectively, you have to set it aside for a time before looking at it again. Give yourself at least a night's rest. If possible, work on something else for a few days. This will give you time to "forget" the draft and approach it more as your readers will. In a few minutes, you will see problems that would have escaped your attention even if you had spent hours revising it right after completing the draft.

How exactly do you revise a draft? Naturally, there is no single way; you should try to develop a technique that works for you.

One tip is to divide the task into several stages. Do not try to catch all the different kinds of errors by reading through the draft once. Make several passes through the draft, looking for a different kind of problem each time. For example, do not try to find spelling errors while you are checking to make sure the headings are clear and parallel.

As an aid in revising, many writers like to work from checklists. Most of the chapters in this book conclude with checklists. The chapter on proposals, for instance, includes a list of questions that you might want to use as a guide. Of course, you should adapt any checklist to the particular needs of your audience and purpose.

Regardless of whether you use checklists, your purpose in revising is to locate and eliminate any problems that will cut down on the effectiveness of your writing. Keep in mind the characteristics of effective technical writing as described in Chapter 1:

1. *Clarity*. Does the writing convey one meaning, and only one meaning, to your readers?
2. *Accuracy*. Are your facts and figures accurate? Is your presentation objective and unbiased?
3. *Accessibility*. Have you structured the document so that your readers can easily find the information they are seeking?
4. *Conciseness*. Have you reduced the length of the document to make it as concise as possible but still clear and comprehensive?
5. *Correctness*. Does your document abide by the conventions of grammar, punctuation, and usage? Are all the words spelled correctly?

Beware of that happy feeling of accomplishment that comes over you when you complete a draft. You will be fooling only yourself if you simply assume—or hope—that the draft is error-free. Many writers would like to think that their drafts do not need any revision. Every good writer knows that isn't true.

After you have revised your draft, give it to someone else to read. This will provide a more objective assessment of your writing. Choose a person who comes close to the profile of your eventual readers; someone more knowledgeable about the subject than your intended audience will understand the document even if it isn't clear and hence will not be sufficiently critical. After your colleague has read the document, find out what he or she thinks about it: strong points, unclear passages, sections that need to be added, deleted, or revised, and so forth. If possible, have your colleague read it as your eventual readers will: if the document is a set of instructions, see if he or she can perform the task.

EXERCISES

1. Choose a topic you are familiar with and interested in (such as some aspect of your academic requirements, a neighborhood concern, or some

issue of public policy). Brainstorm for ten minutes, listing as many items as you can that might be relevant in a report on the topic. Finally, after analyzing your list, arrange the items in an outline, discarding irrelevant items and adding necessary ones that occur to you.

2. For each of the following topics, think of a pattern of development that might be used. Defend each choice in a brief paragraph.

a. a brochure written for purchasers of a gasoline-engine model airplane
b. a report, addressed to the owner of an expensive stereo set, describing the repairs you performed
c. a proposal, addressed to a homeowner, on how to upgrade the energy efficiency of his house

3. The following portions of outlines contain logical flaws. In a sentence each, explain the flaws.

a. I. Advantages of collegiate football
  A. Fosters school spirit
  B. Teaches sportsmanship
II. Increases revenue for college
  A. From alumni gifts
  B. From media coverage

b. A. Effects of new draft law
  1. On Army personnel
  2. On males
  3. On females

c. I. Components of a Personal Computer
  A. Central Processing Unit
  B. External Storage Device
  C. Keyboard
  D. Magnetic Tape
  E. Disks
  F. Diskettes

d. A. Types of Common Screwdrivers
  1. Standard
  2. Phillips
  3. Ratchet-type
  4. Screw-holding tip
  5. Jeweler's
  6. Short-handled

e. A. Types of Health Care Facilities
  1. Hospitals
  2. Nursing Homes
  3. Care at Home
  4. Hospices

4. The following numbered statements have been taken from a formal outline. Reconstruct the outline, using all of these statements.

Writing Situation: to explain to the employees the reasons for instituting a fair system of evaluating job performance

1. The following are the characteristics of the present evaluation system.
2. One of the goals is to decrease employee turnover.
3. Currently, evaluations are performed at irregular intervals.
4. Ethical reasons motivate us to improve our evaluation system.
5. Evaluations should be carried out at regular intervals.
6. What are the purposes of an evaluation system?
7. Economic reasons compel us to improve our evaluation system.
8. The findings should be useful to both management and employees.
9. Following are the characteristics of a fair system of evaluating job performance.
10. The federal government forbids discrimination.
11. The objectives of the evaluation should be clear.
12. The number of legal actions against the city has increased.
13. Another goal of the system is to decrease absenteeism.
14. The criteria of the evaluation are subjective.
15. Why should we institute a fair system of evaluating job performance?
16. Legal reasons compel us to improve our system.
17. The criteria of evaluation should be objective.
18. Another purpose is to reward good job performance and discourage bad job performance.

CHAPTER FIVE

TECHNICAL WRITING STYLE

Perhaps you have heard it said that the best technical writing style is no style at all. This means simply that the readers should not be aware of your presence as a writer. They should not notice that you have a wonderful vocabulary or that your sentences flow beautifully—even if those things are true. In the best technical writing, the readers "notice" nothing. They are aware only of the information being conveyed. The writer fades completely into the background.

This is as it should be. Few people read technical writing for pleasure. Most readers either *must* read it as part of their work or want to keep abreast of new developments in the field. People read technical writing to gather information, not to appreciate the writer's flair. For this reason, experienced writers do not try to be fancy. The old saying has never been more appropriate: *Write to express, not to impress.*

The word *style*, as it is used in this chapter, encompasses word choice, sentence construction, and paragraph structure. Technical writing, like any other kind of writing, requires conscious decisions—about which words or phrases to use, what kinds of sentences to create, and how to turn those sentences into clear and coherent paragraphs.

DETERMINING THE APPROPRIATE STYLISTIC GUIDELINES

Most successful writers agree that the key to effective writing is revision: coming back to a draft and adding, deleting, and changing. Time permitting, an important document might go through four or five drafts before the writer finally has to stop and send it off to be typed. In revising, the writer will make many stylistic changes in an attempt to get closer and closer to the exact meaning he or she wishes to convey.

Some stylistic matters, however, can be determined before you start to write. Learning the "house style" that your organization follows will cut down the time needed for revision.

An organization's stylistic preferences may be defined explicitly in a company style guide that describes everything from how to write numbers to how to write the complimentary close at the end of a letter. In some organizations, an outside style manual, such as *The Chicago Manual of Style*, is the "rule book." (Figure 5-1 shows a page from the *U.S. Government Printing Office Style Manual*.) In many organizations, however, the stylistic preferences are implicit; no style manual exists, but over the course of years a set of unwritten guidelines has evolved. The best way to learn the unwritten house style is to study some letters, memos, and reports in the files and to ask more-experienced coworkers for explanations. Secretaries, in particular, often are valuable sources of information.

The following discussion covers three important stylistic matters about which organizations commonly have clear preferences.

ACTIVE AND PASSIVE VOICE

There are two voices: active and passive. In the active voice, the subject of the sentence performs the action expressed by the verb. In the passive voice, the subject is the recipient of the action. (In the following examples, the subjects are italicized.)

ACTIVE

Brushaw drove the launch vehicle.

PASSIVE

The launch *vehicle* was driven by Brushaw.

ACTIVE

Many *physicists* support the big-bang theory.

PASSIVE

The big-bang *theory* is supported by many physicists.

In most cases, the active voice is preferable to the passive voice. The active voice sentence more clearly emphasizes the actor. In addition, the active voice sentence is shorter, because it does not require a verbal and a prepositional phrase, as the passive voice sentence does. In the second example, for instance, the verb is "support," rather than "is supported," and "by" is unnecessary.

The passive voice, however, is generally more appropriate in four cases:

1. The actor is clear from the context.

EXAMPLE

Students are required to take both writing courses.

The context makes it clear that the college requires that students take both writing courses.

2. The actor is unknown.

EXAMPLE

The comet was first referred to in an ancient Egyptian text.

We don't know *who* referred to the comet.

3. The actor is unimportant

EXAMPLE

The documents were hand-delivered this morning.

It doesn't matter *who* the messenger was.

4. A reference to the actor is embarrassing, dangerous, or in some other way inappropriate.

EXAMPLE

Incorrect data were recorded for the flow rate.

It might be inappropriate to say *who* recorded the incorrect data.

Many people who otherwise take little interest in grammar have strong feelings about the relative merits of active and passive. A generation ago, students were taught that the active voice is inappropriate because it emphasizes the person who does the work rather than the work itself and thus robs the writing of objectivity. In many cases, this idea is valid. Why write, "I analyzed the sample for traces of iodine," when you can say, "The sample was ana-

FIGURE 5-1

From the U.S. Government Printing Office Style Manual

12. NUMERALS

(See also Tabular Work; Leaderwork)

12.1. Most rules for the use of numerals are based on the general principle that the reader comprehends numerals more readily than numerical word expressions, particularly in technical, scientific, or statistical matter. However, for special reasons numbers are spelled out in indicated instances.

12.2. The following rules cover the most common conditions that require a choice between the use of numerals and words. Some of them, however, are based on typographic appearance rather than on the general principle stated above.

12.3. Arabic numerals are generally preferable to Roman numerals.

NUMBERS EXPRESSED IN FIGURES

12.4. A figure is used for a single number of *10* or more with the exception of the first word of the sentence. (See also rules 12.9, 12.23.)

50 ballots	24 horses	about 40 men
10 guns	nearly 10 miles	10 times as large

Numbers and numbers in series

☆ **12.5.** Figures are used in a group of 2 or more numbers, or for related numbers, any one of which is *10* or more. The sentence will be regarded as a unit for the use of figures.

lyzed for traces of iodine"? If there is no ambiguity about who did the analysis, or if it is not necessary to identify who did the analysis, a focus on the action being performed is appropriate.

Supporters of the active voice argue that the passive voice creates a double ambiguity. When you write, "The sample was analyzed for traces of iodine," your reader is not quite sure who did the analysis (you or someone else) or when it was done (as part of the project being described or some time previously). Even though a passive voice sentence can contain all the information found in its active voice counterpart, often the writer omits the actor.

The best approach to the active-passive problem is to recognize how the two voices differ and use them appropriately.

Each of 15 major commodities (9 metal and 6 nonmetal) was in supply.
but Each of nine major commodities (five metal and four nonmetal) was in supply.

Petroleum came from 16 fields, of which 8 were discovered in 1956.
but Petroleum came from nine fields, of which eight were discovered in 1956.

That man has 3 suits, 2 pairs of shoes, and 12 pairs of socks.
but That man has three suits, two pairs of shoes, and four hats.

Of the 13 engine producers, 6 were farm equipment manufacturers, 6 were principally engaged in the production of other types of machinery, and 1 was not classified in the machinery industry.
but Only nine of these were among the large manufacturing companies, and only three were among the largest concerns.

There were three 6-room houses, five 4-room houses, and three 2-room cottages, and they were built by 20 men. (See rule 12.21.)
There were three six-room houses, five four-room houses, and three two-room cottages, and they were built by nine men.
Only 4 companies in the metals group appear on the list, whereas the 1947 census shows at least 4,400 establishments.
but If two columns of sums of money add or subtract one into the other and one carries points and ciphers, the other should also carry points and ciphers.
At the hearing, only one Senator and one Congressman testified.
There are four or five things which can be done.

179

AWKWARD

He lifted the cage door, and a hungry mouse was seen.

BETTER

He lifted the cage door and saw a hungry mouse.

AWKWARD

The new catalyst produced good-quality foam, and a flatter mold was caused by the new chute-opening size.

BETTER

The new catalyst produced good-quality foam, and the new chute-opening size resulted in a flatter mold.

FIRST, SECOND, AND THIRD PERSON

Closely related to the question of voice is the question of person. The term *person* refers to the different forms of the personal pronoun:

FIRST PERSON

I worked . . . , we worked . . .

SECOND PERSON

You worked . . .

THIRD PERSON

He worked . . . , she worked . . . , it worked . . . ,
they worked . . .

Organizations that prefer the active voice generally encourage the use of the first-person pronouns: "We analyzed the rate of flow." Organizations that prefer the passive voice often *prohibit* the use of the first-person pronouns: "The rate of flow was analyzed."

Another question of person that often arises is whether to use the second or the third person in instructions. In some organizations, instructional material—step-by-step procedures—is written in the second person: "You begin by locating the ON/OFF switch." The second person is concise and easy to understand. Other organizations prefer the more formal third person: "The operator begins by locating the ON/OFF switch." Perhaps the most popular version is the second person in the imperative: "Begin by locating the ON/OFF switch." In the imperative, the *you* is implicit. Regardless of the preferred style, however, be consistent in your use of the personal pronoun.

SEXIST LANGUAGE

Sexist language favors one sex at the expense of the other. Although sexist language can shortchange males—as in some writing about female-dominated professions such as nursing—in an overwhelming majority of cases the female is victimized. Common examples include nouns such as *workman* and *chairman* and pronouns as used in the sentence "Each worker is responsible for his work area."

In some organizations, the problem of sexist language is still considered trivial; in internal memos and reports, sexist language is used freely. In most organizations, however, sexist language is a serious matter. Unfortunately, it is not easy to eliminate all gender bias from writing.

A number of male-gender words have no standard nongender substitutes, and there is simply no graceful way to get around the pronoun *he*. However, many organizations have formulated guidelines in an attempt to reduce sexist language.

The relatively simple first step is to eliminate the male-gender words. *Chairman*, for instance, is being replaced by *chairperson* or *chair*. *Firemen* are *firefighters*, *policemen* are *police officers*.

Rewording a sentence to eliminate masculine pronouns is also effective.

SEXIST

The operator should make sure he logs in.

NONSEXIST

The operator should make sure to log in.

In this revision, an infinitive replaces the *he* clause.

NONSEXIST

Operators should make sure they log in.

In this revision, the masculine pronoun is eliminated through a switch from singular to plural.

Notice that sometimes the plural can be unclear:

UNCLEAR

Operators are responsible for their operating manuals.

Does each operator have one operating manual or more than one?

CLEAR

Each operator is responsible for his or her operating manual.

In this revision, "his or her" clarifies the meaning. *He or she* and *his or her* are awkward, especially if overused, but they are at least clear.

In some quarters, the problem of sexist language is side-stepped. The writer simply claims innocence: "The use of the pronoun *he* does not in any way suggest a male bias." Many readers find this kind of approach equivalent to that of a man who enters a crowded elevator, announces that he knows it is rude to smoke, and then proceeds to light up a cheap cigar. Sexism in language is not a trivial matter, although most people would agree that eliminating sexual discrimination regarding salaries, benefits, and promotions deserves a higher priority. Still, sexist language reinforces more active forms of sexual discrimination and thus perpetuates inequality in the whole fabric of our society.

CHOOSING THE RIGHT WORDS AND PHRASES

Choosing the right words and phrases is, of course, as important as choosing the appropriate voice or person or avoiding sexist language. Choosing the best word or phrase to convey what you want to say is a different kind of problem, however; it involves constant, sustained concentration. Your organization's style guide can't help you. You're on your own. Most writers don't worry about word choice when writing a first draft, but for subsequent drafts they are always on the lookout for the better word.

The following discussion includes seven basic guidelines for choosing the right word:

1. Be specific.
2. Avoid unnecessary jargon.
3. Avoid wordy phrases.
4. Avoid clichés.
5. Avoid pompous words.
6. Focus on the "real" subject.
7. Focus on the "real" verb.

BE SPECIFIC Being specific involves using precise words, providing adequate detail, and avoiding ambiguity.

Wherever possible, use the most precise word you can. A Ford Mustang is an automobile, but it is also a vehicle, a machine, and a thing. In describing the Ford Mustang, the word *automobile* is better than *vehicle*, because the less specific word also refers to trains, hot-air balloons, and other means of transport. As the words become more abstract—from *machine* to *thing*, for instance—the chances for misunderstanding increase.

In addition to using the most precise words you can, be sure to provide enough detail. Remember that the reader knows less than you do. What might be perfectly clear to you might be too vague for the reader.

VAGUE

An engine on the plane experienced some difficulties.

What engine? What plane? What difficulties?

CLEAR

The left engine on the Jetson 411 unaccountably lost power during flight.

Avoid ambiguity. That is, don't let the reader wonder which of two meanings you are trying to convey.

AMBIGUOUS

After stirring by hand for ten seconds, add three drops of the iodine mixture to the solution.

Stir the iodine mixture or the solution?

CLEAR

Stir the iodine mixture by hand for ten seconds. Then add three drops to the solution.

CLEAR

Stir the solution by hand for ten seconds. Then add three drops of the iodine mixture.

What should you do if you don't have the specific data? You have two options: to approximate—and clearly tell the reader you are doing so—or to explain why the specific data are unavailable and indicate when they will become available.

VAGUE

The leakage in the fuel system is much greater than we had anticipated.

CLEAR

The leakage in the fuel system is much greater than we had anticipated; we estimate it to be at least five gallons per minute, rather than two.

AVOID UNNECESSARY JARGON

Jargon is shoptalk. The term *fish eyes* means something to workers in the plastics industry that it doesn't mean to everyone else, just as *CAFE* is a meaningful acronym among auto executives. Although jargon is often held up to ridicule, it is a useful and natural kind of communication in its proper sphere. Two baseball pitchers would find it hard to talk to each other about their craft if they couldn't use terms such as *slider* and *curve*.

In one sense, the abuse of jargon is simply a needless corruption of the language. The best current example of this phenomenon is the degree to which computer science terminology has crept into everyday English. In offices, employees are frequently asked to provide "feedback" or told that the coffee machine is "down."

To many people, such words seem dehumanizing; to others, they seem silly. What is a natural and clear expression to one person is often strange or confusing to another. Communication breaks down.

An additional danger of using jargon outside a very limited professional circle is that it sounds condescending to many people, as if the writer is showing off—displaying a level of expertise that excludes most readers. While the readers are concentrating on how much they dislike the writer, they are missing the message.

If some readers are offended by unnecessary jargon, others are intimidated. They feel somehow inadequate or stupid because they do not know what the writer is talking about. When the writer casually tosses in jargon, many readers really *can't* understand what is being said.

If you are addressing a technically knowledgeable audience, feel free to use appropriate jargon. However, an audience that includes managers or the general public will probably have trouble with specialized vocabulary. If your document has separate sections for different audiences—as in the case of a technical report with an executive summary—use jargon accordingly. A glossary (list of definitions) is useful if you suspect that the technical sections will be read by the managers.

AVOID WORDY PHRASES

Wordy phrases weaken technical writing by making it unnecessarily long. Sometimes writers deliberately choose phrases such as "demonstrates a tendency to" rather than "tends to." The long phrase rolls off the tongue easily and appears to carry the weight of scientific truth. But the humble "tends to" says the same thing—and says it better for having done so concisely. The sentence "We can do it" is the concise version of "We possess the capability to achieve it."

Some wordy phrases just pop into writers' minds. We are all so used to hearing *take into consideration* that we don't realize that *consider* gets us there faster. Replacing wordy phrases with concise ones is therefore more difficult than it might seem. Avoiding the temptation to write the long phrase is only half of the solution. The other half is to try to root out the long phrase that has infiltrated the prose unnoticed.

Following are a wordy sentence and a concise translation.

WORDY

I am of the opinion that, in regard to profit achievement, the statistics pertaining to this month will appear to indicate an upward tendency.

CONCISE

I think this month's statistics will show an increase in profits.

A special kind of wordiness to watch out for is unnecessary redundancy, as in *end result*, *any and all*, *each and every*, *completely eliminate*, and *very unique. Be content to say something once. Use "The liquid is green," not "The liquid is green in color."*

REDUNDANT

We initially began our investigative analysis with a sample that was spherical in shape and heavy in weight.

BETTER

We began our analysis with a heavy, spherical sample.

AVOID CLICHÉS

The English writer George Orwell once offered some good advice about writing: "Never use a metaphor, simile or other figure of speech which you are used to seeing in print." Rather than writing, "It's a whole new ball game," write, "The situation has changed completely." Don't write, "I am sure the new manager can cut the mustard"; write, "I am sure the new manager can do his job effectively." The use of clichés suggests laziness.

Sometimes, writers further embarrass themselves by getting their clichés wrong: expressions become so timeworn that users forget what the words mean. The phrase "a new bag of worms" has found its way into print; the producer Sam Goldwyn was famous for such howlers as "An oral agreement isn't worth the paper it's written on." And the phrase "I could care less" often is used when the writer means just the opposite. The best solution to this problem is, of course, not to use clichés.

Following are a cliché-filled sentence and a translation into plain English.

TRITE

Afraid that we were between a rock and a hard place, we decided to throw caution to the winds with a grandstand play that would catch our competition with its pants down.

PLAIN

Afraid that were we in a hopeless situation, we decided on a risky and aggressive move that would surprise our competition.

AVOID POMPOUS WORDS

Writers sometimes try to impress their readers by using pompous words, such as *initiate* for *begin*, *perform* for *do*, and *prioritize* for *rank*. When asked why they use big words where small ones will do, writers say that they want to make sure their readers know they have a strong vocabulary, that they are well educated.

Undoubtedly, pompous words have a role in some kinds of communication. Sports commentator Howard Cosell and conservative columnist William F. Buckley, Jr., both understand the comic potential of pomposity. But in technical writing, plain talk is best. If you know what you're talking about, be direct and simple. Even if you're not so sure of what you're talking about, say it plainly; big words won't fool anyone for more than a few seconds.

Following are a few pompous sentences translated into plain English.

POMPOUS

The purchase of a minicomputer will enhance our record maintenance capabilities.

PLAIN

Buying a minicomputer will help us maintain our records.

POMPOUS

It is the belief of the Accounting Department that the predicament was precipitated by a computational inaccuracy.

PLAIN

The Accounting Department thinks a math error caused the problem.

In the long run, your readers will be impressed by your clarity and accuracy. Don't waste your time thinking up fancy words.

FOCUS ON THE "REAL" SUBJECT

The conceptual or "real" subject of the sentence should also be the sentence's grammatical subject, and it should appear prominently in technical writing. Don't bury the real subject of the sentence in a prepositional phrase following a useless or "limp" grammatical subject. In the following examples, notice how the limp subjects disguise the real subjects. (The grammatical subjects are italicized.)

WEAK

The *use* of this method would eliminate the problem of motor damage.

STRONG

This *method* would eliminate the problem of motor damage.

WEAK

The *presence* of a six-membered lactone ring was detected.

STRONG

A six-membered lactone *ring* was detected.

Another way to make the subject of the sentence prominent is to eliminate grammatical expletives. You can almost always remove expletive constructions—*it is . . .* , *there is . . .* , and *there are . . .*—without changing the meaning of the sentence.

WEAK

There are many problems that must be worked out.

STRONG

Many problems must be worked out.

WEAK

It is with great pleasure that I welcome you to our annual development seminar.

STRONG

With great pleasure I welcome you to our annual development seminar.

STRONG

I am pleased to welcome you to our annual development seminar.

FOCUS ON THE "REAL" VERB

A "real" verb, like a "real" subject, should be prominent in every sentence. Few stylistic problems weaken a sentence more than nominalizing verbs. To nominalize the real verb, you convert it into a noun; you must then supply another, usually weaker, verb to convey your meaning. "To install" becomes "to effect an installation," "to analyze" becomes "to conduct an analysis." Notice how nominalizing the real verbs makes the following sentences both awkward and unnecessarily long. (The nominalized verbs are italicized.)

WEAK

Each *preparation* of the solution is done twice.

STRONG

Each solution is prepared twice.

WEAK

An *investigation* of all possible alternatives was undertaken.

STRONG

All possible alternatives were investigated.

WEAK

Consideration should be given to an acquisition of the properties.

STRONG

We should consider acquiring the properties.

Watch out for the *-tion* endings; they generally signify a nominalized verb that can be eliminated with a little rewriting.

SENTENCE STRUCTURE AND LENGTH

In addition to recognizing word and phrase problems that can weaken a sentence, you should also understand how to use different sentence structures and lengths to make your writing more effective.

TYPES OF SENTENCES

There are four basic types of sentences:

1. simple (one independent clause)

The technician soon discovered the problem.

2. compound (two independent clauses, linked by a semicolon or by a comma and one of the seven common coordinating conjunctions: *and*, *or*, *nor*, *for*, *so*, *yet*, and *but*)

The technician soon discovered the problem, but he found it difficult to solve.

3. complex (one independent clause and at least one dependent clause, linked by a subordinating conjunction)

Although the technician soon discovered the problem, he found it difficult to solve.

4. compound-complex (at least two independent clauses and at least one dependent clause)

Although the technician soon discovered the problem, he found it difficult to solve, and he finally determined that he needed some additional parts.

Most commonly used in technical writing is the simple sentence, because it is clear and direct. However, a series of three or four simple sentences can bore and distract the reader. The main shortcoming of the simple sentence is that it can communicate only one basic idea, because it is made up of only one independent clause.

Compound and complex sentences communicate more sophisticated ideas. The compound sentence works on the principle of coordination: the two halves of the sentence are roughly equivalent in importance. The complex sentence uses subordination: one half of the sentence is less important than the other.

Compound-complex sentences are useful in communicating very complicated ideas, but their length makes them unsuitable for most kinds of technical writing.

One common construction deserves special mention: the compound sentence linked by the coordinating conjunction *and*. In many cases, this construction is a lazy way of linking two thoughts that could be linked more securely. For instance, the sentence "Enrollment is up, and the school is planning new course offerings" represents a weak use of the *and* conjunction, because it doesn't show the cause-effect relationship between the two clauses. A stronger link would be *so*: "Enrollment is up, so the school is planning new course offerings." A complex sentence would sharpen the relationship even more: "Because enrollment is up, the school is planning new course offerings." Another example of the weak *and* conjunction is the sentence "The wires were inspected, and none was found to be damaged." In this case, a simple sentence would be more effective: "The inspection of the wires revealed that none was damaged" or "None of the inspected wires was damaged."

SENTENCE LENGTH

Although it would be artificial and distracting to keep a count of the number of words in your sentences, sometimes sentence length can work against effective communication. In revising a draft, you might want to compute an average sentence length for a page of writing.

There are no firm guidelines covering appropriate sentence length. In general, a length of 15 to 20 words is effective for technical writing. A succession of 10-word sentences would be abrupt and choppy; a series of 35-word sentences would probably be too demanding.

The best approach to determining an effective sentence length is to consider the writing situation—the audience and the purpose.

AUDIENCE

The more the readers know about the subject, the more easily they will be able to handle longer sentences.

PURPOSE

If you are writing a set of instructions or some other kind of information that your reader will be working from directly, short sentences are more effective. In addition, you should further decrease sentence length to emphasize a particularly important point.

The following four examples show how the same information can be conveyed in sentences of varying type and length.

The ruling on dumping was expected to have widespread implications for the chemical industry. However, problems in interpreting the ruling have led to protracted legal battles. As a result, the government has not yet won a single conviction.

This example contains three simple sentences, of 14, 12, and 12 words. This version would be suitable for most general audiences.

The ruling on dumping was expected to have widespread implications for the chemical industry, but problems in interpreting the ruling have led to protracted legal battles. As a result, the government has not yet won a single conviction.

In this version, the first two simple sentences have been combined into a compound sentence—two independent clauses linked by a coordinating conjunction—of 26 words. Although the material has remained the same, this version is more difficult than the previous one because it asks the readers to remember more information as they read the longer first sentence.

The ruling on dumping was expected to have widespread implications for the chemical industry. Because problems in interpreting the ruling have led to protracted legal battles, however, the government has not yet won a single conviction.

Here, the second and third sentences of the first version have been combined in a complex sentence. The original second sentence is now a dependent clause; the original third sentence is now an independent clause. This version, of 14 and 22 words, is about as difficult as the second version.

Although the ruling on dumping was expected to have widespread implications for the chemical industry, problems in interpreting the ruling have led to protracted legal battles, and the government has not yet won a single conviction.

In this version, the three original sentences have been combined into a 36-word compound-complex sentence: a dependent clause and two independent clauses. This is the most sophisticated and most demanding version of the information.

A NOTE ON READABILITY FORMULAS

Readability formulas are mathematical techniques used to determine how "readable" a piece of writing is; that is, how difficult it is for someone to read and understand it. More than a dozen different readability formulas exist, and they are being used increasingly by government agencies and private businesses in an attempt to improve writing effectiveness. For this reason, you should become familiar with the concept behind them.

Most readability formulas are based on the idea that short words and sentences are easier to understand than long ones. One of the more popular formulas, Robert Gunning's "Fog Index," works like this:

1. Find the average number of words per sentence, using a 100-word passage.
2. Find the number of "difficult words" in that same passage. "Difficult words" are words of three or more syllables, except for proper names, combinations of simple words (such as *manpower*), and verbs whose third syllable is *-es* or *-ed* (such as *contracted*).
3. Add the average sentence length and the number of "difficult words."
4. Multiply this sum by 0.4.

The figure obtained from this procedure represents the approximate grade level that the reader must have reached in order to understand the writing.

Average number of words per sentence:	13.9
Number of difficult words:	16
	$29.9 \times 0.4 = 12$

The average twelfth-grade student could understand the writing.

Readability formulas are easy to use, and it is appealing to think you can be objective in assessing your writing, but unfortunately they have not been proven to work. They simply have not been shown to reflect accurately how difficult it is to read a piece of writing (Selzer 1981; Battison and Goswami 1981). The problem with readability formulas is that they attempt to evaluate words on paper—without considering the reader. It is possible to measure how well a specific person understands a specific writing sample. But words on paper don't communicate until somebody reads them, and everyone is different. A "difficult word" for a lawyer might not be difficult for a biologist, and vice versa. And someone who is interested in the subject being discussed will understand more than the reluctant reader will.

Readability formulas offer a false sense of security by suggesting that short words and sentences will make writing easy to read. Things are not so simple. Good writing has to be well thought out and carefully structured. The sentences have to be clear, and the vocabulary has to be appropriate for the readers. *Then* the writing will be readable. Nonetheless, many organizations use readability formulas, and you should be familiar with them.

USING HEADINGS AND LISTS

Much of the preceding discussion of style applies to any kind of nonfiction writing. Headings and lists, although not unique to technical writing, are a major stylistic feature of reports, memos, and letters. Although headings—and to a lesser extent, lists—might at first appear to be mechanical elements that need be considered only in the final stages of writing, they are in fact fundamental, and their use determines content as well as form.

HEADINGS

The main purpose of a heading is to announce the subject of the discussion that follows it: the word HEADINGS, for instance, tells you that this discussion will cover the subject of headings. For the writer, the use of headings eliminates the need to announce the subject in a sentence such as "Let us now turn to the subject of headings."

A second important purpose of a heading is to clarify for the reader the hierarchical relationships within the document. This text, for example, has four main parts, each of which is subdivided into chapters. Each chapter, in turn, is subdivided into major units (such as **USING HEADINGS AND LISTS**). Most of the major units are subdivided again into smaller units (such as HEADINGS). Understanding the level of importance of the various discussion components helps the reader concentrate on the discussion itself.

To signify different hierarchical levels in headings in typewritten manuscript, use capitalization, underlining, and indentation. Capital (uppercase) letters are more emphatic than small (lowercase) letters; therefore, an all-capitals heading (PROCEDURE) signals a more important category of information than an initial-capital heading (Procedure). Similarly, underlined headings (PROCEDURE) are more emphatic than nonunderlined headings (PROCEDURE). Finally, a heading centered horizontally on the page is more emphatic than one that begins at the left margin. And a heading that begins at the left margin is more emphatic than one that is run into a paragraph.

Figure 5–2 shows how capitalization, underlining, and indentation can be used to create headings that show different levels of hierarchy.

FIGURE 5–2

A Four-Level Heading System

FIRST LEVEL	REPAIRING ASPHALT STREETS
SECOND LEVEL	INTRODUCTION
	This manual is intended ....
	....
SECOND LEVEL	TYPES OF DEFECTS
	Three basic types of defects in asphalt streets ....
	....
THIRD LEVEL	Cracks
	A crack is defined as ....
	....
FOURTH LEVEL	Identifying Cracks. To identify a crack, ...
	....
FOURTH LEVEL	Repairing Cracks. The first step ....
	....
THIRD LEVEL	Grade Depressions
	A grade depression is defined as ....
	....

FOURTH LEVEL Identifying Grade Depressions. To identify a grade depression, ...

....

FOURTH LEVEL Repairing Grade Depressions. The first step ....

....

THIRD LEVEL Potholes

A pothole is defined as ....

....

FOURTH LEVEL Identifying Potholes. To identify a pothole, ...

....

FOURTH LEVEL Repairing Potholes. The first step ....

....

To further reinforce the hierarchical levels, you can add a numbering scheme, such as the traditional outline system or the decimal system:

```
I.        1.
  A.        1.1
    1.        1.1.1
    2.        1.1.2
  B.        1.2
    1.        1.2.1
    2.        1.2.2
      a.        1.2.1.1
      b.        1.2.1.2
        (1)       1.2.1.1.1
        (2)       1.2.1.1.2
```

Notice that the decimal system can easily accommodate a large number of hierarchical levels—an advantage when you are writing complicated technical documents.

When you make up the headings, keep in mind that you are not restricted to simple phrases. Your assessment of the writing situation might indicate that a heading such as "What Is Wrong with the Present System?" would be more effective than one such as "Problem." Although general terms such as *problem*, *results*, and *conclusions* are well known by engineers and scientists, general readers and upper-level managers might have an easier time with more-informative phrases or questions.

LISTS

Like headings, lists let you manipulate the placement of words on the page to improve the effectiveness of the communication.

Many sentences in technical writing are long and complicated:

We recommend that more work on heat-exchanger performance be done with a larger variety of different fuels at the same temperature, with similar fuels at different temperatures, and with special fuels such as diesel fuel and shale-oil-derived fuels.

The difficulty in this sentence is that the readers cannot concentrate on the information because they must worry about remembering all the *with* phrases following "done." If they could "see" how many phrases they had to remember, their job would be easier.

Revised as a list, the sentence is easier to follow:

We recommend that more work on heat-exchanger performance be done:
1. with a larger variety of different fuels at the same temperature
2. with similar fuels at different temperatures
3. with special fuels such as diesel fuels and shale-oil-derived fuels

In this version, the placement of the words on the page reinforces the meaning. The readers can easily see that the sentence contains three items in a series. And the fact that each item begins at the same left margin helps, too.

Make sure the items in the list are presented in a parallel structure. (See Appendix A for a discussion of parallelism.)

NONPARALLEL

Here is the schedule we plan to follow:
1. construction of the preliminary proposal
2. do library research
3. interview with the Bemco vice-president
4. first draft
5. revision of the first draft
6. after we get your approval, typing of the final draft

PARALLEL

Here is the schedule we plan to follow:
1. write the preliminary proposal
2. do library research
3. interview the Bemco vice-president
4. write the first draft
5. revise the first draft
6. type the final draft, after we receive your approval

In this example, the original version of the list is sloppy, a mixture of noun phrases (items 1, 3, 4, and 5), a verb phrase (item 2), and a participial phrase preceded by a dependent clause (item 6). The revision uses parallel verb phrases and deemphasizes the dependent clause in item 6 by placing it after the verb phrase.

Note that reports, memos, and letters do not have to look "formal," with traditional sentences and paragraphs covering the whole page. Headings and lists make writing easier to read and understand.

PARAGRAPH STRUCTURE AND LENGTH

A paragraph can be defined as a group of sentences (or sometimes a single sentence) that is complete and self-sufficient but that also contributes to a larger discussion. The challenge of creating an effective paragraph in technical writing is to make sure, first, that all of the sentences clearly and directly substantiate one main point, and second, that the whole paragraph follows logically from the material that precedes it. Readers tend to pause between paragraphs (not between sentences) to digest the information given in one paragraph and link it with that given in the previous paragraphs. For this reason, the paragraph is the key unit of composition. Readers might forgive or at least overlook a slightly fuzzy sentence. But if they can't figure out what a paragraph says or why it appears where it does, the communication process is likely to break down completely.

PARAGRAPH STRUCTURE

Too often in technical writing, paragraphs look as if they were written for the writer, not the reader. They start off with a number of details: about who worked on the problem before and what equipment or procedure they used; about the ups and downs of the project, the successes and setbacks; about specifications, dimensions, and computations. The paragraph winds its way down the page until, finally, the writer concludes: "No problems were found."

This structure—moving from the particular details to the general statement—accurately reflects the way the writer carried out the activity he or she is describing, but it makes the paragraph difficult to follow. As you put a paragraph together, focus on your readers' needs. Do they want to "experience" your writing, to regret your disappointments and to celebrate your successes? Probably not. They just want to find out what you have to say.

Help your readers. Put the point—the topic sentence—up front. Technical writing should be clear and easy to read, not full of suspense. If a paragraph describes a test you performed on a piece of equipment, include the result in your first sentence: "The point-to-point continuity test on Cabinet 3 revealed no problems." Then go on to explain the details. If the paragraph describes a complicated idea, start with an overview: "Mitosis occurs in five stages: (1) interphase, (2) prophase, (3) metaphase, (4) anaphase, and (5) telophase." Then describe each stage. In other words, put the "bottom line" on top.

Notice, for instance, how difficult the following paragraph is, because the writer structured the discussion in the same order she performed her calculations:

> Our estimates are based on our generating power during eight months of the year and purchasing it the other four. Based on the 1982 purchased power rate of $0.034/KW (January through April cost data) inflating at 8% annually, and a constant coal cost of $45–50, the projected 1985 savings resulting from a conversion to coal would be $225,000.

Putting the bottom line on top makes the paragraph much easier to read. Notice how the writer adds a numbered list after the topic sentence.

> The projected 1985 savings resulting from a conversion to coal are $225,000. This estimate is based on three assumptions: (1) that we will be generating power during eight months of the year and purchasing it the other four, (2) that power rates inflate at 8% from the 1982 figure of $0.034/KW (January through April cost data), and (3) that coal costs remain constant at $45–50.

The topic sentence in technical writing functions just as it does in any other kind of writing: it summarizes or forecasts the main point of the paragraph.

Why don't writers automatically begin with the topic sentence if that helps the readers understand the paragraph? One reason is that it is easier to present the events in their natural order, without first having to decide what the single most important point is. Perhaps a more common reason is that writers feel uncomfortable "exposing" the topic sentence at the start of a paragraph. Most people who do technical writing were trained in science or engineering or technology; they were taught that they must not reach a conclusion before obtaining sufficient evidence to back it up. Putting the topic sentence up top *looks* like stating a conclusion without "proving" it, even though the proof follows the topic sentence directly in the rest of the paragraph. Beginning the paragraph with the sentence "It was concluded that human error caused the overflow" somehow seems risky. It isn't. The real risk is that you might frustrate or bore your readers by making them hunt for the topic sentence.

After the topic sentence comes the support. The purpose of the support is to make the topic sentence clear and convincing. Sometimes a few explanatory details can provide all the support needed. In the paragraph about estimated fuel savings presented earlier, for example, the writer simply fills in the assumptions she used in making her calculation: the current energy rates, the inflation rate, and so forth. Sometimes, however, the support must carry a heavier load: it has to clarify a difficult thought or defend a controversial one.

Because every paragraph is unique, it is impossible to define the exact function of the support. In general, however, the support fulfills one of the following roles:

1. to define a key term or idea included in the topic sentence
2. to provide examples or illustrations of the situation described in the topic sentence
3. to identify factors that led to the situation described in the topic sentence
4. to define implications of the situation described in the topic sentence
5. to defend the assertion made in the topic sentence

The techniques used in developing the support include those used in most nonfiction writing: definition, comparison and contrast, classification and partition, and causal analysis. These techniques are described fully in Part II.

PARAGRAPH LENGTH

How long should a paragraph of technical writing be? In general, a length of 75 to 125 words will provide enough space for a topic sentence and four or five supporting sentences. Long paragraphs are more difficult to read than short paragraphs, for the simple reason that the readers have to concentrate longer. A second factor that makes shorter paragraphs preferable is that many readers are intimidated by unbroken stretches of type. Some readers actually will skip over long paragraphs.

Don't let an arbitrary guideline about length take precedence over your analysis of the writing situation. Often you will need to write very brief paragraphs. You might need only one or two sentences—to introduce a graphic aid, for example. A transitional paragraph—one that links two other paragraphs—also is likely to be quite short. If a brief paragraph fulfills its function, let it be. Do not combine two ideas in one paragraph in order to achieve a minimum word count.

While it is confusing to include more than one basic idea in a paragraph, the concept of unity can be violated in the other direction. Often you will find it necessary to divide one idea into two or more paragraphs. A complex idea that would require 200 or 300 words probably should not be squeezed into one paragraph.

The following example shows how a writer addressing a general audience divided one long paragraph into two (Marion 1974: 82–83).

> If we could find a way to make efficient use of solar energy, we would have a continuing "free" supply of energy which would not degrade our environment and which would lift at least a portion of the burden on our nonrenewable fuel supplies. The primary problem associated with utilizing solar energy is that the energy is spread thinly over the Earth and is variable due to local weather conditions and the regular day-night cycle. In the relatively cloudless desert regions of the southwestern United States, for example, the rate at which solar energy reaches the Earth's surface during the 6 to 8 hours around mid-day is about 0.8kW/m^2. The energy absorbed per square meter per year amounts to about 2000 kWh.
>
> The second problem is that only a small fraction of the absorbed solar energy can actually be converted into electrical energy. Estimates of the conversion efficiency for proposed systems are about 10%. That is, the annual absorbed solar energy per square meter on the surface of the Earth represents about 200 kWh of electrical energy under favorable weather conditions.

A strict approach to paragraphing would have required one paragraph, not two, because all of the information presented provides support for the topic sentence that begins the first paragraph. Many readers, in fact, could easily understand a one-paragraph version. However, the writer found a logical place to create a second paragraph and thereby increased the effectiveness of his communication.

Another writer might have approached the problem differently, making each "problem with solar energy" a separate paragraph.

> If we could find a way to make efficient use of solar energy, we would have a continuing "free" supply of energy which would not degrade our environment and which would lift at least a portion of the burden on our nonrenewable fuel supplies.
>
> The primary problem associated..
> ....
> ....
>
> The second problem...
> ....
> ....

The original topic sentence becomes a transitional paragraph that leads clearly and logically into the two explanatory paragraphs.

MAINTAINING COHERENCE WITHIN AND BETWEEN PARAGRAPHS

After you have blocked out the main structure of the paragraph—the topic sentence and the support—make sure the paragraph is coherent. In a coherent paragraph, thoughts are linked together logically and clearly. If the paragraph moves smoothly from sentence to sentence, emphasize the coherence by adding transitional words and phrases, repeating key words, and using demonstratives.

Maintaining coherence *between* paragraphs is the same process as maintaining coherence *within* paragraphs: place the transitional device as close as possible to the beginning of the second element. For example, the link between two sentences within a paragraph should be near the start of the second sentence:

> The new embossing machine was found to be defective. *However*, the warranty on the machine will cover replacement costs.

The link between the two paragraphs should be near the start of the second paragraph:

> The complete system would be too expensive for us to purchase now...
> ....
> ....

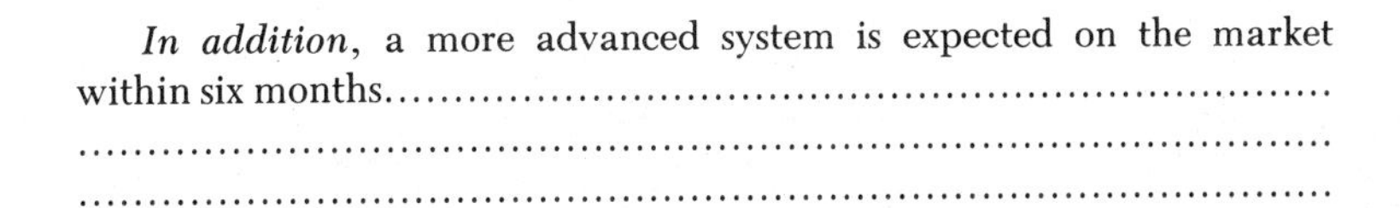

In addition, a more advanced system is expected on the market within six months..
....
....

Transitional words and phrases help the reader understand a discussion by pointing out the direction the thoughts are following. Here is a list of the most common logical relationships between two thoughts and some of the common transitions that express those relationships:

RELATIONSHIP	TRANSITIONS
addition	also, and, finally, first (second, etc.), furthermore, in addition, likewise, moreover, similarly
comparison	in the same way, likewise, similarly
contrast	although, but, however, nevertheless, on the other hand, yet
illustration	for example, for instance, in other words
cause-effect	as a result, because, consequently, hence, so, therefore, thus
time or space	above, around, earlier, later, next, to the right (left, west, etc.), soon, then
summary or conclusion	at last, finally, in conclusion, to conclude, to summarize

In the following examples, the first versions contain no transitional words and phrases. Notice how much clearer the second versions are.

WEAK

Neurons are not the only kind of cell in the brain. Blood cells supply oxygen and nutrients.

IMPROVED

Neurons are not the only kind of cell in the brain. *For example*, blood cells supply oxygen and nutrients.

WEAK

The project was originally expected to cost $300,000. The final cost was $450,000.

IMPROVED

The project was originally expected to cost $300,000. *However*, the final cost was $450,000.

WEAK

The manatee population of Florida has been stricken by an unknown disease. Marine biologists from across the nation have come to Florida to assist in manatee-disease research.

IMPROVED

The manatee population of Florida has been stricken by an unknown disease. *As a result*, marine biologists from across the nation have come to Florida to assist in manatee-disease research.

Repetition of key words—generally, nouns—helps the reader follow the discussion. Notice in the following example how the first version can be confusing.

UNCLEAR

For months the project leaders carefully planned their research. The cost of the work was estimated to be over $200,000. (*What is the* work, *the planning or the research?*)

CLEAR

For months the project leaders carefully planned their *research*. The cost of the *research* was estimated to be over $200,000.

Out of a misguided desire to be "interesting," some writers keep changing their important terms. *Plankton* becomes *miniature seaweed*, then *the ocean's fast food*. Leave this kind of word game to TV sportscasters; technical writing must be clear.

In addition to transitional words and phrases and repetition of key phrases, demonstratives—*this*, *that*, *these*, and *those*—can help the writer maintain the coherence of a discussion by linking ideas securely. Demonstratives should in almost all cases serve as adjectives rather than as pronouns. In the following examples, notice that a demonstrative pronoun by itself can be confusing.

UNCLEAR

New screening techniques are being developed to combat viral infections. These are the subject of a new research effort in California.

What is being studied in California, new screening techniques or viral infections?

CLEAR

New screening techniques are being developed to combat viral infections. *These techniques* are the subject of a new research effort in California.

UNCLEAR

The task force could not complete its study of the mine accident. This was the subject of a scathing editorial in the union newsletter.

What was the subject of the editorial, the mine accident or the task force's inability to complete its study of the accident?

CLEAR

The task force failed to complete its study of the mine accident. *This failure* was the subject of a scathing editorial in the union newsletter.

Even when the context is clear, a demonstrative pronoun refers the reader to an earlier idea and therefore interrupts the reader's progress.

INTERRUPTIVE

The law firm advised that the company initiate proceedings. This resulted in the company's search for a second legal opinion.

FLUID

The law firm advised that the company initiate proceedings. *This advice* resulted in the company's search for a second legal opinion.

Transitional words and phrases, repetition, and demonstratives cannot *give* your writing coherence: they can only help the reader to appreciate the coherence that already exists. Your job is, first, to make sure your writing is coherent and, second, to highlight that coherence.

TURNING A "WRITER'S PARAGRAPH" INTO A "READER'S PARAGRAPH"

The best way to demonstrate what has been said in this discussion is to take a weak paragraph and improve it. The paragraph that follows is an excerpt from a status report written by a branch manager of the utility company mentioned in Chapter 1. The subject of the paragraph is how the writer decided on a method to increase the company's business within his particular branch.

There were two principal alternatives considered for improving Montana Branch. The first alternative was to drill and equip additional sources of supply with sufficient capacity to provide for the present and projected system deficiencies. The second alternative was to provide for said deficiencies through a combination of additional sources of supply and a storage facility. Unfortunately, ground water studies which were conducted in the Southeast Montana area by the consulting firm of Smith and Jones indicated that although ground water is available within this general area of our system, it is limited as to quantity, and considerable separation between well sources is necessary in order to avoid interference between said sources. This being the case, it becomes necessary to utilize the sources that are available or that can be developed in the most efficient manner, which means operating them in conjunction with a storage facility. In this way, the sources only have to be capable of providing for the average demand on a maximum day, and the storage facility can be utilized to provide for the peaking requirements plus fire protection. Consequently, the second alternative as mentioned hereinabove was determined to be the more desirable alternative.

First, let's be fair. The paragraph has been taken out of its context—a 17-page report—and was never meant to stand alone on a page. Also, it was not written for the general reader, but for an executive of the water company—someone who, in this case, is technically knowledgeable in the writer's field. Still, an outsider's analysis of an essentially private communication can at least isolate the weaknesses.

The most important element of a paragraph is the topic sentence, so you look for it in the usual place: the beginning. At first glance, the topic sentence looks good. The statement that two principal alternatives were considered appears to function effectively as an introduction to the rest of the paragraph. You assume that the paragraph will define the two alternatives.

The more you think about the topic sentence, however, the less good it looks. The problem is suggested by the word *principal*. Most complex decisions eventually come down to a choice between the two best alternatives; why bother labeling the two best alternatives "principal"? The writer almost sounds as if he is congratulating himself for weeding out the undesirable alternatives. (Further, the sentence begins with the expletive "there are" and is written in the passive voice; the result is a weak topic sentence.)

Structured as it is, the paragraph focuses on the process of selecting an alternative rather than on the choice itself. Perhaps the writer thinks his readers are curious about how he does his job; perhaps he wants to suggest that he's done his best, so that if it turns out that he made the wrong choice, at least he cannot be criticized for negligence; or perhaps the writer is unintentionally recounting the scientific method by describing how he examined the two alternatives and finally came to a decision.

For whatever reasons, the writer has built a lot of suspense into the paragraph. It proceeds like a horse race, with first one alternative gaining, then the other. The problem with this strategy is that the readers end up worrying about the outcome and can't concentrate on the specific reasons why one choice prevailed over the other. Quite likely, the only thing you will remember about the paragraph is the final emphatic statement: that the writer decided to go ahead with the second alternative. And given the complexity of the paragraph, it would be easy to forget what that alternative is.

Without too much trouble the paragraph can be rewritten so that it is much easier to understand. A careful topic sentence—one that defines the choice of a technical solution to the problem—is a constructive start. Then the writer should elaborate on the necessary details of the solution. Only after he has finished his discussion of the solution should he define and explain the alternative that was rejected. Using this structure, the writer clearly emphasizes what he *did* decide to do, putting what he decided *not* to do in a subordinate position.

Here is the writer's paragraph translated into a reader's paragraph:

> We found that the best way to improve the Montana branch would be to add a storage facility to our existing supply sources. Currently, we can handle the average demand on a maximum day; the storage facility will enable us to meet peaking requirements and fire-protection needs. In conducting our investigation, we considered developing new supply sources with sufficient capacity to meet current and future needs. This alternative was rejected, however, when our consultants (Smith and Jones) did ground water studies that revealed that insufficient ground water is available and that the new wells would have to be located too far apart if they were not to interfere with each other.

One obvious advantage of the revision is that it is about half as long as the original. The structure of the original version is largely responsible for its length: the writer announced that two alternatives were considered; defined them; explained why the first one was impractical and why it led to the second alternative; and, finally, announced the choice of the second alternative. A direct structure eliminates all this zigzagging and clarifies the paragraph.

The only possible objection to the streamlined version is that it is *too* clear, that it leaves the writer vulnerable in case his decision turns out to have been wrong. The response to this objection is that if the decision doesn't work out, the writer will be responsible no matter which way he described it, and that he will end up in more trouble if a supervisor has to investigate in order to find out who made the decision. Good writing is the best bet under any circumstances.

EXERCISES

1. The following sentences are vague. Revise the sentences to substitute specific information for the vague elements. Make up any reasonable details.

   a. The results won't be available for a while.
   b. The fire in the laboratory caused extensive damage.
   c. Analysis using a highlighting fluid revealed an abnormality in the tissue culture.

2. Consider voice (active or passive) in the following sentences. If a sentence would be more effective in the other voice, revise the sentence. Be prepared to defend your choices.

   a. The proposal was submitted to the Planning Commission by Dr. Hendrick.
   b. A study of the three available flood-control systems was approved by the Planning Commission.
   c. On Tuesday, employee John Pawley drove the one-millionth Plymouth Reliant off the assembly line.

3. In the following sentences, the real subjects are buried in prepositional phrases or obscured by expletives. Revise the sentences so that the real subjects appear prominently.

   a. The creation of the Energy Task Force will decrease our overhead costs.
   b. It is on the basis of recent research that I recommend the newer system.
   c. There is the need for new personnel to learn to use the computer.
   d. The completion of the new causeway will enable 40,000 cars to cross the ravine each day.
   e. There has been a decrease in the number of students enrolled in two of our training sessions.
   f. The use of point-of-purchase video presentations has resulted in a dramatic increase in business.

4. In the following sentences, unnecessary nominalization has obscured the real verb. Revise the sentences to focus on the real verb.

   a. Pollution constitutes a threat to the Wilson Wildlife Reserve.
   b. The switch from our current system to the microfilm can be accomplished in about two weeks.
   c. The construction of each unit will be done by three men.
   d. The ability to expand both the working memory and the mass storage must be possible.

5. The following sentences contain wordy phrases. Revise the sentences to make them more direct.

   a. As far as experimentation is concerned, much work has to be done on animals.

b. The analysis should require a period of three months.
c. The second forecasting indicator used is that of energy demand.
d. In Kevin McCarthy's article "Computerized Tax Returns," he argues that within a few years no firms will be computing returns "by pencil."

6. The following sentences contain clichés. Revise the sentences to eliminate the clichés.

a. I would like to thank each and every one of you.
b. With our backs to the wall, we decided to drop back and punt.
c. If we are to survive this difficult period, we are going to have to keep our ears to the ground and our noses to the grindstone.

7. The following sentences contain pompous words. Revise the sentences to eliminate the pomposity.

a. This state-of-the-art beverage procurement module is to be utilized by the personnel associated with the Marketing Department.
b. It is indeed a not unsupportable inference that we have been unsuccessful in our attempt to forward the proposal to the proper agency in advance of the mandated date by which such proposals must be in receipt.
c. This system will facilitate a reduction in employee time in reference to filing materials.
d. Our aspiration is the expedition of the research timetable.

8. The following sentences contain sexist language. Revise the sentences to eliminate the sexism.

a. Each doctor is asked to make sure he follows the standard procedure for handling Medicare forms.
b. Policemen are required to live in the city in which they work.
c. Two of the university's distinguished professors—Prof. Henry Larson and Ms. Anita Sebastian—have been elected to the editorial board of *Modern Chemistry*.

9. In each of the following compound sentences, the coordinating conjunction *and* links two clauses. Revise the sentences to make the link between the two clauses stronger.

a. We inteviewed George Karney, president of Whelk Industries, and he said that our suggestions seem very interesting.
b. This new schedule will give the employees more freedom, and they will be more productive.
c. A section of the upper layer of the tissue is seen in Figure 3W, and it is identical to the original sample.

10. The following sentences might be too long for some readers. Break each one into two or more sentences. If appropriate, add transitional words and phrases or other coherence devices.

a. In the event that we get the contract, we must be ready by June 1 with the necessary personnel and equipment to get the job done, so

with this end in mind a staff meeting, which all group managers are expected to attend, is scheduled for February 12.
b. Once we get the results of the stress tests on the 125-Z fiberglass mix, we will have a better idea where we stand in terms of our time constraints, because if it isn't suitable we will really have to hurry to find and test a replacement by the Phase I deadline.
c. Although we had a frank discussion with Becker's legal staff, we were not able to get them to discuss specifics on what they would be looking for in an out-of-court settlement, but they gave us a strong impression that they would rather not take the matter to court.

11. The following examples contain choppy and abrupt sentences. Combine sentences to create a smoother prose style.

a. I need a figure on the surrender value of a policy. The policy number is A6423146. Can you get me this figure by tomorrow?
b. There are advantages to having your tax return prepared by a professional. There are also disadvantages. One of the advantages is that it saves you time. One of the disadvantages is that it costs you money.
c. We didn't get the results we anticipated. The program obviously contains an error. Please ask Paul Davis to go through the program.

12. The information contained in each of the following sentences could be conveyed more effectively in a list. Rewrite each sentence in the form of a list.

a. The freezer system used now is inefficient in several ways: the chef cannot buy in bulk or take advantage of special sales, there is a high rate of spoilage because the temperature is not uniform, and the staff wastes time buying provisions every day.
b. The causes of burnout can be studied from three areas: physiological—the roles of sleep, diet, and physical fatigue; psychological—the roles of guilt, fear, jealousy, and frustration; environmental—the roles of the physical surroundings at home and at work.
c. There are many problems with the on-line registration system currently used at Dickerson. Firstly, lists of closed sections cannot be updated as often as necessary. Secondly, students who want to register in a closed section must be assigned to a special terminal. Thirdly, the computer staff is not trained to handle the student problems. Fourthly, the Computer Center's own terminals cannot be used on the system; therefore, the university has to rent fifteen extra terminals to handle registration.

13. Provide a topic sentence for each of the following paragraphs.

a. ____

____.
All service centers that provide gas and electric services in the tri-county area must register with the TUC, which is empowered to carry out unannounced inspections periodically. Additionally, all service centers must adhere to the TUC's Fair Deal Regulations, a set of stan-

dards that encompasses every phase of the service-center operations. The Fair Deal Regulations are meant to guarantee that all centers adhere to the same standards of prompt, courteous, and safe work at a fair price.

b. ____

____.

The reason for this difference is that a larger percentage of engineers working in small firms may be expected to hold high-level positions. In firms with fewer then twenty engineers, for example, the median income was $31,200. In firms of twenty to two hundred engineers, the median income was $28,345. For the largest firms, the median was $25,600.

14. Develop the following topic sentences into full paragraphs.

a. Job candidates should not automatically choose the company that offers the highest salary.
b. Every college student should learn at least the fundamentals of computer science.
c. The one college course I most regret not having taken is ________.
d. Sometimes two instructors offer contradictory advice about how to solve the same kind of problem.

15. In this exercise, several paragraphs have been grouped together into one long paragraph. Re-create the separate paragraphs by marking where the breaks would appear.

IN SEARCH OF ROBOT VISION

During the past decade, much research and development has been aimed at simulating sensory functions. Industrial-robot manufacturers are using the general-purpose computer as part of the control system and are implementing software that allows robots to accept sensory data. General Motors has participated in, and has encouraged, these developments. More specifically, GM has invested considerable effort in a machine-vision program, which has resulted in development of a fully integrated production prototype machine-vision/industrial-robot system that adaptively reacts to visually sensed changes in workpiece type and placement. Machine vision is obtained by connecting a computer to a special electric camera and by developing programs to interpret the camera data for part recognition and inspection or for determining position and orientation for robot control. This visual sense will enable robots to sense and react to their environment. The vision systems can be positioned to observe the robots and workpieces, or they can be mounted on the robot itself for feedback control. A major goal of the GM vision research program was to determine whether the technology could be developed to go beyond blocks and simple shapes and to deal with real production parts and conditions. Such devices as automatic welding robots were already at work in our plants, and our research people recognized that they would be many times more useful if they could be equipped with vision and the ability to adaptively adjust their actions to varying job requirements. A complementary effort was exploring the requirements of real-time robot control. In an early experiment, a 128- by 128-element solid-state camera was connected to a computer running a simple program to find the center of white objects on a black conveyor belt. The center coordinates were transmitted to a computer that had been substituted for the standard control hardware of a Unimate 2000 robot. The robot could pick up white plastic cups that it sensed visually as they moved by on a black conveyor. Although the vision system could not yet reliably sense real parts under production conditions, we were able to get the robot control ready to accept data from the more capable vision system under development. The new vision approach used the displacement of a projected line of light to detect the part regardless of color or reflectivity, and it demonstrated the feasibility of visually guided robot handling of real parts on a moving conveyor. We are developing a prototype system for production testing and evaluation. (Based on Beecher and Dewar 1979)

16. In the following paragraphs, transitional words and phrases have been removed. Add an appropriate transition in each blank space. Where necessary, add punctuation.

a. As you know, the current regulation requires the use of conduit for all cable extending more than 18″ from the cable tray to the piece of equipment. ________ conduit is becoming increasingly expensive: up 17% in the last year alone. ________ we would like to determine whether the NRC would grant us any flexibility in its conduit regulations. Could we ________ run cable without conduit for lengths up to 3′ in low-risk situations such as wall-mounted cable or low-traffic areas? We realize ________ that conduit will always remain necessary in high-risk situations. The cable specifications for the Unit Two report to the NRC are due in less than two months; ________ we would appreciate a quick reply to our request, as this matter will seriously affect our materials budget.

b. Acrylonitrile is a substance used in the production of many plastics and synthetic fibers. ________ when fed to laboratory rats in large doses, it produces malignant tumors. ________ to reduce the risk to industrial workers who have frequent contact with the substance, OSHA last year placed a limit on the amount of acrylonitrile permitted in factory air. The regulation will prevent about seven cancer deaths a year and cost industry $24.3 million, or $3.5 million for each life saved. Many executives in the chemical industry have serious questions about the regulation. They wonder ________ if the expense is worth it? Might not the money be spent more effectively elsewhere? Such questions are ________ difficult to deal with, for they run counter to religious and moral tenets, which hold that human life cannot be discussed in terms of material values. It is understandable ________ that OSHA has avoided the economic issues in the acrylonitrile debate by saying its mandate is to promote the health and safety of workers, and that cost is irrelevant. (Based on "What price safety?" 1979)

17. The following paragraphs are poorly organized and developed. Rewrite the paragraphs, making sure to include in each a clear topic sentence, adequate support, and effective transitions.

a. The cask containing the spent nuclear fuel is qualified to sustain a 30-foot fall to an unyielding base. If the cask is allowed to pierce a floor after a 30-foot fall, we can safely assume that the cask is not qualified to maintain its integrity. This could result in spent fuel being released from the cask, resulting in high radioactive exposure to personnel.

b. The use of a single-phase test source results in partial energization of the polarizing and operating circuits of relay units other than the one under test. It is necessary that these "unfaulted" relay units be disabled to assure that only the unit being tested is providing information regarding operation or nonoperation. The disabling is accomplished by disengaging printed circuit cards from the card sockets in the relay logic unit. The following table lists the cards to be disengaged by card address in the logic unit under test with respect to STF position.

c. The results indicate that the plaques are not chemically homogeneous: the percentage of NILOC varied among the various sections. This fact represents a problem with determinations such as were conducted in this study. Since only a portion of each plaque submitted was analyzed, differences in the amount of materials extracted could be solely a function of the portion of the plaque examined. Another problem in these determinations was the suspended materials present in the samples. It is not known if any of the compounds of interest occluded onto or interacted with the precipitate. For these reasons, the results of the tests should be considered as having a ±10% error.

REFERENCES

Battison, R., and D. Goswami. 1981. Clear writing today. *Journal of Business Communication* 18, no. 4: 5–16.

Beecher, R. C., and R. Dewar. 1979. Robot trends at General Motors. *American Machinist* (August):74.

Marion, J. B. 1974. *Energy in perspective*. New York: Academic.

Selzer, J. 1981. Readability is a four-letter word. *Journal of Business Communication* 18, no. 4: 23–34.

What price safety? The "zero-risk" debate. 1979. *Dun's Review* (September): 49.

PART TWO

TECHNIQUES OF TECHNICAL WRITING

CHAPTER SIX

DEFINITIONS

No technique is more basic to technical writing than that of definition. Because technical writing requires clear, objective communication of factual information, you will often have to define objects, processes, and ideas.

The world of business and industry depends on clear and effective definitions. Without written definitions, the working world would be chaotic. For example, suppose that you learn at a job interview that the potential employer pays tuition and expenses for its employees' job-related education. That's good news, of course, if you are planning to continue your education. But until you study the employee-benefits manual, you will not know with any certainty just what the company will pay for. Who, for instance, is an *employee*? You would think an employee would be anybody who works for and is paid by the company. But you might find that for the purposes of the manual, an employee is someone who has worked for the company in a full-time (35 hours-per-week) capacity for at least six uninterrupted months. A number of other terms would have to be defined in the description of tuition benefits. What, for example, is *tuition*? Does the company's definition include incidental laboratory or student fees? What is *job-related education*? Does a course on methods of dealing with stress qualify under the company's definition? What, in fact, constitutes *educa-*

tion? All of these terms and many others must be defined in order for the employees to understand their rights and responsibilities.

Definitions play a major role, therefore, in communicating policies and standards "for the record." When a company wants to purchase some air-conditioning equipment, it might require that the supplier provide equipment that has been certified by a professional organization of air-conditioner manufacturers. The organization's definitions of acceptable standards of safety, reliability, and efficiency will provide some assurance that the equipment is of high quality.

Definitions, of course, have many uses outside legal or contractual contexts. Two such uses occur very frequently. Definitions can help the writer clarify a description of a new technology or a new development in a technical field. When a new animal species is discovered, for instance, it is named and defined. When a new laboratory procedure is devised, it is defined and then described in an article printed in a technical journal. The other basic use of definitions is to help a specialist communicate with a less knowledgeable audience. A manual that explains how to tune up a car will include definitions of several parts and tools. A researcher at a manufacturing company will use definitions in describing a new product to the sales staff.

Definitions, then, are crucial in all kinds of technical writing, from brief letters and memos to technical reports, manuals, and journal articles. All kinds of readers, from the layperson to the expert, need effective definitions to carry out their jobs every day.

Definitions, like every other technique in technical writing, require thought and planning. Before you can include a definition in a document you are writing, you must carry out three steps:

1. Analyze the writing situation.
2. Determine the kind of definition that will be appropriate.
3. Decide where to place the definition.

THE WRITING SITUATION

The first step in writing effective definitions is to analyze the writing situation: the audience and purpose of your document.

Unless you know whom you are addressing and how much that audience already knows about the subject, you cannot know which terms need to be defined or the kind of definition to write. A group of physicists wouldn't need a definition of the word *en-*

tropy, but a group of lawyers might. Builders know what a Molly screw is, but some insurance agents might not. If you are aware of your audience's background and knowledge, you can easily devise effective informal definitions. For example, if you are describing a cassette deck to a group of readers who understand automobiles, you can use a familiar analogy: "The PAUSE button is the brake pedal of the cassette deck."

Keep in mind, too, your purpose in writing. If you want to give your readers only a basic understanding of a concept—say, time-sharing vacation resorts—a brief, informal definition will usually be sufficient. However, if you want your readers to understand an object, process, or concept thoroughly and be able to carry out tasks based on what you have written, then a more formal and elaborate definition is called for. For example, a definition of a "Class 2 Alert" written for operators at a nuclear power plant will have to be comprehensive, specific, and precise.

TYPES OF DEFINITIONS

The preceding analysis of the writing situation has suggested that definitions can be short or long, informal or formal. Three basic types of definitions can be isolated:

1. parenthetical
2. sentence
3. extended

PARENTHETICAL DEFINITION

A parenthetical definition is a brief clarification placed unobtrusively within a sentence. Sometimes a parenthetical definition is a mere word or phrase:

The crane is located on the starboard (right) side of the ship.

Summit Books announced its desire to create a new colophon (emblem or trademark).

United Engineering is seeking to purchase the equity stock (common stock) of Minnesota Textiles.

A parenthetical definition can also take the form of a longer explanatory phrase or clause:

Motorboating is not forbidden in the Jamesport Estuary, the portion of the bay that meets the mouth of the Jamesport River.

The divers soon discovered the kentledge, the pig-iron ballast.

Before the metal is plated, it is immersed in the pickle, the acid bath that removes scales and oxides from the surface.

Parenthetical definitions are not, of course, authoritative. They serve mainly as quick and convenient ways of introducing new terms to the readers. Because parenthetical definition is particularly common in writing addressed to general readers, make sure that the definition itself is clear. You have gained nothing if your readers don't understand your clarification:

Next, check for blight on the epicotyl, the stem portion above the cotyledons.

If your readers are botanists, this parenthetical definition will be clear (although it might be unnecessary, for if they know the meaning of *cotyledons*, they are likely to know *epicotyl*). However, if you are addressing the general reader, this definition will merely be frustrating.

SENTENCE DEFINITION

A sentence definition is a one- or two-sentence clarification. It is more formal than the parenthetical definition. Usually, the sentence definition follows a standard pattern: the item to be defined (the *species*) is placed within a category of similar items (the *genus*) and then distinguished from the other items (by the *differentia*):

SPECIES	=	GENUS	+	DIFFERENTIA
A flip flop	is	a circuit		containing active elements that are capable of assuming either one of two stable states at any given time.
An electrophorus	is	an instrument		used to generate static electricity.
Hypnoanalysis	is	a psychoanalytical technique		in which hypnosis is used to elicit unconscious information from a patient.
A Bunsen burner	is	a small laboratory burner		consisting of a vertical metal tube connected to a gas source.
An electron microscope	is	a microscope		that uses electrons rather than visible light to produce magnified images.

Sentence definitions are useful when your readers require a more formal or more informative clarification than parenthetical definition can provide. Sentence definitions are often used to establish a working definition for a particular document: "In this report, the term *electron microscope* will be used to refer to any microscope that uses electrons rather than visible light to produce magnified images."

When you write sentence definitions, keep several points in mind.

First, be as specific as you can in writing the *genus* and the *differentia*. Remember, your purpose is not merely to provide some information about the item you are defining; you are trying to distinguish it from all other similar items. If you write, "A Bunsen burner is a burner that consists of a vertical metal tube connected to a gas source," the imprecise *genus*—"a burner"—defeats the purpose of your definition: there are many types of large-scale burners that use vertical metal tubes connected to gas sources. If you write, "Hypnoanalysis is a psychoanalytical technique used to elicit unconscious information from a patient," the imprecise *differentia*—"used to elicit . . ."—ruins the definition: there are many psychoanalytical techniques used to elicit a patient's unconscious information. If more than one *species* is described by your definition, you have to sharpen either the *genus* or the *differentia*, or both.

Second, avoid writing circular definitions: that is, definitions that merely repeat the key words of the *species* (the ones being defined) in the *genus* or *differentia*. In "A required course is a course that is required," what does "required" mean? Required of whom, by whom? The word is never defined.

Similarly, "A balloon mortgage is a mortgage that balloons" is useless. Note, however, that you can use *some* kinds of words in the *species* as well as in the *genus* and *differentia*; in the definition of *electron microscope* given earlier, for example, the word *microscope* is repeated. Here, *microscope* is not the "difficult" part of the species; readers know what a microscope is. The purpose of defining *electron microscope* is to clarify the *electron* part of the term.

Third, be sure the *genus* contains a noun or a noun phrase rather than a phrase beginning with *when*, *what*, or *where*.

INCORRECT

A brazier is what is used to . . .

CORRECT

A brazier is a metal pan used to . . .

INCORRECT

An electron microscope is when a microscope . . .

CORRECT

An electron microscope is a microscope that . . .

INCORRECT

Hypnoanalysis is where hypnosis is used . . .

CORRECT

Hypnoanalysis is a psychoanalytical technique in which . . .

EXTENDED DEFINITION

An extended definition is a long (one- or several-paragraph), detailed clarification of an object, process, or idea. Often an extended definition begins with a sentence definition, which is then elaborated. For instance, the sentence definition "An electrophorus is an instrument used to generate static electricity" tells you the basic function of the device, but it leaves many questions unanswered: How does it work? What does it look like? An extended definition would answer these and other questions.

Extended definitions are useful, naturally, when you want to give your readers a reasonably complete understanding of the item or concept. And the more complicated or more abstract the term being defined, the greater the need for an extended definition.

There is no one way to "extend" a definition. Your analysis of the audience and purpose of the communication will help to indicate which method to use. In fact, an extended definition will sometimes employ several different methods. Often, however, one of the following techniques will work effectively:

1. exemplification
2. analysis
3. principle of operation
4. comparison and contrast
5. negation
6. etymology

Exemplification, using an example (or examples) to clarify an object or idea, is particularly useful in making an abstract term easy to understand. The following paragraph is an extended definition of the psychological defense mechanism called *conversion* (Wilson 1964: 84).

A third mechanism of psychological defense, "conversion," is found in hysteria. Here the conflict is converted into the symptom of a physical illness. In a case of conversion made famous by Freud, a young woman went out for a long walk with her brother-in-law, with whom she had fallen in love. Later, on learning that her sister lay gravely ill, she hurried to her bedside. She arrived too late and her sister was dead. The young woman's grief was accompanied by sharp pain in her legs. The pain kept recurring without any apparent physical cause. Freud's explanation was that she felt guilty because she desired the husband for herself, and unconsciously converted her repressed feelings into an imaginary physical ailment. The pain struck her in the legs because she unconsciously connected her feelings for the husband with the walk they had taken together. The ailment symbolically represented both the unconscious wish and a penance for the feelings of guilt which it engendered.

Notice that the first two sentences in this paragraph are essentially a sentence definition that might be paraphrased as follows: "Conversion is a mechanism of psychological defense by which the conflict is converted into the symptoms of a physical illness."

This extended definition is effective because the writer has chosen a clear and interesting (and therefore memorable) example of the subject he is describing. No other examples are necessary in this case. If conversion were a more difficult concept to describe, an additional example might be useful.

Analysis is the process of dividing a thing or idea into smaller parts so that the reader can more easily understand it.

A load-distributing hitch is a trailer hitch that is designed to distribute the hitch load to all axles of the tow vehicle and the trailer. The crucial component of the load-distributing hitch is the set of spring bars that attaches to the trailer. For a complete understanding of the load-distributing hitch, however, the following other components should be explained first:

1. the shank
2. the ball mount
3. the sway control
4. the frame bracket

The shank is the metal bar that is attached to the frame of the tow vehicle . . .

This extended definition of a load-distributing hitch uses analysis as its method of development. The hitch is divided into its major components, each of which is then defined and described.

The *principle of operation* is an effective way to develop an extended definition, especially that of an object or process. The following extended definition of a thermal jet engine is based on the mechanism's principle of operation.

A thermal jet engine is a jet-propulsion device that uses air, along with the combustion of a fuel, to produce the propulsion. In operation, the thermal jet engine draws in air, increases its pressure, heats it by combustion of the fuel, and finally ejects the heated air (and the combustion gases). The increased velocity of the ejected mixture determines the thrust: the greater the difference in velocity between air entering and leaving the unit, the greater the thrust.

Notice that this extended definition begins with a sentence definition.

Comparison and contrast is another useful technique for developing an extended definition. With this technique, the writer discusses similarities or differences between the item being defined and an item with which the readers are more familiar. The following definition of a bit brace begins by comparing and contrasting it to the more common power drill.

A bit brace is a manual tool used to drill holes. Cranked by hand, it can theoretically turn a bit to bore a hole in any material that a power drill can bore. Like a power drill, a bit brace can accept any number of different sizes and shapes of bits. The principal differences between a bit brace and a power drill are:

1. A bit brace drills much more slowly.
2. A bit brace is a manual tool, and so it can be used where no electricity is available.
3. A bit brace makes almost no noise in use.

The bit brace consists of the following parts. . . .

A special kind of contrast is sometimes called *negation* or *negative statement*. Negation is the technique of clarifying a term by distinguishing it from a different term with which the reader might have confused it.

An ambulatory patient is *not* a patient who must be moved by ambulance. On the contrary, an ambulatory patient is one who can walk without assistance from another person. . . .

Negation is rarely the only technique used in an extended definition. In fact, negation is used most often in a sentence or two at the start of a definition: for after you state what the item is *not*, you still have to define what it *is*.

Etymology, the history of a word, is often a useful and interesting way to develop a definition.

No-fault auto insurance, as the name suggests, is a type of insurance that ignores who or what caused the accident. When the accident damage is less than a specific amount set by the state, the people involved in the accident may not use legal means to determine who was at fault. What this means to the driver is that . . .

The word *mortgage* was originally a compound of *mort* (dead) and *gage* (pledge). The meaning of the word has not changed substantially since its origin in Old French. A mortgage is still a pledge that is "dead" upon either the payment of the loan or the forfeiture of the collateral and payment from the proceeds of its sale.

Etymology, like negation, is rarely used alone in technical writing, but it is an effective way to introduce an extended definition.

The following extended definition of fission is excerpted from a discussion of nuclear reactors (Nero 1979: 3–4).

The nuclear power plants of this century depend on a particular type of nuclear reaction, fission, for the generation of heat. Fission is the splitting of a heavy nucleus, the center of an atom such as uranium, into two or more principal fragments, as well as lighter pieces, such as neutrons. In principle, fission may occur spontaneously, but in nuclear reactors this splitting is induced by the interaction of a neutron with a fissionable nucleus. Neutrons are, in fact, one of the two basic components of nuclei (the other is the proton) and, as noted, they are released during fission, thereby becoming available to induce subsequent fission events. Under suitable conditions, a "chain" reaction of fission events may be sustained. The energy released from the fission reactions provides the heat, part of which is ultimately converted into electricity. In present-day nuclear power plants, this heat is removed from the nuclear fuel by water that is pumped past the rods containing the fuel. Other fluids may be used as a coolant, but in every case this coolant delivers the heat, either directly or indirectly, to the electrical generating part of the plant. This mode of heat transfer does not differ in principle from that used in plants that depend on chemical reactions, including the burning of coal, oil, or gas.

The first two sentences of this paragraph contain what is in effect a sentence definition: "Fission is a type of nuclear reaction in which a heavy nucleus, the center of an atom such as uranium, is split. . . ." The writer extends this definition by explaining how the process of fission occurs in a nuclear power plant: how fission is induced, how it is sustained, and how the heat produced by the fission is removed.

PLACEMENT OF DEFINITIONS

In many cases, the writer does not need to "decide" where to place a definition. In writing your first draft, for instance, you may realize that most of your readers will not be familiar with a term you want to use. If you can easily provide a parenthetical definition that will satisfy your readers' needs, simply do so.

Often, however, in assessing the writing situation before beginning the draft, you will conclude that one or more complicated terms will have to be introduced, and that your readers will need more detailed and comprehensive clarifications, perhaps sentence definitions and extended definitions. In these cases, you should plan—at least tentatively—where you are going to place them in the document.

Definitions can be placed in four different locations:

1. in the text
2. in footnotes
3. in a glossary
4. in an appendix

The text itself can accommodate any of the three kinds of definitions. Parenthetical definitions, because they are brief and unobtrusive, are almost always included in the text; even if a reader already knows what the term means, the slight interruption will not be annoying. Sentence definitions are often placed within the text. If you want all of your readers to see your definition or you suspect that many of them need the clarification, the text is the appropriate location. Keep in mind, however, that unnecessary sentence definitions can easily become bothersome to your readers. Extended definitions are rarely placed within the text, because of their length. The obvious exception, of course, is the extended definition of a term that is central to the discussion; a discussion of recent changes in workmen's compensation insurance will likely begin with an extended definition of that kind of insurance.

Footnotes are a logical location for an occasional sentence definition or extended definition. The reader who needs the definition can find it easily at the bottom of the page; the reader who doesn't need it is not interrupted. Footnotes are difficult to type, however, and they make the page look choppy. So if you are going to need more than one footnote for every two or three pages, consider creating a glossary.

A glossary, which is simply an alphabetized list of definitions, can handle sentence definitions and extended definitions of fewer than three or four paragraphs. A glossary can be placed at the start of a document (such as after the executive summary in a report) or at the end, preceding the appendixes. The principal advantage of a glossary is that it provides a convenient collection of definitions that otherwise might clutter up the document. For more information on how to set up a glossary, see Chapter 11.

An appendix is an appropriate place to put an extended definition, one of a page or longer. A definition of this length would be cumbersome in a glossary or in a footnote, and—unless it explains a crucial term—too distracting in the text. Because the definition is an appendix in the document, it will be listed in the table of contents. In addition, it can be referred to—with the appropriate page number—in the text.

WRITER'S CHECKLIST

This checklist covers parenthetical, sentence, and extended definitions.

1. Are all terms that need definitions defined? ____

2. Are the parenthetical definitions
   a. Appropriate for the audience? ____
   b. Clear? ____

3. Does each sentence definition
   a. Contain a sufficiently specific *genus* and *differentia*? ____
   b. Avoid circular definition? ____
   c. Contain a noun in the genus? ____

4. Are the extended definitions developed logically and clearly? ____

5. Are the definitions placed in the most useful location for the readers? ____

EXERCISES

1. Add parenthetical definitions of the italicized terms in the following sentences.

   a. Reluctantly, he decided to *drop* the physics course.
   b. Last week the computer was *down*.

c. The culture was studied *in vitro*.

d. The tire plant's managers hope they do not have to *lay off* any more employees.

e. The word processor comes complete with a *printer*.

2. Write a sentence definition for each of the following terms:

a. a computer chip
b. a digital watch
c. a job interview
d. a cassette deck
e. flextime

3. Write an extended definition for each of the following terms:

a. flextime
b. binding arbitration
c. energy
d. an academic major (don't focus on any particular major; define what a major is)
e. quality control

4. Revise any of the following sentence definitions that need revision.

a. Dropping a course is when you leave the class.
b. A thermometer measures temperature.
c. The spark plugs are the things that ignite the air-gas mixture in a cylinder.
d. Double parking is where you park next to another car.
e. A strike is when the employees stop working.

REFERENCES

Nero, A. V., Jr. 1979. *A guidebook to nuclear reactors.* Berkeley: Univ. of Calif. Press.

Wilson, J. R. 1964. *The mind.* New York: Time, Inc.

CHAPTER SEVEN

ANALYSIS

Different writing situations demand different strategies. Often you will have to describe for your readers two pieces of equipment, both of which are designed to satisfy a need within your organization. Comparing and contrasting them would be a logical way to structure your discussion. Or you might want to create categories, to help your readers make sense of a large number of items. In describing photocopy machines, for example, you might want to classify them by size (desk-top models and floor models) or by features (those that collate and those that do not). Often in technical writing you will have to describe causality—what caused some occurrence, or what will be the effect of some occurrence. For instance, you might have to predict how your company's business will be affected by the opening of a competing retail store nearby. Or you might be asked to investigate what caused a recent decline in sales of one of your company's products.

The term *analysis*, as used in this chapter, refers to these three basic techniques of writing:

1. comparison and contrast
2. classification and partition
3. cause and effect

COMPARISON AND CONTRAST

Much technical writing is structured according to the technique of comparison and contrast—the same technique by which most of us learned as children. What is a zebra? A zebra looks like a horse (comparison), but it has stripes (contrast). In technical writing, comparison-and-contrast discussions are developed much more fully, of course, but the technique is essentially the same.

Comparison and contrast is a common writing technique in the working world because organizations constantly make decisions. Should we purchase the model A computer, or the B or the C? Should we hire person X or person Y? Should we carry out a comprehensive analysis of this potential site for a new plant, or would a more limited analysis be preferable? Probably the most common kind of report in technical writing is a "recommendation report," in which the writer recommends a course of action. Almost always, this kind of report requires a substantial comparison-and-contrast section.

The first step in comparing and contrasting two or more items is to establish a basis (or several bases) on which to evaluate the items. If you have only one basis, comparing and contrasting the items is a simple task. Next semester you might need to take a course that is offered at ten o'clock on Mondays, Wednesdays, and Fridays. Your one basis of comparison and contrast is the time at which it is offered—nothing else matters.

Almost always, however, you will have several bases (or criteria) by which to carry out your comparison and contrast. For example, you might need a three-credit science course that meets at ten o'clock on Mondays, Wednesdays, and Fridays. In this case, you have three criteria: number of credits, general subject area, and schedule. When you are comparing and contrasting items for a study on the job, you are likely to have six or eight or even more criteria. For a recommendation report on which kind of computer your company should buy, for example, your comparison and contrast would probably include all of the following criteria:

1. availability of software
2. amount and type of storage capacity
3. ease of operation
4. reliability
5. availability of peripherals
6. accessibility of maintenance and service personnel
7. initial costs and maintenance costs

The more criteria you use in your comparison and contrast, the more precise your analysis will be. If you want to buy a used car but your only criterion is price, you don't have to ponder whether to buy the $1,000 car or the $2,000 car. However, when you add more and more criteria—such as age, reliability, and features—you will probably conclude that the more expensive car is "better"—according to the analysis.

Choose sufficient and appropriate criteria when you compare and contrast items. Don't overlook an important criterion. A beautifully written comparison and contrast of office computers is of little value if you have neglected price. If your company can spend no more than $35,000 and you carry out a detailed comparison and contrast of models that cost more than $50,000, you will have wasted everyone's time. Be careful, too, that the criteria you choose are sensible. If, for instance, your company has a large, empty storeroom whose temperature is just right for the computer, don't use size as one of your criteria: there is no reason to spend more money for a compact unit if a larger model would be perfectly acceptable.

Once you have chosen your criteria and carried out your research, choose a structure for writing the comparison and contrast. The two structures from which to choose are called *whole-by-whole* and *part-by-part*. In the whole-by-whole structure, the first item is discussed, then the second, and so on:

Item 1
  Criterion A
  Criterion B
  Criterion C
Item 2
  Criterion A
  Criterion B
  Criterion C
Item 3
  Criterion A
  Criterion B
  Criterion C

In the part-by-part structure, each criterion is discussed separately:

Criterion A
  Item 1
  Item 2
  Item 3

Criterion B
  Item 1
  Item 2
  Item 3
Criterion C
  Item 1
  Item 2
  Item 3

The whole-by-whole structure, because it focuses on the individual items, is easy for your audience to read and understand. If your purpose is to give a profile of each of the items you are comparing and contrasting, the whole-by-whole structure is probably more effective. If, however, you want to focus on the criteria themselves, the part-by-part structure is more effective. Your readers will find it easier to study your discussion of the memory of three computers you are evaluating, for instance, if memory is discussed in one place in the report, rather than in three places.

The following comparison of passive and active solar-energy systems employs the whole-by-whole structure.

> A passive solar-energy system is one that uses the greenhouse effect: the sun shines through the windows of the house, and the heat is retained in the house. A new house designed for a passive system is generally situated so that its long dimension is on the east-west plane. Most of its glass is on the southern exposure; the north has few windows. Often, a greenhouse is added to the southern exposure.
>
> An active solar-energy system consists of solar collectors mounted on a wall or the roof and a delivery system to carry the heat to the different rooms of the house or to a storage system. An active system can be incorporated easily into most new or existing houses, because almost every house has an appropriate site for a collector.

Notice in this example that each paragraph has a two-part structure. First, each type of solar energy system is described. Next, the siting requirements are discussed. Of course, the whole-by-whole pattern can be expanded beyond a few paragraphs. A detailed discussion of active and passive solar-energy systems, for instance, might require several chapters of a book.

The following two paragraphs, taken from a longer discussion of the two beetles that transmit Dutch elm disease in the United States (Strel and Lanier 1981), exemplify a comparison-and-contrast passage that uses the part-by-part structure:

> European and native bark beetles are similar in size (between two and 3.5 millimeters in length) but dissimilar in appearance. The European

species is shiny and distinctly two-toned, the wing covers are dark reddish brown and the thorax is black. The native beetle is a drab grayish brown and has a rough texture. The immature stages of the two kinds of beetle can easily be distinguished by the gallery systems in which they are found. The female European beetle bores its egg gallery along the grain of the wood, and the tunnels made by the larvae radiate from the egg gallery, resulting in an elliptical pattern. The egg gallery of the native beetle crosses the grain, and the larvae tend to bore along the grain, creating a butterfly-shaped pattern.

The native elm bark beetle inhabits the entire natural range of elms in North America, but it is apparently absent in the Rocky Mountain states and westward, where elms were introduced in comparatively recent times. The European beetle is now found wherever elms grow in North America except for some northern areas where its over-wintering larvae are destroyed by temperatures lower than about minus 25 degrees Fahrenheit.

In the first paragraph, the two beetles are compared and contrasted according to the criteria of size and appearance and then of gallery systems. In the second paragraph, the beetles are compared and contrasted according to the criterion of location of habitation.

CLASSIFICATION AND PARTITION

CLASSIFICATION

Classification (sometimes called classification and division) is the process of establishing categories, of grouping items that share certain characteristics. Classification also works in reverse: sometimes you will wish to break categories down into smaller units. To return to the terms used in the discussion of definition, a *genus* contains several *species*: the species can be thought of as belonging to the larger genus, or class; the class can be divided into individual species, or items, on the basis of *differentiae*. Consider the computer science major whose courses for the semester include operating systems, file structures, discrete structures, American literature since 1860, and physical education. These courses all belong to the general class *college courses* and to the more specific class *courses that this student is taking this semester*. However, you might wish to divide this more specific class into two smaller classes, such as *major requirements* and *electives*. The first subclass would include operating systems, file structures, and discrete structures; the second would include American literature since 1860 and physical education. Accordingly, classification can cover a range of divisions. The process involves both the creation of larger categories so that similar items can be understood as a group and the division of larger groups into individual items. Although these exercises would appear to be opposites, they are in fact the same, as they explore the same relations between items and groups of items.

You classify when you define; you also classify when you outline. When you select the information in your notes that is relevant to your topic, you are creating a class—relevant information—to which the selected items belong. Looking at the process from another angle, you are dividing the bulk of your information into two classes, relevant information and irrelevant information. You then subclassify the relevant information. In this way you can organize an understandable discussion, regardless of whether you are producing a report, a memo, a letter, or any other kind of technical document.

PARTITION

Partition, the process of separating a unit into its components, shares some ground with classification and follows many of the same rules. However, partition is best thought of as an independent technique that is used frequently in technical writing. For example, if you wished to discuss a stereo system, you might partition it into the following components: amplifier, tuner, turntable, and speakers. Each component is separate; the only relation among them is that together they form a stereo system. Similarly, you cannot have a complete stereo system that lacks any of these components. Each component can of course be partitioned further. Partition underlies descriptions of mechanisms (see Chapter 8) and descriptions of processes (see Chapter 9).

SOME GUIDELINES FOR CLASSIFICATION AND PARTITION

The most important concept to keep in mind when you are classifying or partitioning is that you are establishing a variety of levels. As you know from outlining, you must avoid confusing major and minor categories (faulty coordination). An item that is out of place can destroy any system that you are employing. Consider the following attempt to classify keyboard instruments:

pianos
organs
harpsichords
synthesizers
concert grand pianos

Clearly, "concert grand pianos" is a subcategory of pianos and does not belong in this system of classification. The following partition of a small sailboat is similarly faulty:

hull
sails
rudder
centerboard
mast
mainsail
jib

"Mainsail" and "jib" are subcategories of "sails" and represent a level of partition beyond that defined by the rest of this list.

The following guidelines should help you to avoid this kind of error:

1. Use only one basis at at time.
2. Choose a basis consistent with your audience and purpose.
3. Avoid overlap.
4. Be inclusive.
5. Arrange the categories in a logical sequence.

1. *Use only one basis of classification or partition at a time.* If you are classifying fans according to movement, do not include another basis of classification:

oscillating
stationary
inexpensive

"Inexpensive" is inappropriate here because it has a different basis of classification—price—from the other categories. Also, do not mix classification and partition. If you are partitioning a fan into its major components, do not include a stray basis of classification:

motor
blade
casing
metal parts

"Metal parts" is inappropriate here because motor, blade, and casing are partitions based on function; "metal parts," as opposed to "plastic parts," might be a useful classification, but it does not belong in this partition.

2. *Choose a basis of classification or partition that is consistent with your audience and purpose.* Never lose sight of your audience—their backgrounds, skills, and reasons for reading—and your purpose. If you were writing a brochure for hikers in a particular state park, you would probably classify the local snakes according to whether they are poisonous or nonpoisonous, not according to size, color, or any other basis. If you were writing a manual describing do-it-yourself maintenance for a particular car, you would most likely partition the car into its component parts so that the owner could locate specific discussions.

3. *Avoid overlap.* Make sure that no single item could logically be placed in more than one category of your classification or, in partition, that no listed component includes another listed component. Overlapping generally results from changing the basis of classification or the level at which you are partitioning a unit. In the following classification of bicycles, for instance, the writer introduces a new basis of classification that results in overlapping categories:

racing bikes
touring bikes
ten-speed bikes

The first two categories here have use as their basis; the third category is based on number of speeds. The third category is illogical, because a particular ten-speed bike could be a touring bike or a racing bike. In the following partition of a guitar, the writer changes the level of his focus:

body
bridge
neck
frets

The first three items here are basic parts of a guitar. Frets, as part of the neck, represent a further level of partition.

4. *Be inclusive.* When classifying or partitioning, be sure to include all the categories. For example, a classification of music according to type would be incomplete if it included popular and

classical music, but not jazz. A partition of an automobile by major systems would be incomplete if it included the electrical, fuel, and drive systems, but not the cooling system.

If your purpose or audience requires that you omit a category, tell your readers what you are doing. If, for instance, you are writing in a classification passage about recent sales statistics of General Motors, Ford, and Chrysler, don't use "American car manufacturers" as a classifying tag for the three companies. Although all three belong to the larger class "American car manufacturers," there are a number of other American car manufacturers. Rather, classify General Motors, Ford, and Chrysler as "the big three" or refer to them by name.

5. *Arrange the categories in a logical sequence.* Once you have established your categories and subcategories of classification or partition, arrange them according to some reasonable plan: time (first to last), space (top to bottom), importance (most to least), and so forth. For example, a classification of record turntables might begin with the simplest kind and proceed to the most sophisticated. See Chapter 4 for a discussion of patterns of development.

Following are several examples of classification and partition.

The first example is taken from a description of background sound systems used in industry.

> The various background sound systems available for industrial use can be classified according to whether the client has any control over the content of the programs. Piped-in phone line systems and FM radio systems are examples of those services that are not controlled by the client. In the piped-in phone line system, a special phone hook-up brings in the music, which is programmed entirely by the vendor. The FM radio system uses a special receiver to edit out news and other nonmusic broadcasts. Sound systems that can be controlled by the client include on-the-premise systems and customized external systems. An on-the-premise system is any arrangement in which a person or persons at the job site selects and broadcasts recorded music. A customized external system lets the client choose from a number of different programs or, for an extra fee, stipulate what the vendor is to broadcast.

This classification is based on who controls the music being broadcast. The various music systems could easily be classified according to other bases: for example, according to the source of the music (FM radio and nonradio); or according to cost (expensive and inexpensive systems); or according to where the music physically originates (on the job site or off).

The next example, from a discussion of the industrial uses of microorganisms (Demain and Solomon 1981), shows how a writer can use classification and subclassification effectively. (The paragraphs have been numbered for convenience.)

1 There are four classes of industrially important microorganisms: yeasts, molds, single-cell bacteria and actinomycetes. The yeasts and the molds are more highly developed; together they constitute fungi. Organisms of this type are eukaryotic, that is, their cells, like the cells of plants and animals, have a membrane-enclosed nucleus and more than one chromosome; they also contain organelles such as mitochondria (the tiny sausage-shaped bodies that are responsible for the cell's main energy supply). The single-cell bacteria and the actinomycetes, in contrast, are prokaryotic: they have no nuclear membrane or mitochondria, and they have only one chromosome. In addition the cells of prokaryotes are typically much smaller than those of eukaryotes. In spite of these basic biological differences there is a superficial resemblance between the molds and the actinomycetes in that both are filamentous: they grow as a branched system of threadlike hyphae rather than as single cells. The yeasts and the bacteria, on the other hand, are unicellular under normal conditions.

2 The commercially important products of these microoganisms fall into four major categories: (1) the microbial cells themselves; (2) the large molecules, such as enzymes, that they synthesize; (3) their primary metabolic products (compounds essential to their growth), and (4) their secondary metabolic products (compounds not required for their growth). In general both the primary and the secondary metabolites of commercial interest have a fairly low molecular weight: less than 1,500 daltons, compared with the molecular weight of an enzyme, which can range from 10,000 to several million daltons.

3 Microbial cells have two main commercial applications. The first is as a source of protein, primarily for animal feed. In its commonest form this product is referred to as single-cell protein, although in fact it usually includes the entire microbial cell, the major component of which is protein.

4 Microbial cells are also used to carry out biological conversions, processes in which a compound is changed into a structurally related compound by means of one or more enzymes supplied by the cells. Biological conversions, also known as microbial transformations, can be accomplished with growing cells, nongrowing cells, spores or even dried cells. Microorganisms, which can carry out almost every kind of chemical reaction, have many advantages over chemical reagents. For example, many nonbiological chemical reactions call for a considerable input of energy to heat or cool the reaction vessel; in addition they are generally conducted in solvents and require inorganic catalysts, both of which may be pollutants. Finally, many nonbiological chemical reactions yield unwanted by-products that must be removed in a separate purification step.

5 Unlike most nonbiological chemical reactions, biological conversions proceed at biological temperatures with water as the solvent. The cells

can often be immobilized on a supporting structure for continuous processing. Another valuable asset of biological conversions is their specificity: one enzyme usually catalyzes only one kind of reaction at a specific site on the substrate molecule. The enzyme can also be made to select one isomer, or molecular form of a compound, in a mixture of forms to produce a single isomer of the product. These characteristics account for the high yields typical of biological conversions, which can reach 100 percent.

Paragraph 1 classifies microorganisms that have industrial applications (and then uses comparison and contrast to describe each class). Paragraph 2 classifies the important products of the four microorganisms. Paragraph 3 classifies the two important commercial applications of the microbial cells (the first of the four products classified in paragraph 2) and defines the first one: as a source of protein. The second application, as an agent of biological conversions, is the subject of paragraphs 4 and 5 (which are structured so as to contrast with each other).

Following is an example of partition.

Preventing computer abuse requires a three-pronged attack. First, issue a policy statement that defines specifically what can and cannot be done with the computer. This in itself won't prevent most computer abuse, which is deliberate and planned, but it will at least establish guidelines under which abusers who are caught can be prosecuted. Second, investigate potential employees carefully. A middle manager requires a more substantial investigation than a keypunch clerk, of course. A number of firms specialize in investigating departmental personnel; their fees range from $25 to $800, depending on the detail required. Third, form an information security department to prevent, detect, and deal with computer abuse. Again, a number of consulting companies specialize in instructing clients how to set up this kind of department.

This paragraph partitions the process of guarding against computer abuse into three phases, thus making the process easier to understand and remember.

For more examples of partition, see Chapters 8 and 9, which cover descriptions of mechanisms and processes.

CAUSE AND EFFECT

Technical writing often involves cause-and-effect reasoning. Sometimes you will reason forward: cause to effect. If we raise the price of a particular product we manufacture (cause), what will happen to our sales (effect)? The government has rules that we may not use a particular chemical in our production process (cause); what will we have to do to keep the production process running smoothly (effect)? Sometimes you will reason backward: from effect to cause. Productivity went down by 6% in the last quarter (effect); what factors led to this decrease (causes)? The federal government has decided that used-car dealers are not required to tell potential customers about the cars' defects (result); why did the federal government reach this decision (causes)?

Causality, the use of cause-and-effect reasoning, is therefore the way to answer the following two questions:

What will be the effect(s) of X?

What caused X?

Causality requires an understanding of the two basic forms of reasoning: inductive and deductive.

In inductive reasoning, the writer moves from the specific to the general by adding up a series of facts and coming to a conclusion. If a laboratory rat dies after being injected with a chemical, you might suspect that the chemical caused the death, but you couldn't be sure. But if a significant number of rats—say, 20—all reacted in the same way to exactly the same treatment, you could be reasonably sure that the chemical was responsible. Inductive reasoning is the bedrock of the modern scientific method, the basis of trial-and-error experimentation. However, it can be challenged, and often is.

Generally, challenges to inductive reasoning are aimed not at the validity of the reasoning itself, but at its usefulness. For example, defenders of the chemical saccharin assert that in experiments mice were given a quantity of the chemical several hundred times greater than a human could possibly ingest and, therefore, that the experiments did not demonstrate that saccharin is harmful to humans. Sometimes, the very nature of inductive reasoning is challenged. Tobacco lobbyists assert that the scientific community has not "proven" that cigarette smoking is harmful to the smoker. The lobbyists do not dispute the data that link cigarette smoking with cancer, heart disease, and a number of other disorders; they just don't accept the surgeon general's inductive conclusion: that smoking has been found to be hazardous to your health. Inductive reasoning does produce only circumstantial evidence, but arguments that it doesn't "prove" anything are usually considered self-serving.

In deductive reasoning, the writer moves from the general to the specific. The common illustration of deductive reasoning is the syllogism, in which a minor premise is added to a major premise to yield a conclusion:

MAJOR PREMISE

All men are mortal.

MINOR PREMISE

Dworski is a man.

CONCLUSION

Dworski is mortal.

Notice that syllogisms cannot be restructured arbitrarily:

Dworski is mortal.

All men are mortal.

Dworski is a man.

Dworski could be a fish or a carrot. Similarly:

Rain makes the streets wet.

The streets are wet.

It rained.

Maybe a water main burst.

Like inductive reasoning, deductive reasoning can be challenged. In most cases, the challenge is based on a dispute over the validity of the major premise. Those who believe the Bible is a scientifically valid document might create a syllogism such as this:

The Bible is scientifically accurate.

The Bible says that God created all of the animals as fixed forms.

God created all of the animals as fixed forms.

Obviously, they would find it difficult to agree with an evolutionist, who would not accept that the Bible is a scientifically accurate document.

The relationship between inductive and deductive reasoning should by now be clear: inductive reasoning creates the major premises used in deductive reasoning. Mischievous people will infer from this relationship that it is impossible ever to know anything—that when we toss a ball into the air, we can't count on its coming down (the results of all those other ball tosses were just coincidence). If gravity is just a rumor, all deductions based on the existence of gravity are nonsense.

These games can be fun, but they hide the fact that establishing causal links is a serious task that requires great care. When you use inductive logic, you must be sure to abide by the principles of research in your field: if the specifics you collect are inaccurate or insufficient, the conclusion you derive from them might be invalid. And when you use deductive logic, you must make sure your major premise is clear and indisputable (at least within your professional community) and that your syllogistic reasoning is correct.

When you use causality, be careful to avoid three basic logical errors:

1. inadequate sampling
2. *post hoc, ergo propter hoc*
3. begging the question

1. *Inadequate sampling.* Don't draw conclusions about causality unless you have an adequate sample of evidence. It would be illogical to draw the following conclusion:

> The new Gull is an unreliable car. Two of my friends own Gulls, and both have had reliability problems.

The writer would have to study a much larger sample and compare the findings to those for other cars in the Gull's class before he or she could reach any valid conclusions.

2. *Post hoc, ergo propter hoc.* This phrase means "After this, therefore, because of this." The fact that *A* precedes *B* does not mean that *A* caused *B*. The following statement is illogical:

> There must be something wrong with the new circuit breaker in the office. Ever since we had it installed, the air conditioners haven't worked right.

The loss of electricity *might* be caused by a problem in the circuit breaker; on the other hand, it might be caused by inadequate wiring or a problem in one of the air conditioners.

3. *Begging the question.* To beg the question is to assume the truth of what you are trying to prove. The following statement is illogical:

The Matrix 411 printer is the best on the market because it is better than the other printers.

The writer here states that the Matrix 411 is the best because it is the best; he does not show why he thinks it is the best.

The following passage, from a discussion of why mammals survived but dinosaurs did not (Jastrow 1977: 104–105), illustrates a very effective use of cause-and-effect reasoning.

Effect
Cause no. 1

Why were the ancestral mammals brainier than the dinosaurs? Probably because they were the underdogs during the rule of the reptiles, and the pressures under which they lived put a high value on intelligence in the struggle for survival. These little animals must have lived in a state of constant anxiety—keeping out of sight during the day, searching for food at night under difficult conditions, and always outnumbered by their enemies. They were Lilliputians in the land of Brobdingnag. Small and physically vulnerable, they had to live by their wits.

Cause no. 2

The nocturnal habits of the early mammal may have contributed to his relatively large brain size in another way. The ruling reptiles, active during the day, depended mainly on a keen sense of sight; but the mammal, who moved about in the dark much of the time, must have depended as much on the sense of smell, and on hearing as well. Probably the noses and ears of the early mammals were very sensitive, as they are in modern mammals such as the dog. Dogs live in a world of smells and sounds, and, accordingly, a dog's brain has large brain centers devoted solely to the interpretation of these signals. In the early mammals, the parts of the brain concerned with the interpretation of strange smells and sounds also must have been quite large in comparison to their size in the brain of the dinosaur.

Effect
Cause

Intelligence is a more complex trait than muscular strength, or speed, or other purely physical qualities. How does a trait as subtle as this evolve in a group of animals? Probably the increased intelligence of the early mammals evolved in the same way as their coats of fur and other bodily changes. In each generation, the mammals slightly more intelligent than the rest were more likely to survive, while those less intelligent were likely to become the victims of the rapacious dinosaurs. From generation to generation, these circumstances increased the number of the more intelligent and decreased the number of the less intelligent, so that the average intelligence of the entire population steadily improved, and their brains grew in size. Again the changes were imperceptible from one generation to the next, but over the course of many millions of generations the pressures of a hostile world created an alert and relatively large-brained animal.

Notice that in this discussion, which is addressed to a general audience, the author has sequenced the two causal relationships just as he has structured the cause and effect within each relationship. Each relationship begins with a question that states the effect and reasons backward to the cause. Similarly, the first causal relationship—the fact that ancestral mammals were more intelligent than the dinosaurs—is the "effect" of the second causal relationship—that natural selection "caused" the greater intelligence among the mammals.

WRITER'S CHECKLIST

The following checklist covers comparison and contrast, classification and partition, and cause and effect.

COMPARISON AND CONTRAST

1. Does the comparison-and-contrast passage
   - **a.** Include the necessary criteria? _____
   - **b.** Employ an appropriate structure? _____

CLASSIFICATION AND PARTITION

1. Does the classification or partition passage
   - **a.** Use only one basis at a time? _____
   - **b.** Use a basis that is consistent with the audience and purpose? _____
   - **c.** Avoid overlap? _____
   - **d.** Include all of the appropriate categories? _____
   - **e.** Arrange the categories in a logical sequence? _____

CAUSE AND EFFECT

1. Does the cause-and-effect passage
   - **a.** Arrange the items in a logical sequence? _____
   - **b.** Avoid inadequate sampling? _____
   - **c.** Avoid *post hoc, ergo propter hoc* reasoning? _____
   - **d.** Avoid begging the question? _____

EXERCISES

1. Write a 500-word comparison-and-contrast paper on one of the following topics. Be sure to indicate a clear purpose for your paper.

a. lecture classes and recitation classes
b. the record reviews in *Stereo Review* and *Rolling Stone*
c. manual transmission and automatic transmission automobiles
d. black-and-white and color photography

2. Rewrite the following comparison-and-contrast paragraph so that it contains a clear topic sentence.

> Both the "Gorilla" and the "Hunter" welding shields can be worn without a hard hat and allow the welder to move freely in a congested area. Neither provides head protection. The "Gorilla" is a leather hood. It is uncomfortable because it fogs up easily and absorbs unpleasant odors. Most welders prefer the "Hunter" because it does not fog up and because, being made of plastic, it does not absorb odors. Inspectors have found that welders wear it even when not required to do so. The result has been that sometimes it is difficult to find them. No problem is found in locating the "Gorilla" shields.

3. Write a 500-word classification paper on one of the following topics:

a. foreign cars
b. smoke alarms
c. college courses
d. personal computers
e. insulation used in buildings
f. cameras

4. Write a 500-word partition paper about a piece of equipment or machinery you are familiar with, or about one of the following topics.

a. a student organization on your campus
b. a tape cassette
c. a portable radio
d. a bicycle
e. an electric clock

5. Rewrite the following classification paragraph to improve its structure, development, and writing style.

> There are three basic types of tape decks that are available. The first type is the reel to reel. The reel to reel is the best type of tape deck on the market today, simply because they have been designed to be more versatile than the other types. The next type of tape deck is the eight-track tape player. This type of unit uses a cartridge to make tape loading easier. However, this type of unit sacrifices the quality of sound for ease of operation. The next type of tape deck is the cassette. This type of unit is perhaps the least likely choice for hi-fi since it uses a slower tape speed, which in turn sacrifices quality. In both eight-track and cassette tape decks, quality has been sacrificed for ease of operation. However, the amount of sound quality sacrificed will not be able to be

detected by the human ear, because of its lack of sensitivity. Therefore, these types of units are competitive on today's market.

6. Write a 500-word causality paper—reasoning either forward or backward—on one of the following topics.

a. the number of women serving in the military
b. the price of gasoline
c. the emphasis on achieving high grades in college
d. the prospects for employment in your field

7. In one paragraph each, describe the logical flaws in the following items.

a. The election couldn't have been fair—I don't know anyone who voted for the winner.
b. 1. Endangered animals should be protected.
   2. Alligators are plentiful.
   3. Alligators should not be protected.
c. Environmentalists ought to be ignored because we don't need any more cranks in this world.
d. 1. Fish swim.
   2. Bill swims.
   3. Bill is a fish.
e. Since the introduction of cola drinks at the start of this century, cancer has become the second greatest killer in the United States. Cola drinks should be outlawed.

REFERENCES

Demain, A. L., and N. A. Solomon. 1981. Industrial microbiology. *Scientific American* 245, no. 3 (September): 69–71.

Jastrow, R. 1977. *Until the sun dies.* New York: Norton.

Strel, G. A., and G. N. Lanier. 1981. Dutch elm disease. *Scientific American* 245, no. 2 (August): 60.

CHAPTER EIGHT

MECHANISM DESCRIPTIONS

A mechanism description might be defined as a verbal representation of an object or device. This definition, however, requires a little elaboration. First, a mechanism description often is not purely verbal; generally, graphic aids are included to help the readers understand the verbal description. And second, the words *object* and *device* are so broad as to be almost meaningless. In one sense, a paper clip is an object, and so is a nuclear power plant. For the purposes of this discussion, however, *mechanism* will be used to refer to objects or devices with several identifiable parts that work together as a system. Given this qualification, a paper clip, because it is so simple and does not consist of several parts, would not be considered a mechanism. A nuclear power plant, however, would be thought of as a mechanism—despite its immense complexity—because it can be described (very generally) as a system consisting of several different "parts." Each of these parts—such as the cooling system—is of course made up of dozens of different, smaller mechanisms.

Most of the mechanisms described in technical writing are either pieces of equipment (generators, transformers, carburetors, and so forth) or tools (hydraulic lifts, radial-arm saws, centrifuges, and the like). Mechanism descriptions are used in almost every kind of technical writing. A writer wants to persuade his readers to authorize the purchase of some equipment; part of his proposal will be a mechanism description. An engineer is trying to describe to her research-and-development department the features she would like incorporated in a piece of equipment that is now being designed; she will include a mechanism description in her report. An engineer is trying to describe to the sales staff how a product works so that they can advertise it effectively; he will include a mechanism description. Mechanism descriptions are used frequently by companies communicating with their clients and customers; operating instructions, for instance, often include mechanism descriptions.

Notice that a mechanism description rarely appears as a separate document. Almost always, it is part of a larger report. For example, a maintenance manual for a boiler system might begin with a mechanism description, to help the reader understand how the system operates. However, the ability to write an effective mechanism description is so important in technical writing that it is discussed here as a separate technique.

THE WRITING SITUATION

Before you begin to write a mechanism description as part of a longer document, consider carefully how the audience and purpose of the document will affect the way you describe the mechanism.

What does the audience already know about the general subject? If, for example, you are to describe an electron microscope, you first have to know whether your readers understand what a microscope is. The audience will also determine your use of technical vocabulary as well as your sentence and paragraph length. Another audience-related factor is the use of graphic aids. Less knowledgeable readers rely on simple illustrations, but they might have trouble understanding sophisticated schematics or decision charts.

What are you trying to accomplish in the mechanism description? If you want your readers to understand how a basic computer works, you will want to write a general description: that is, a description that applies to several different varieties of the computer. If, however, you want to enable your readers to understand the intricacies of a particular model of computer, you will want to write a particular description. A general description of a typewriter will use classification to point out that there are manual, electric, and electronic typewriters and then go on to a fundamental partition of typical models of each, describing the basic parts. A particular description of a Smith Corona Model 2100 will describe only that one typewriter and most likely include a highly detailed partition of that model. An understanding of your purpose will determine every aspect of the description, including length and amount of detail and number and type of graphic aids included.

THE STRUCTURE OF THE MECHANISM DESCRIPTION

Most mechanism descriptions have a three-part structure:

1. a general introduction that tells the reader what the mechanism is and what it does (definition/classification)
2. a part-by-part description of the mechanism (partition)
3. a conclusion that summarizes the description and tells how the parts work together

THE GENERAL INTRODUCTION

The general introduction provides the basic information that your readers will need in order to understand the detailed description that follows. In writing the introduction, answer these five questions about the mechanism:

1. What is it?
2. What is its function?
3. What does it look like?
4. How does it work?
5. What are its principal parts?

1. *What is the mechanism?* Generally, the best way to answer this question is to provide a sentence definition (see Chapter 6): "An electron microscope is a microscope that uses electrons rather than visible light to produce magnified images." Then elaborate if necessary.

2. *What is the function of the mechanism?* State clearly what the mechanism does: "Electron microscopes are used to magnify objects that are smaller than the wavelengths of visible light." Often, the function of a mechanism is implied in its sentence definition: "A hydrometer is a sealed, graduated tube, weighted at one

end, that sinks in a fluid to a depth that indicates the specific gravity of the fluid."

3. *What does the mechanism look like?* Include a photograph or drawing if possible (see Chapter 10). If not, use an analogy or a comparison to a mechanism that would be familiar to your readers: "The cassette that encloses the tape is a plastic shell, about the size of a deck of cards." Mention the material, texture, color, and the like, if relevant.

4. *How does the mechanism work?* In a few sentences, define the operating principle of the mechanism:

A tire pressure gauge is essentially a calibrated rod fitted snugly within an open-ended metal cylinder. When the cylinder end is placed on the tire nozzle, the pressure of the air escaping from the tire into the cylinder pushes against the rod. The greater the pressure, the farther the rod is pushed.

5. *What are the principal parts of the mechanism?* Few mechanisms described in technical writing are so simple that you could discuss all of the parts. And, in fact, you rarely will need to describe them all. Some parts will be too complicated or too unimportant to be mentioned; others will already be sufficiently understood by your readers. A description of a bicycle would not mention the dozens of nuts and bolts that hold the mechanism together, nor would it mention the bicycle seat. The focus of the description would be the chain, the pedals, the wheels, and the frame.

At this point in the description, the principal parts should merely be mentioned; the detailed partition comes in the next section. The parts can be named in a sentence or (if there are many parts) in a list. Make sure that you have named the parts in the order in which they will be described in the body of the description.

The information provided in the introduction generally follows this five-part pattern. However, don't feel you have to answer each question in order, or in separate sentences. If your readers' needs or the nature of the mechanism suggests a different sequence—or different questions—adjust the introduction accordingly.

THE PART-BY-PART DESCRIPTION

The body of the mechanism description—the part-by-part description—is essentially like the introduction in that each major part is treated as if it were itself a mechanism. That is, in writing the body you define what each part is, then describe its function, operating principle, and appearance. The discussion of the appearance should include shape, dimension, material, and physical details such as texture and color (if essential). For some descriptions, other qualities, such as weight or hardness, might also be appropriate. If a part has any important subparts, describe them in the same way. A mechanism description therefore resembles a map with a series of detailed insets. A description of a stereo set would include the turntable as one of its parts. The description of the turntable, in turn, would include the tonearm as one of *its* parts. And finally, the description of the tone arm would include the cartridge as one of *its* parts. This process of ever-increasing specificity would continue as required by the complexity of the mechanism and the needs of the readers.

In planning to partition the mechanism into its parts, don't forget to discuss them in a logical sequence. The most common structure reflects the way the mechanism works or is used. In a stereo set, for instance, the "sound" begins at the turntable, travels into the amplifier, and then flows out through the speakers. Another common sequence is based on the physical structure of the mechanism: from top to bottom, outside to inside, and so forth. A third sequence is to move from more-important to less-important parts. (Be careful when you use this sequence: Will your readers understand and agree with your assessment of "importance"?) Most mechanism descriptions could be organized in a number of different ways. Just make sure you consciously choose a pattern; otherwise you might puzzle and frustrate your readers.

THE CONCLUSION

Mechanism descriptions generally do not require elaborate or long conclusions. A brief conclusion is necessary, however, if only to summarize the description and prevent the readers from placing excessive emphasis on the part discussed last in the part-by-part description. A common technique for accomplishing these goals is to describe briefly how the different parts interact as the mechanism performs its function. The conclusion of a description of a telephone, for example, might include a paragraph such as the following:

> When the phone is taken off the hook, a current flows through the carbon granules. The intensity of the speaker's voice causes a greater or lesser movement of the diaphragm and thus a greater or lesser intensity in the current flowing through the carbon granules. The phone receiving the call converts the electrical waves back to sound waves by means of an electromagnet and a diaphragm. The varying intensity of the current transmitted by the phone line alters the strength of the current in the elec-

tromagnet, which in turn changes the position of the diaphragm. The movement of the diaphragm reproduces the speaker's sound waves.

The following general description of a record turntable is addressed to the lay reader (Isganitis 1977). Marginal comments have been added to the description.

GENERAL DESCRIPTION OF A RECORD TURNTABLE

INTRODUCTION

Because the turntable is well known, the writer uses an informal parenthetical definition rather than a formal definition. The definition concentrates on the function rather than the nature of the mechanism.

The music that comes out of the speakers of a record player is encoded in the grooves of the phonograph record. The first step in getting the music out of the record and into the air is performed by the record turntable, the component that "reads" the grooves of the record and passes the encoded signal on to be amplified and finally transmitted.

Figure 1 shows the four basic components of the record turntable.

The writer describes the operating principle of the mechanism.

The operating principle of the turntable is simple. The motor and the drive system rotate the platter, on which the record rests. As the record rotates, the tonearm picks up the recorded signals and transmits them to the amplifier.

THE COMPONENTS

The writer describes the role of the motor in the operation of the whole mechanism. The mechanism description is general, not particular.

THE MOTOR The function of the motor is to provide a source of power to turn the platter at an accurate and consistent speed. Speed accuracy is important in preventing "wow and flutter," the short-term fluctuations that give the recorded music a harsh and muddy sound.

Although several different kinds of motors are used in turntables, the most common kind is the AC synchronous motor, which operates at a fixed speed determined by the frequency of the AC voltage from the

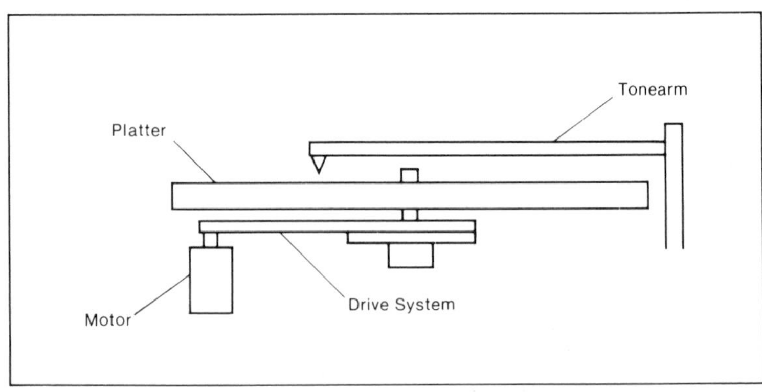

FIGURE 1

Basic Components of a Turntable

power line. A good turntable will have a speed accuracy better than ± 0.10%.

The description of the drive system focuses on its function rather than its nature. Because the writer is going to describe each of the three types of drive systems, he introduces them in a list.

THE DRIVE SYSTEM The drive system transmits the power from the motor to the platter. A good drive system will not introduce any mechanical vibrations that cause "rumble," a humming sound.

Three basic drive systems are used in turntables:

1. belt-drive
2. idler-rim drive
3. direct-drive

The writer uses sketches to clarify his descriptions of each of the three types of drive systems.

After describing each drive system, the writer briefly points out its advantages and (if applicable) its disadvantages.

In the belt-drive system (see Figure 2), a pulley is mounted on the end of the motor shaft. This pulley drives a flexible rubber belt coupled to the underside of the platter. The advantage of the belt-drive system is that the belt acts as a mechanical damper, reducing the rumble and the wow and flutter. However, the belt eventually deteriorates, varying the speed at which the platter turns.

In the idler-rim system (see Figure 3), the motor shaft is linked to the platter by a hard rubber wheel. Idler-rim drive is a sturdy system, but because the platter is directly linked to the motor, rumble is relatively high.

In the direct-drive system (see Figure 4), the platter rests directly on the motor shaft. The direct-drive is powered by a DC servocontrolled motor, which can be operated at the exact platter speed required: 33⅓ or 45 RPM. Direct-drive requires no linkage, and thus it produces the least rumble and wow and flutter. Direct-drive is a sturdy and reliable system, but it is expensive.

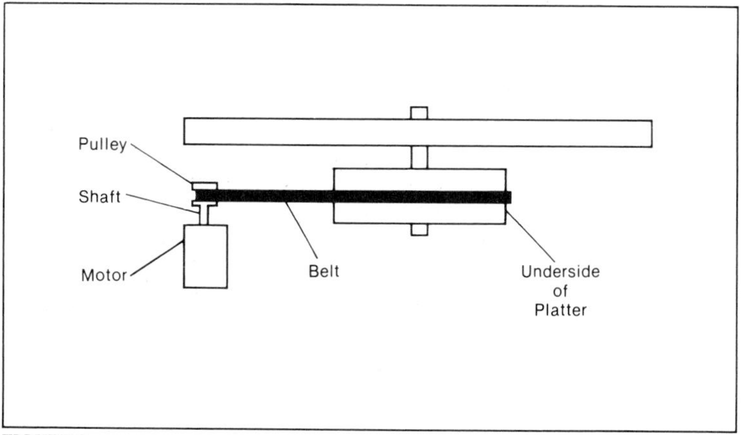

FIGURE 2
Belt-drive System

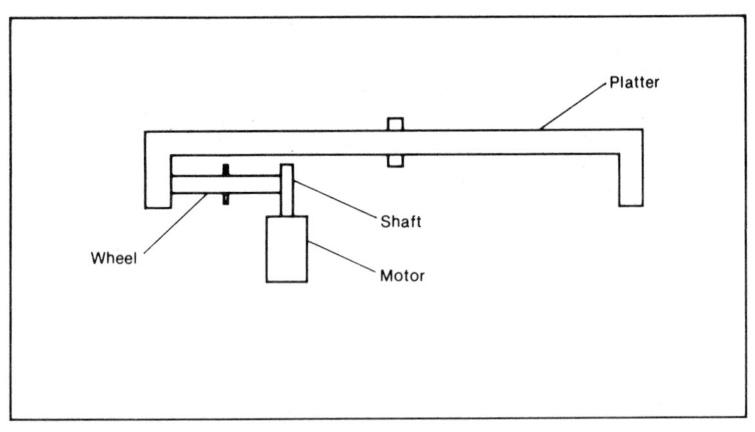

FIGURE 3
Idler-rim Drive System

The platter requires only a brief discussion because it is a relatively simple component.

THE PLATTER The platter is the machined metallic disc (sometimes covered with a soft material such as felt) on which the record rests while it is being played. There is little variation in the quality of different platters.

The sentence definition of the tonearm mentions its function.

THE TONEARM The tonearm houses the cartridge, the removable case that contains the stylus (the "needle") and the electronic circuitry, which converts the mechanical vibrations of the record grooves into electrical signals. The cartridge is the crucial component of the turntable because it actually touches the grooves of the record, thus affecting not only the sound quality but also the record wear.

The design of every kind of tonearm is unique. However, the operating characteristics of a tonearm can be measured. Four factors are generally considered in describing the performance of a tonearm:

1. tracking force
2. pivot friction

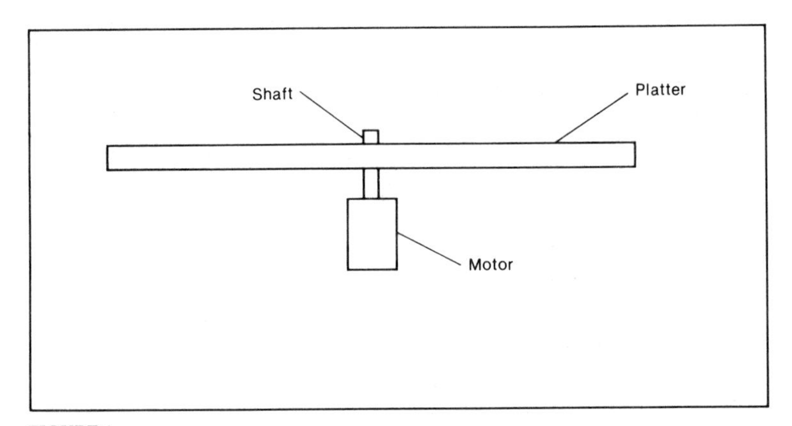

FIGURE 4
Direct-drive System

3. tracking error
4. tonearm resonance

"Tracking force" is defined effectively in a brief extended definition.

Tracking force is the force that the stylus and tonearm exert upon the record being played. The amount of tracking force required by a particular cartridge depends on the quality of its stylus construction. High-quality cartridges require low tracking forces from ½ to 1½ grams. A tonearm should therefore be capable of operating at low tracking forces so that high-quality cartridges may be used.

"Pivot friction" is also defined effectively.

Pivot friction is horizontal and vertical friction in the pivot of the tonearm. Excessive pivot friction requires higher tracking forces to prevent the stylus from jumping the record groove. This in turn increases record and stylus wear.

"Tracking error" is not defined clearly. The reader does not know what tracking error is until almost the end of the paragraph.

Tracking error (or tangent error) results when the cartridge is not held exactly tangent to the record circumference, as shown in Figure 5. The tracking error figure for a tonearm is a measure of the error in the angle of the cartridge with respect to the exact tangent. Depending on the arm geometry, tracking error will vary from point to point along the record radius. When tracking error occurs, the stylus does not ride properly in the record groove, distorting the sound and damaging the record grooves. The maximum value of tracking error for a high-quality tonearm is approximately 5 degrees at the outer edge of a 12-inch record.

"Tonearm resonance" could also be defined more clearly.

Tonearm resonance occurs when the tonearm encounters a signal (or vibration) of a frequency equal to its own resonant frequency. At this resonant frequency, the arm undergoes abnormally large vibrations, resulting in mistracking of the record groove by the stylus. Resonance is a common problem in many mechanical devices and is not easily avoided. However, proper design can limit the range of frequencies over which resonance will occur. For tonearms, the ideal range for resonant frequencies is between 10 and 15 Hz, because it is below the audible frequency range (20 to 20,000 Hz).

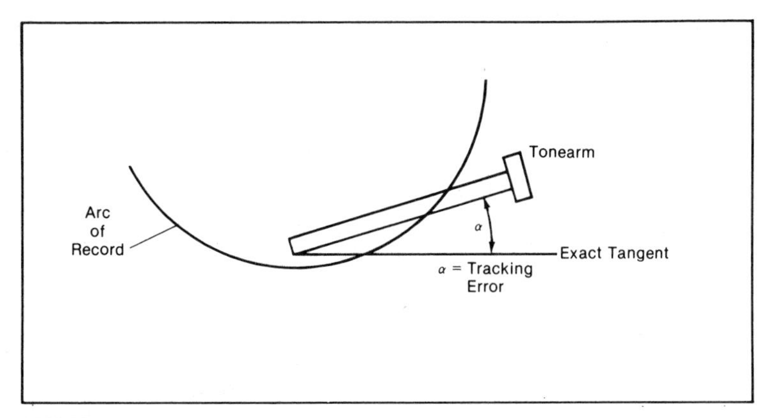

FIGURE 5
Illustration of Tracking Error (Top View)

CONCLUSION

Here the writer summarizes his description by showing how the various parts of the turntable work together.

Turntables can cost from about thirty dollars to more than ten times that amount. They range from single-play models, which require that the operator lift the tonearm and place it on the record, to sophisticated systems that the operator can program to search out particular songs without playing the entire side of the record. However, the basic principle of all turntables is the same: a motor produces power that rotates the platter. The cartridge picks up and deciphers the signals embedded in the grooves of the record that is resting on the platter. The signal is then sent along to the other components—the preamplifier, amplifier, and speakers—to produce audible sound.

Following is a description of a solar heating system called the MODEL-TEA (Temple and Adams 1980: 10–19). Marginal comments have been added to the description.

The writers use a process description (see Chapter 9) to introduce the three basic components of the system.

The MODEL-TEA system consists of a solar collector, delivery and control system, and storage. The collector receives energy from the sun, which heats a black metal plate within the collector. Air is blown through the collector, becomes hot, and then is delivered to the living space of the house. The control system automatically governs the air flow according to the demand for heat and available sun. For large systems, a storage bin is provided. Then, when the collector is providing more heat than the living space requires, the surplus heat is sent to storage to be held in reserve for use at night or during periods of cloudy weather.

The writers compare and contrast their system with the more familiar commercial systems, thus clarifying the introduction.

This general relationship of subsystems is similar to that of most commercial active solar systems. All solar systems use either a liquid (water or anti-freeze) or air to carry heat from the collector to the living space. Air was chosen as the heat-carrying medium for the MODEL-TEA system. Appendix B contains a complete discussion of this choice and of other decisions made in the design of this system. The most important difference between the MODEL-TEA and commercially available systems is that the MODEL-TEA is actually built on-site, fully integrated into the building structure, thus providing a tremendous savings in cost.

1.1 COLLECTOR

A sentence definition leads into a further division of the component into subcomponents.

The MODEL-TEA collector is an air-type, flat-plate collector designed for installation on either pitched roofs or vertical walls in new or existing buildings. . . . The major components of the collector are a glass cover system, a black-ribbed metal absorber plate, Thermo-ply building sheathing, and sheet-metal manifold pans located between the outer studs or rafters.

A process description is used again, this time to describe how the component operates.

Sunlight passes through the glazing cover and is converted to heat on the black absorber plate. The glazing cover reduces losses to the outdoor air and the collector plate heats up. Air from the room or rock bin is ducted into the collector manifold pan. . . . Air passes through the manifold openings in the sheathing and travels the length of the collector in the horizontal channels formed between the raised portion of the ribbed absorber and the sheathing. Heat is transferred from the absorber to the air, and heated air then passes through manifold openings into a manifold pan at the opposite end of the collector, and is returned to the room or a rock bin through air ducts. A blower is required to move the air through the system. Other components found in a complete system include rock bin storage, thermostatic controllers, and motorized dampers.

The process of creating the collector is described as a process.

The construction of the MODEL-TEA collector is straightforward. The roof (or wall) is framed in the usual manner, horizontal blocking is added, and manifold blocking is installed. Sheet metal manifold pans are fastened between the rafters (or studs) at each end of the collector. . . . For collectors longer than 26 feet, manifold pans are also located between the rafters on each side of the center rafter, as explained in Chapter 4. Caulking is applied generously all around the upper edge of the manifold pans. The Thermo-ply sheathing is attached to the roof (or wall), and all seams are carefully caulked. Manifold slots are cut in the sheathing. Blocking is fastened around the perimeter of the Thermo-ply.

The absorber plate is 8-inch ribbed, industrial aluminum siding, painted flat black. It is installed over the sheathing, extending to the outer manifold blocking. Then the ends are sealed with EPDM (ethylene propylene diene monomer) rubber end closure strips and a continuous caulk bead, and the top and bottom edges are caulked. Battens are fastened to the absorber plate. The glazing system is then attached: a single layer of glass on the roof version, and double glass on the wall collector. The collector is completed by installing flashing and aluminum glazing bars.

1.2 AIR-HANDLING AND CONTROL SYSTEMS

The first sentence indicates the function of the component and divides it into subcomponents.

The air-handling system consists of fans, ductwork, automatic dampers, and controls which operate to deliver the heated air. If the collector is small, the house can usually receive all the heated air directly from the collector, eliminating the need for a storage bin. Then the required air-handling system is very simple: it consists of one fan, with supply and return ductwork between the collector and the living space. Whenever there is sufficient sunshine and the living space is below the overheating temperature, the controls turn on the fan.

Here the writers introduce the three air-handling modes that they will describe in more detail.

Each of the three modes is described in the same way: first, the writers state when the mode is used, then they describe how the mode operates.

A larger collector can usually deliver more heat to the living space in the daytime than is necessary. In these systems, a storage rock bin is used to hold the surplus heat for a later time, such as at night, when it is needed. This type of full-scale installation is characterized by three air-handling modes: collector to house, collector to storage, and storage to house (Figures 1 to 3). When the collector is producing useful heat and the house concurrently requires heat, the first mode is activated. Air from the house is drawn through the collector, heated, and then delivered back to the house. If the collector heat is not sufficient to totally fulfill the house need, the house temperature continues to drop, and the second stage of the thermostat activates the auxiliary heating system. The collector-to-house mode continues to operate as long as there is both a heating

Figures 1 through 3 are clear and easy to understand.

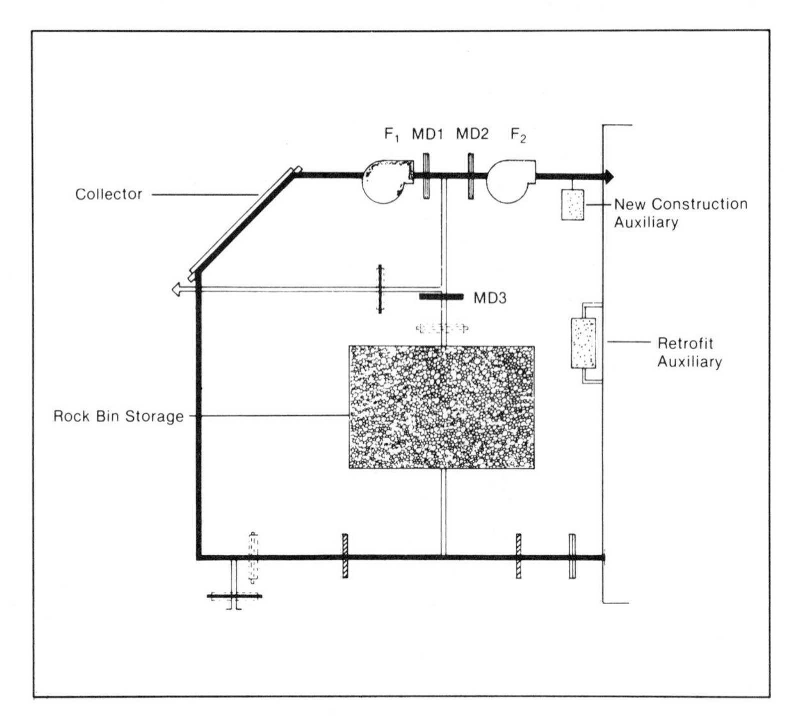

FIGURE 1
Collector-to-house Mode

need and the collector air temperature remains above a minimum set temperature.

The second mode, collector to storage, is engaged whenever the collector is warmer (by 20°F) than the cool side of the rock bin, and the house does not require heat. Air from the cool side of the rock bin is drawn through the collector, heated, and then returned to the hot side of the rock bin. If the collector air temperature drops to within approximately 4° of the cool side of the rock bin, this mode is halted.

The third mode, storage to house, operates when the house requires heat, but the collector is not hot enough to turn on. Air is drawn from the hot side of the rock bin, delivered to the house, and returned to the rock bin's cool side. If this heat is insufficient to meet the house need, the house temperature continues to drop, and the second stage of the thermostat activates the auxiliary heating system. The storage-to-house mode continues until either the house demand is satisfied or, in the case of retrofit, the air being delivered from the rock bin drops below 85°F.

Here the writers describe the hardware used in the three-mode air-handling system.

The three-mode air-handling system uses two fans and three on/off dampers. It is assembled on-site from standard components and is suitable for either new construction or retrofitting to existing buildings. A water coil can be placed in the ductwork so the system may provide do-

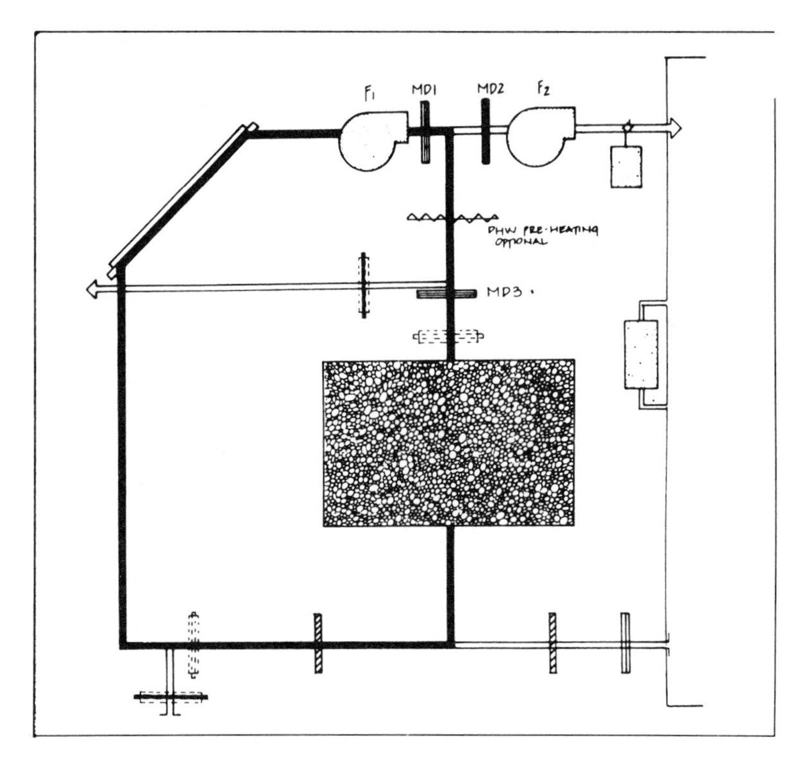

FIGURE 2
Collector-to-storage Mode

mestic hot water (DHW) in addition to space heating. The cost-effectiveness of the system is thereby increased, since it can provide useful service throughout the year.

T.E.A. also developed an innovative air-handling system which is suitable only for new construction. This is an even lower cost system, obtained by careful integration with the building design and the use of a different type of rock bin. This system is described in detail in Chapter 8.

1.3 STORAGE

The principle of operation of the storage system is described.

Installations with sufficiently small collector areas (less than 200 square feet for retrofit and less than 150 square feet for new super-insulated houses), do not require storage. For larger collector areas, storage is essential to achieving efficient use of the collected heat. Storage for the MODEL-TEA system consists of an insulated box, or bin, filled with small rocks. The rocks have sufficient spaces (voids) between them so that air can be blown through the bin between duct connections on the two ends. As heated air from the collector is blown through the bin, heat is transferred from the air to the rocks, and the cooled air then returns to

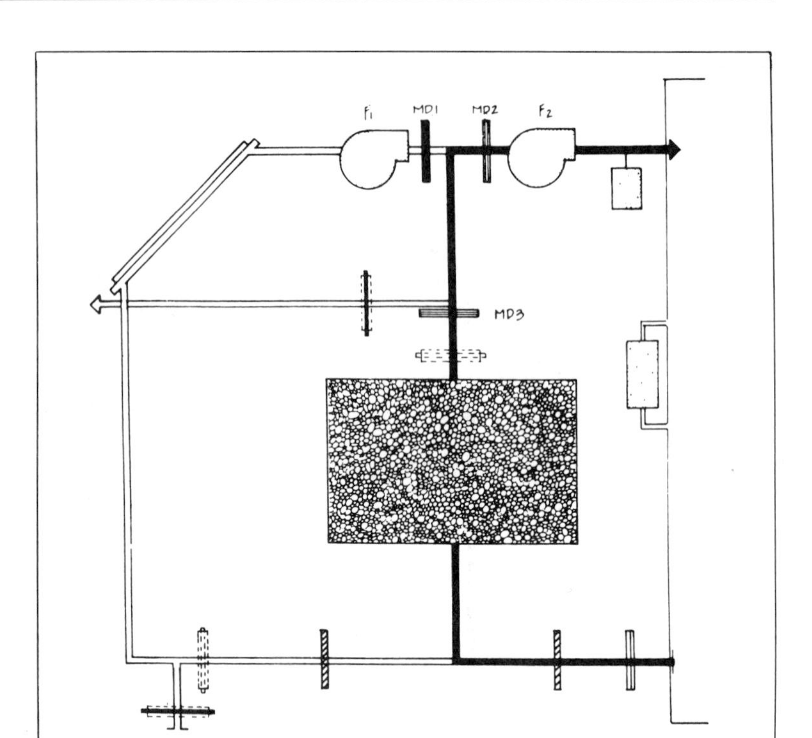

FIGURE 3
Storage-to-house Mode

the collector to be reheated. There is good heat transfer from the air to the rocks, due to the large surface-to-volume ratio of the rocks. In addition, rocks have the capacity to store a large amount of heat with only a moderate rise in temperature.

In the storage-to-house mode, air flows through the rock bin in the opposite direction. Cool house air is drawn into the cool side of the bin, heated by the rocks as it passes, and delivered from the hot side back to the house. Since air flow direction reverses, depending upon whether the rock bin is being charged or discharged, this type of storage is known as a two-way rock bin. Thermal stratification, the fact that the rock bin usually maintains a temperature difference between the two sides, is due to the relatively poor thermal conductivity of the rock-void combination.

The mechanism description has no conclusion. The writers describe each of the components and systems in greater detail in later chapters of the book.

The MODEL-TEA rock bin has a U-shaped air-flow path which allows both inlet and outlet duct connections to be made at the top of the bin, and limits the overall height. These are important features when retrofitting rock bins into existing basements where space may be tight. Rocks can usually be dumped into the bin on a chute through a basement window. For houses on slabs, the rock bin must be integrated into a space on the first floor. The rock bin is constructed on site from standard framing lumber, and lined with plywood and gypsum board. Joints are carefully caulked and taped to prevent air leakage and the bin is fully insulated. Chapter 8 contains a description of the one-way rock bed used in T.E.A.'s innovative system, suitable only for new construction.

WRITER'S CHECKLIST

1. Does the introduction to the mechanism description
   - a. Define the mechanism? _____
   - b. Identify its function? _____
   - c. Describe its appearance? _____
   - d. Describe its principle of operation? _____
   - e. List its principal parts? _____
2. Does the part-by-part description
   - a. Answer, for each of the major parts, the questions listed in item 1? _____
   - b. Describe each part in the sequence in which it was listed in the introduction? _____
   - c. Include illustrations of each of the major parts of the mechanism? _____
3. Does the conclusion
   - a. Summarize the major points made in the part-by-part description? _____
   - b. Include, if appropriate, a description of the mechanism performing its function? _____

EXERCISES

1. Write a 500-word description of one of the following mechanisms or of a piece of equipment used in your field. Be sure to specify your audience and indicate the type of description (general or particular) you are writing. Include appropriate graphic aids.

   a. a carburetor
   b. a locking bicycle rack
   c. a deadbolt lock
   d. a folding card table
   e. a lawn mower
   f. a photocopy machine
   g. a cooling tower
   h. a jet engine
   i. a telescope
   j. an ammeter
   k. a television set

l. an automobile jack
m. a speaker (in a stereo set)
n. a refrigerator
o. a computer
p. a cigarette lighter

2. In an essay of 250 to 350 words, evaluate the effectiveness of one of the following mechanism descriptions.

DESCRIPTION OF A CAMERA

A camera is really just a box that does not admit light. On one end is the lens, on the other is the film. The lens focuses an image of what it sees on the film. As seen in Figure 1, this image is upside down.

How does the image become reversed again? The film is made of silver halide, which undergoes a chemical change when it is exposed to light—the more light, the greater the change. During the process of developing the film, the change is emphasized further. When a positive is made from this negative, the image comes out correct.

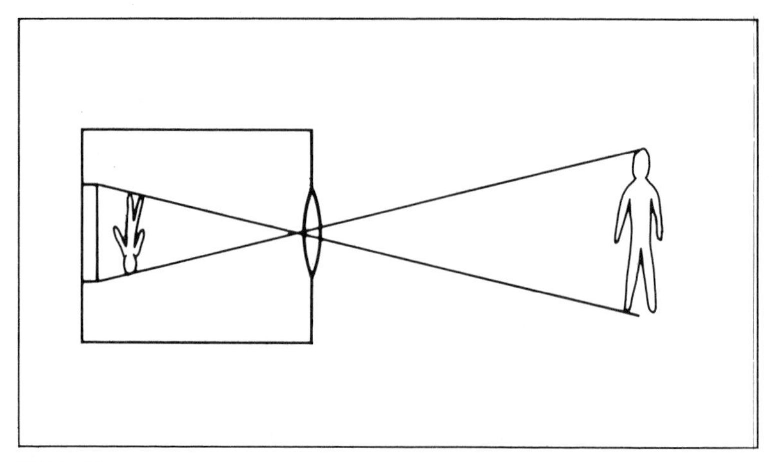

FIGURE 1

SPARK PLUGS

A spark plug is a stationary pair of points inside the combustion chamber. One of the electrodes is connected to ground; the other is insulated. When high voltage flows from the coil to the insulated electrode, it will jump across the gap to the grounded electrode. This is the spark. The center electrode is insulated by the porcelain that surrounds it. It is sensitive to oil deposits. If enough oil collects on the porcelain, the current will flow down the outside of the porcelain without jumping the gap. This is called a dead spark plug, and if enough of them are dead the engine won't start. For this reason, it is necessary to clean the plugs periodically.

3. Rewrite one of the mechanism descriptions in Exercise 2.

REFERENCES

Isganitis, E. J. 1977. Report.

Temple, P. L., and J. A. Adams. 1980. *Solar heating: A construction manual.* Radnor, Pa.: Chilton.

CHAPTER NINE

PROCESS DESCRIPTIONS AND INSTRUCTIONS

This chapter covers two closely related writing techniques: process descriptions and instructions. Both of these techniques involve telling the reader about activities that take place over time. The difference lies in the audience's reason for reading. A process description is a verbal representation of how something happens or how something is done. The readers of a process description want to understand the process, *but they don't want actually to perform the process.* A set of instructions, on the other hand, is a step-by-step guide intended to enable the readers to perform the process.

Process descriptions can be written about activities that people have nothing to do with, such as how the Earth was formed and how cells reproduce, but they are also written about complex activities, such as how an automobile is built, that involve human participation. Some activities can be discussed in either process descriptions or instructions. A process description of making wine focuses on the chemical reactions that transform the fruit juice and other ingredients into the alcoholic beverage. A set of instructions on the same subject focuses on the actual activities that a person must carry out to make wine. In other words, a process description answers the question "How is wine made?" A set of instructions answers the question "How do I go about making wine?"

PROCESS DESCRIPTIONS

Process descriptions are common in technical writing because often readers who are not directly involved in performing a process nevertheless need to understand the process. If, for example, a new law is enacted that places limits on the amount of heated water that a nuclear power plant may discharge into a river, the plant managers have to understand how water is discharged. If the plant is in violation of the law, engineers have to devise alternative solutions—ways to reduce the quantity or temperature of the discharge—and describe them to their supervisors so that a decision can be reached on which alternative to implement. Or consider a staff accountant who performs an audit of one of her company's branches and then writes up a report describing the process and its results. Her readers will review her report to determine whether she performed the audit properly and to learn her findings.

As this last example shows, a process description is not necessarily a separate document. In fact, a process description most often is a preliminary section of a report that reaches conclusions or makes recommendations that the readers can act on. Like mechanism descriptions, however, process descriptions are written so often and are so important that they are discussed here as a separate technique.

THE WRITING SITUATION

Before you write a process description—whether it is to be a separate document or part of a longer one—consider carefully how the audience and purpose will affect the way you describe the process.

What does the audience already know about the process? If, for example, you are to describe how the use of industrial robots will affect the process of manufacturing cars, you first have to know whether your readers already understand the current process. In addition, you need to know whether they understand the basics of robotics. The audience will determine your use of technical vocabulary and your sentence and paragraph length. Also, the audience will affect the number and kinds of graphic aids you use. The less your readers know about your subject, the more graphic aids you will want to use. However, less knowledgeable readers might have trouble understanding complicated figures and tables.

What are you trying to accomplish in the description? Do you want to write a general description, one that encompasses several different varieties of the process? An example of a general description would be a report on how steel is made. Or do you want to write a particular description, one that applies to only one version of the process? An example of a particular description would be a report on how the Johnstown plant makes steel. An understanding of your purpose will affect every aspect of the description, including length and amount of detail and number and type of graphic aids to include.

THE STRUCTURE OF THE PROCESS DESCRIPTION

The structure of the process description is essentially the same as that of the mechanism description. The only real difference is that "steps" replace "parts": the process is partitioned into a reasonable (usually chronological) sequence of steps. Most process descriptions contain the following three components:

1. a general introduction that tells the reader what the process is and what it is used for
2. a step-by-step description of the process
3. a conclusion that summarizes the description and tells how the steps work together

THE GENERAL INTRODUCTION The general introduction gives your readers the basic information they will need to understand the detailed description that follows. In writing the introduction, answer these six questions about the process:

1. What is the process?
2. What is its function?
3. Where and when does it take place?
4. Who or what performs it?
5. How does it work?
6. What are its principal steps?

1. *What is the process?* Generally, the best way to answer this question is to provide a sentence definition (see Chapter 6): "Debugging the program is the process of identifying and eliminating any errors within the program." Then elaborate if necessary.

2. *What is the function of the process?* State clearly the function of the process: "The central purpose of performing a census is to obtain up-to-date population figures on which to base legislative redistricting and revenue-sharing revisions." Make sure that your explanation of the function of the process is clear to your readers; if you are unsure whether they already know the function, state it anyway. Few things are more frustrating for your readers than not knowing why the process should be performed.

3. *Where and when does the process take place?* State clearly the location and occasion for the process, to help your readers understand it: "The stream is stocked at the hatchery in the first week of March each year." These details can generally be added simply and easily. Again, omit reference to these facts only if you are certain your readers already know them.

4. *Who or what performs the process?* Most processes are performed by people, by natural forces or machinery, or by some combination of the two. In most cases, you do not need to state explicitly that, for example, the young trout are released into the stream by a person; the context makes that point clear. In fact, much of the description usually is written in the passive voice: "The water temperature is then measured." Do not assume, however, that your readers already know what agent performs the process, or even that they understand you when you have identified the agent. Someone who is not knowledgeable about computers, for instance, might not know whether a compiler is a person or a thing. The term *word processor* often refers ambiguously to a piece of equipment and to the person who operates it. Confusion at this early stage of the process description can ruin the effectiveness of the writing.

5. *How does the process work?* In a few sentences, define the principle or theory of operation of the process:

> The four-treatment lawn-spray plan is based on the theory that the most effective way to promote a healthy lawn is to apply different treatments at crucial times during the growing season. The first two treatments—in spring and early summer—consist of quick-acting organic fertilizers and weed- and insect-control chemicals. The late summer treatment contains postemergence weed-control and insect-control chemicals. The last treatment—in the fall—uses long-feeding fertilizers to encourage root growth over the winter.

6. *What are the principal steps of the process?* The principal steps of the process should be named, in one or several sentences or in a list. Make sure you have named the steps in the order they will be described in the body of the description. The principal steps in changing an automobile tire, for instance, include jacking up the car, replacing the old tire with the new one, and lowering the car back to the ground. The process of changing a tire also includes some secondary steps—such as placing blocks behind the tires to prevent the car from moving once it is jacked up, and assembling the jack—that are explained or referred to at the appropriate points in the description.

The information provided in the introduction to a process description generally follows this pattern. However, don't feel you have to answer each question in order, or in separate sentences. If your readers' needs or the nature of the process suggest a different sequence—or different questions—adjust the introduction accordingly.

THE STEP-BY-STEP DESCRIPTION The body of the process description—the step-by-step description—is essentially like the introduction, in that it treats each major step as if it were a process. Of course, you do not repeat your answer to the question about who or what performs the action unless a new agent performs the action at a particular step of the process. The other principal questions—what the step is, what its function is, and when, where, and how it occurs—should be answered. In addition, if the step has any important substeps that the reader will need to know in order to understand the process, they should be explained clearly.

The structure of the step-by-step description should be chronological: discuss the initial step first and then discuss each succeeding step in the order in which it occurs in the process. If the process is a closed system and hence has no "first" step—such as the cycle of evaporation and condensation—explain clearly to your readers that the process is cyclical. Then simply begin with any principal step.

Although the structure of the step-by-step description should be chronological, don't present the steps as if they are individual occurrences that have nothing to do with one another. In many cases, one step leads to another causally (see Chapter 7 for a discussion of cause and effect). In the operation of a four-cycle gasoline engine, for instance, each step sets up the conditions under which the next step can occur. In the compression cycle, the cylinder travels upward in the piston, compressing the mixture of fuel and air. This compressed mixture is ignited by the spark from the spark plug in the next step, the power cycle. Your readers will find it easier to understand and remember your description if you clearly explain the causality in addition to the chronology.

A word about tense. The steps should be discussed in the present tense, unless of course you are writing about a process that occurred in the historical past. For example, a description of how the Earth was formed would be written in the past tense: "The molten material condensed. . . ." However, a description of how steel is made would be written in the present tense: "The molten material is then poured into. . . ."

THE CONCLUSION Process descriptions usually do not require long conclusions. If the description itself is brief—less than a few pages—a short paragraph summarizing the principal steps is all that is needed. For longer descriptions, a discussion of the importance or implications of the process might be appropriate.

Following is the conclusion from a description of the four-cycle gasoline engine in operation:

In the intake stroke, the piston moves down, drawing the air-fuel mixture into the cylinder from the carburetor. This mixture is compressed as the piston goes up in the compression stroke. In the power stroke, a spark from the spark plug ignites the mixture, which burns rapidly, forcing the piston down. In the exhaust stroke, the piston moves up, expelling the burned gases.

Following is a general description of how a computer program is developed. The description is adapted from an introductory text on data processing (Martin L. Harris, *Introduction to Data Processing*, 2nd edition. Copyright © 1979. Reprinted by permission of John Wiley & Sons, Inc.). All the words and concepts in the description have already been introduced in earlier chapters of the text. Marginal notes have been added.

HOW A COMPUTER PROGRAM IS DEVELOPED

INTRODUCTION

This brief introduction defines *computer program*. The definition—the second sentence—explains the function of the program and identifies the person who develops it.

A computer is capable of carrying out a tremendous number of operations quickly, but it must be told what to do—and in what order. A computer program is a set of instructions, written by a programmer, that tells the computer what operations to perform. Developing a program involves six basic steps:

1. Analyzing the problem
2. Developing the flow charts
3. Writing the program in a coded form
4. Compiling the program
5. Debugging the program
6. Documenting the program

STEPS

The first step—analyzing the problem—is divided into several substeps.

Analyzing the Problem. In analyzing the problem, the programmer must know what input medium is to be used—cards, tape, or the like. He must know the order in which data fields are arranged on the records and how big these fields are, because records frequently contain more data than is required for any one job.

The *process component* of the program describes what operations are to be performed on the data. Here the programmer decides the order in which to arrange these operations and the points at which logical decisions (branches) have to be made.

Finally, the programmer analyzes the output requirements. Is the output to be punched into cards, written on magnetic tape, or printed as a report? If it is to be printed, what are the headings for the report, and where should each output field be placed?

It is essential that all these questions be asked, and answered, before proceeding with the next step. In obtaining the answers, the programmer will talk not only with the people who designed the system but also with the people who will be using the output. He will talk to the latter primarily to determine the form in which the output data would be most useful to the user.

The chronological structure of the process description is apparent here.

Developing the Flow Charts. Once the programmer has the necessary information about the function of the program, he draws a *flow chart*, a

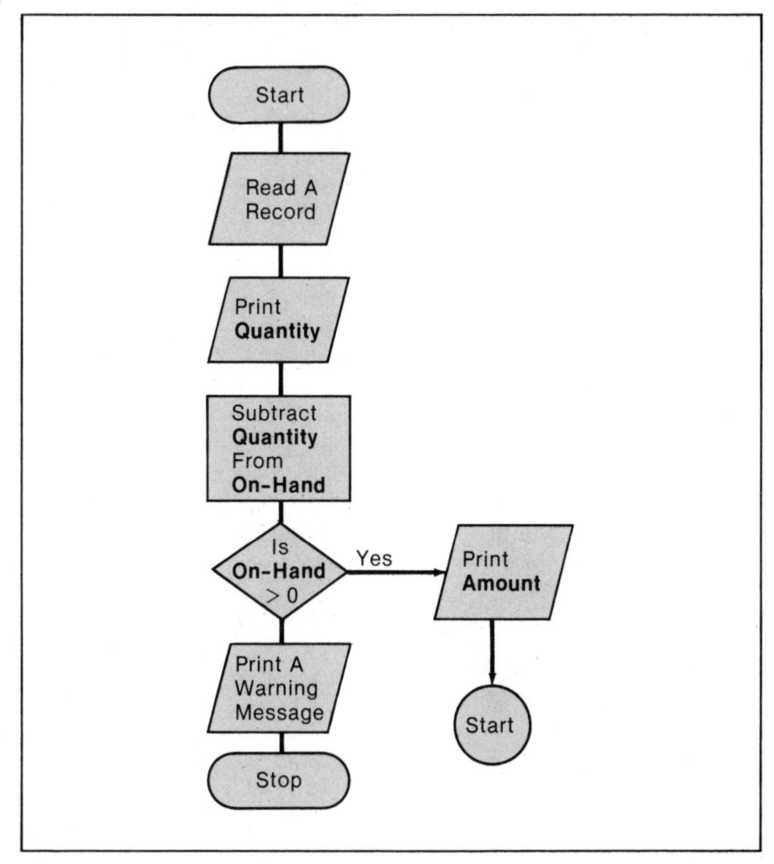

FIGURE 1
A Simple Flow Chart

picture of the program and how it operates (see Figure 1). The different shaped boxes on the flow chart represent different types of operation. Flow charts are useful because they show, in order, the operations to be performed on the data.

The writer explains the causality as well as the chronology of the steps and substeps.

Once the flow chart is drawn, the general accuracy of the proposed program can be checked. The programmer can take some sample data and "pretend" he is the computer system; that is, he does to the data exactly what the flow chart tells him to do. The final output should be exactly what he expected. If it isn't, assuming he didn't make a careless mistake in handling the data, his flow chart contains some incorrect instructions and must be changed. Getting the correct output doesn't guarantee that the completed program will be perfect, since there are many tasks yet to be done, but this simple check will reveal any logical flaws in the programmer's planning so far.

The description of this step contains several parenthetical definitions.

Writing the Program in a Coded Form. After the flow chart is completed, a series of detailed instructions has to be written from it. These instructions are called the *source program* and are usually written in a *symbolic language.* The particular symbolic language chosen for the source program will depend on many factors—two main considerations being the type of job being programmed and the type of computer being used. A symbolic language is one that is relatively easily understood by a human mind but that cannot be understood by a computer. There are many symbolic languages, among them BASIC, FORTRAN, COBOL, ALGOL, PL/1, and RPG.

When the source program has been written, it is punched into cards or stored on magnetic tape or magnetic disks. In this way it becomes available for use when needed.

The first three sentences in this step contain two parenthetical definitions.

Compiling the Program. Next, the source program is translated, or *compiled,* into *machine language,* a language that the computer can understand. Because the source program may still have some errors in it, *the compiler*—a translating program—also produces a printed listing of the source program and a list of the errors in it. These error messages are used by the programmer at the next step of the process, when he is *debugging* his program (that is, when he is finding and removing the errors). Error messages indicate places where the programmer broke some rules of the language used. For example, the English sentence "To the other side of the room go" is not correct according to the rules of English grammar. A teacher might point out the error by saying "the verb is in the wrong place." (The teacher, of course, can decide what the sentence means, but a computer cannot second guess an instruction.)

The writer uses an analogy to clarify his explanation.

The errors detected by the compiler are called *compilation time diagnostics.* They indicate the instruction in which an error occurs and the type of error. All the errors indicated by the compilation time diagnostics must be corrected before the system can process data.

Debugging the Program. When all the compilation time errors have been corrected, the program must then be tested with some data. As the computer system *executes* the program (works through it instruction by instruction), it may find errors in the input data. For example, the input device might be instructed to read a certain field that is supposed to contain numeric data but finds that it contains alphabetic data. This will cause the computer to stop executing the object program and to print some kind of error message. These messages are called *execution time diagnostics*. Another execution time error can be caused by faulty transfer of control instructions. It is possible to put the system into an *infinite loop*. That is, it will continue executing the same sequence, over and over again, because the last instruction sends the system back to the first instruction in the sequence. This means that, in theory, the computer would continue to execute those instructions forever. In practice, however, some time limit is placed on the execution of the program, and if it is not finished before time is up, the system automatically stops the execution and prints a diagnostic message saying why it stopped. Beginning programmers often write infinite loops into their programs.

Documenting the Program. The final step in the preparation of the program comes after it is stored and free of errors; the program is then *documented*. Program documentation refers to the gathering in one place of all relevant information used in the preparation of the program. All job descriptions, flow charts, the written symbolic code, and so on are gathered in one place. If the program develops an error at a later date or has to be modified in some way, this documentation makes it easy for another programmer to see what has been done and why.

A *run manual* is also prepared. This manual contains all the instructions that the computer operator requires in order to run the program—what input files may have to be loaded, what type of output is being prepared, and so forth.

CONCLUSION

The conclusion summarizes the major steps of the process.

Every computer program, regardless of how simple or complicated, was probably developed in the same way. After the data-processing problem is analyzed, the programmer outlines the program in a flow chart and then translates it into a symbolic language. A special program called a *compiler* then translates the program into language that can be understood by the computer. Using clues provided by the compiler, the programmer debugs the program, and once it tests out satisfactorily, he writes the documentation that will be used for reference and instruction guides.

INSTRUCTIONS

A set of instructions is a process description that is written to help the reader perform a specific task. Instructions are central to technical writing, and you are likely to be asked to write them often in your career. An effective set of instructions looks easy to write, but it isn't. You have to make sure not only that your readers will be able to understand and follow your directions easily, but also that in performing the task they won't damage any equipment or, more important, injure themselves or other people.

THE WRITING SITUATION

The purpose of a set of instructions is, of course, obvious and requires little thought. The question of audience, however, is difficult and subtle. In fact, if a set of instructions is ineffective, chances are that the writer inaccurately assessed the audience. Among the most common examples are assembly instructions for a backyard swing or a bicycle. The average consumer has developed a skeptical attitude toward products that announce "Some Assembly Required" on the carton. Assembling the product is easy for the person who wrote the instructions but often difficult for the consumer—because the writer assumes incorrectly that the reader knows the technical vocabulary and can anticipate problems. But most people don't know what a self-locking washer is. And most people don't know that they are wasting their time by tightening all the nuts securely before assembling the whole swing set.

Before you start to write a set of instructions, think carefully about your audience's background and skill level. If you are writing to people who are experienced in the field, you can use the technical vocabulary and make any reasonable assumptions about their knowledge. But if you are addressing people who are unfamiliar with the field, define your terminology and explain your instructions in more detail. Don't be content to write, "Make sure the tires are rotated properly." Define proper rotation and describe how to achieve it.

The best way to make sure you have assessed your audience effectively is to find someone whose background is similar to that of the people to whom you are writing. Give this person a draft of your instructions and watch as he or she tries to perform the tasks. This process will give you valuable information about how clear your instructions are.

One other aspect of audience analysis should be mentioned. Inexperienced writers assume that their readers will read through the instructions carefully before beginning the first step. They won't. Most people will merely glance at the whole set of instructions to see if anything catches their eye before they get started. For this reason, you should organize the instructions in a strict chronological sequence. For example, if you want your readers to make sure the power is off before they throw a switch, don't write:

> Turn the switch to READY. Before you do this, make sure the power is off.

Instead, write:

First, make sure the power is off. Then, turn the switch to READY.

This concept of chronology applies in particular to safety concerns. If a safety precaution applies to the whole procedure, state it emphatically at the start of the instructions.

WARNING:
TO PREVENT SERIOUS EYE INJURY,
SAFETY GOGGLES MUST BE WORN WHEN YOU WELD!

However, if the warning applies only to a particular step in the instructions, insert the warning before that step.

THE STRUCTURE OF THE SET OF INSTRUCTIONS

The structure of the set of instructions is essentially the same as that of a process description. The only real difference is that the conclusion is less a summary than an explanation of how to make sure the reader has followed the instructions correctly. Most sets of instructions contain the following three components:

1. a general introduction that prepares the reader for carrying out the instructions
2. a step-by-step description of the instructions
3. a conclusion

THE GENERAL INTRODUCTION The general introduction gives the readers the preliminary information they will need to follow the instructions easily and safely. In writing the introduction, answer these three questions about the process:

1. Why should this task be carried out?
2. What safety measures or other information about carrying out the instructions should the reader understand?
3. What tools and equipment will be needed?

Why should this task be carried out? Often, the answer to this question is obvious and should not be stated. For example, the purchaser of a backyard barbeque grill does not need an explanation of why it should be assembled. Sometimes, however, the readers need to be told why they should carry out the task. Many preventive maintenance chores—such as changing radiator antifreeze every two years—fall into this category.

What safety measures or other concerns should the reader understand? In addition to the safety measures that apply to the whole task, state any tips that will make your reader's job easier. For example:

NOTE: For ease of assembly, leave all nuts loose.
Give only 3 or 4 complete turns on bolt threads.

What tools and equipment will be needed? The list of necessary tools and equipment is usually included in the introduction so that the readers do not have to interrupt their work to hunt down another tool.

THE STEP-BY-STEP INSTRUCTIONS The step-by-step instructions are essentially like the body of the process description. There are, however, two differences.

First, the instructions should always be numbered. Each step should define a single task that the reader can carry out easily, without having to refer back to the instructions. Don't overload the step:

1. Mix one part of the cement with one part water, using the trowel. When the mixture is a thick consistency without any lumps bigger than a marble, place a strip of about 1″ high and 1″ wide along the face of the brick.

On the other hand, don't make the step so simple that the reader will be annoyed:

1. Pick up the trowel.

Second, the instructions should always be stated in the imperative mood: "Attach the red wire . . ." The imperative is more direct and economical than the indicative mood ("You should attach the red wire . . ." or "The operator should attach the red wire . . ."). Make sure your sentences are grammatically parallel. Avoid the passive voice ("The red wire is attached . . ."), because it can be ambiguous: Is the red wire already attached?

Keep the instructions simple and direct. However, do not omit the articles (*a*, *an*, *the*) to save space. Omitting the articles makes the instructions hard to read and, sometimes, unclear. In the sentence "Locate midpoint and draw line," for example, the reader cannot tell if "draw line" is a noun ("the draw line") or a verb and its object ("draw the line").

Be sure to include graphic aids in your step-by-step instructions. In appropriate cases, each step should be accompanied by a photograph or diagram that shows what the reader is supposed to do. Some kinds of activities—such as adding two drops of a reagent to a mixture—do not need illustration. However, steps that require manipulating physical objects—such as adjusting a chain to a specified tension—can be clarified substantially by graphic aids.

THE CONCLUSION Instructions generally do not require conclusions—except for brief statements about how to turn on equipment that has just been assembled. Often, however, a troubleshooter's checklist is included. This checklist, usually in the form of a table, identifies and solves common problems associated with the mechanism or process described in the instructions.

Following is a portion of the troubleshooter's guide included in the operating instructions of a lawnmower.

PROBLEM	CAUSE	CORRECTION
Mower does not start.	1. Out of gas.	1. Fill gas tank.
	2. "Stale" gas.	2. Drain tank and refill with fresh gas.
	3. Spark plug wire disconnected from spark plug.	3. Connect wire to plug.
Mower loses power.	1. Grass too high.	1. Set mower in "higher-cut" position.
	2. Dirty air cleaner.	2. Replace air cleaner.
	3. Buildup of grass, leaves, and trash.	3. Disconnect spark plug wire, attach to retainer post, and clean underside of mower housing.

Following is a set of instructions on how to change motor oil. The instructions were written by an oil company and distributed by retail stores that sell the oil.

The first section of the instructions ("Here's what you get when you change your own oil") highlights the advantages of doing the

Here's what you get when you change your own oil

Changing your own oil is a simple maintenance job, but one of the most important for your car. Here's what you get:

- **Satisfaction**—you have the satisfaction of doing an important job with your own two hands, and the fun of a job well done.
- **Convenience**—you pick your own good time to do it—no more tying up your car all day or making an appointment.
- **Assurance**—you'll know the right oil is in the right place at the right time.
- **Savings**—do-it-yourselfers always save. Follow the steps outlined here . . . and in your car owner's manual.

What's needed to start

Consult your owner's manual to find out how many quarts of oil are needed and the "weight" of the oil recommended.

Get a new oil filter for the make, model and year of your car. Follow manufacturer's instructions carefully on how to thread and mount filter.

You'll need an oil filter wrench to remove the old filter, and a wrench to fit your car's oil drain plug.

A bayonet spout or other opener will be needed for opening and pouring from oil cans.

The last item is a wide, low pan that will fit under your car to catch draining oil.

Safety first

- Double check your manual for specific safety measures during oil change.
- Place chocks against wheels to prevent car rolling.
- Never use a jack to change oil.
- Don't permit anyone inside car while changing oil.
- Be careful of hot, draining oil.
- Double check to make sure oil filter is seated properly.

An oil change in five easy steps

1. Position car in driveway or similar safe and convenient location. Run engine up to operating temperature, then turn off.

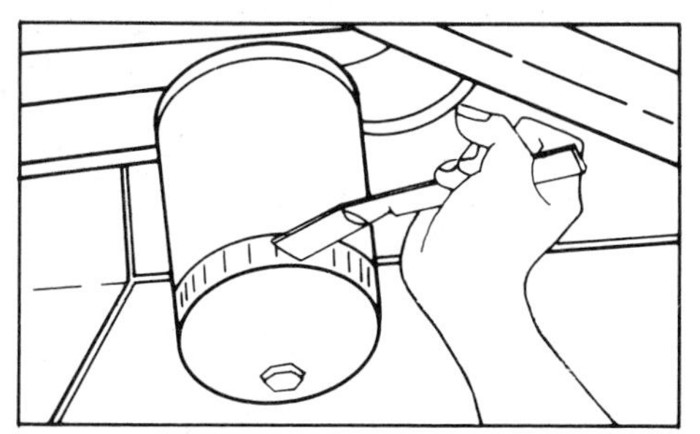

2. Place drain pan under car beneath oil filter. Remove old oil filter with filter wrench. Let oil from old filter drip in pan. Note: turn oil filter *counterclockwise* when removing with filter wrench.

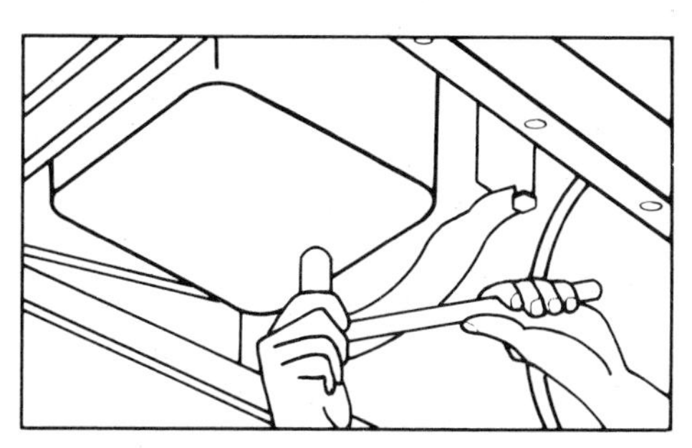

3. Place drain pan under oil pan drain plug. (Note: drain plug located on oil pan, which is lowest part of engine under car. Be careful not to loosen transmission drain plug, which is located on assembly just back of engine). Loosen oil drain plug with wrench. Make last few turns by hand, so plug does not fall in pan. Clean plug so it's ready for replacement after oil has drained. *Be careful not to get hot oil on your hands.*

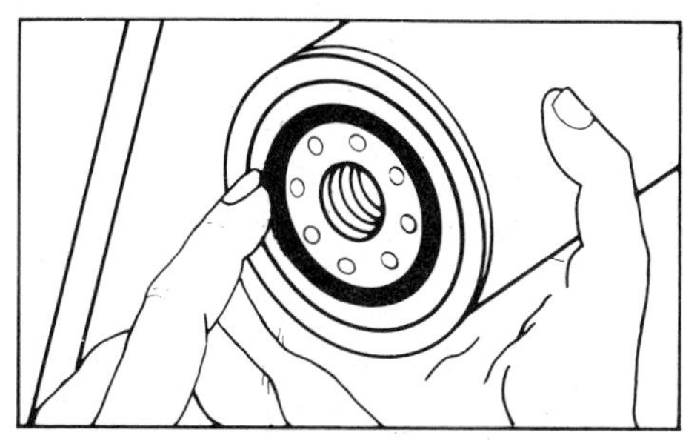

4. Give engine enough time to drain fully. Replace and tighten oil drain plug. Install new oil filter. Place some oil on your fingertip and oil the filter seal to assure a leak-free fit when new filter is seated. Tighten new filter *only by hand.*

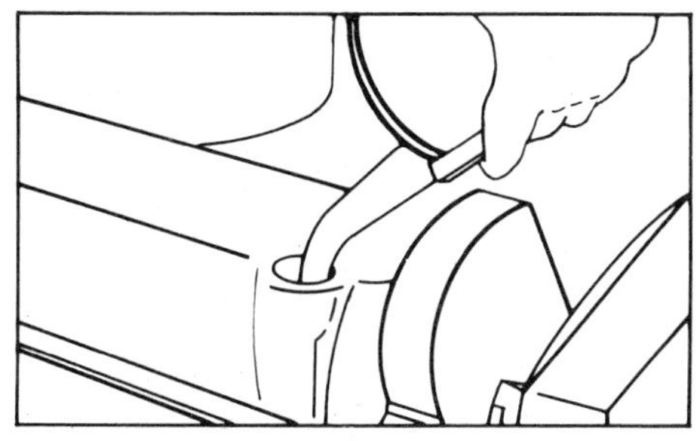

5. Use bayonet spout to open each can of new oil. See owner's manual for oil fill location. Add new oil. Check oil pan drain plug to make sure it is properly seated and sealed. Be sure there is no oil leak around filter. Add all but one quart and run engine a few minutes to circulate oil into filter. Check oil dipstick. If needed, add last quart of oil. Replace oil filter cap.

job yourself rather than having a mechanic do it. The process of changing motor oil does not need to be explained or justified to the readers; therefore, the brochure gets right to the advantages of doing it yourself.

The next section ("What's needed to start") advises readers to consult the car's owner's manual to determine the quantity and weight of oil needed and to learn how to install the new oil filter. The tools and equipment necessary for the job are also listed.

The third section ("Safety First") is effective, except for two of the six points. "Never use a jack to change oil" is an unclear bit of advice. The point the writers wish to make is that the reader should not jack up the car and then get under it—the jack can fail. The final point in the safety list—"Double check to make sure oil filter is seated properly"—is good advice, but it has nothing to do with safety. The comment should appear in the discussion of the steps themselves, rather than in the safety list.

The step-by-step instructions ("An oil change in five easy steps") are brief and generally clear. However, the following steps could be improved:

STEP 1:

The reader of this brochure is likely not to know what the "operating temperature" of the car is. Many cars do not have temperature gauges. The instructions should say how long most cars take to achieve operating temperature. Also, a safety warning should be included: don't run the engine in a closed garage.

STEP 2:

The instruction should explain why a filter wrench is necessary. Without the explanation, many readers will not bother getting the tool. In addition, the last sentence of Step 2 is unclear: "Note: turn oil filter counterclockwise when removing with filter wrench." This sentence could be interpreted as meaning the filter should be turned *clockwise* if a filter wrench is not used.

STEP 3:

The phrasing of the note—"drain plug located on oil pan . . ." is unclear because the writer omitted the articles. The note should read "the drain plug is located on the oil pan. . . ." The note in the brochure could be interpreted as meaning "drain the plug located on the oil pan. . . ." In addition, the warning about hot oil should *precede* the instruction on how to loosen the plug.

STEP 4:

The writer should specify how long draining the old oil will take.

Not reprinted here is the back of the brochure, which gives some helpful advice on how to dispose of the old oil and, also, some advertisements for the manufacturer's products.

WRITER'S CHECKLIST

This checklist covers process descriptions and instructions.

PROCESS DESCRIPTIONS

1. Does the introduction to the process description
   a. Define the process? _____
   b. Identify its function? _____
   c. Identify where and when the process takes place? _____
   d. Identify who or what performs it? _____
   e. Describe how the process works? _____
   f. List its principal steps? _____
2. Does the step-by-step description
   a. Answer, for each of the major steps, the questions listed in item 1? _____
   b. Discuss the steps in chronological order or in some other logical sequence? _____
   c. Make clear the causal relationships among the steps? _____
   d. Include appropriate graphic aids? _____
3. Does the conclusion
   a. Summarize the major points made in the step-by-step description? _____
   b. Discuss, if appropriate, the importance or implications of the process? _____

INSTRUCTIONS

1. Does the introduction to the set of instructions
   a. State the purpose of carrying out the task? _____
   b. Describe safety measures or other concerns that the readers should understand? _____
   c. List necessary tools and equipment? _____
2. Are the step-by-step instructions
   a. Numbered? _____
   b. Expressed in the imperative mood? _____
   c. Simple and direct? _____
3. Are appropriate graphic aids included? _____

4. Does the conclusion

  a. Include any necessary follow-up advice? _____

  b. Include, if appropriate, a troubleshooter's guide? _____

EXERCISES

1. Write a 500-word description of one of the following processes or a similar process with which you are familiar. Be sure to indicate your audience and the type of description (general or particular) you are writing. Include appropriate graphic aids.

- **a.** how steel is made
- **b.** how an audit is conducted
- **c.** how a nuclear power plant works
- **d.** how a bill becomes a law
- **e.** how a suspension bridge is constructed
- **f.** how a microscope operates
- **g.** how we hear
- **h.** how a dry battery operates
- **i.** how a baseball player becomes a free agent
- **j.** how cells reproduce

2. In an essay of 250 to 300 words, evaluate the effectiveness of one of the following process descriptions.

THE FOUR-STROKE POWER CYCLE

The power to drive an engine comes from the four-stroke cycle.

Intake Stroke: Inside the cylinder, the piston moves down, creating a vacuum that draws in the air/fuel mixture through the intake valve.

Compression Stroke: The intake valve shuts and the cylinder moves back up, compressing the air/fuel mixture in what is called the combustion chamber. The purpose of compression is to increase the power of the explosion that will occur in the next stroke.

Power Stroke: The spark across the electrodes of the spark plug ignites the air/fuel mixture, pushing the piston down. This in turn moves the wheels of the car.

Exhaust Stroke: The burned gases escape through the exhaust valve as the piston moves up again.

THE PROCESS OF XEROGRAPHY

Xerography—"dry writing"—is the process used in most photocopy machines. Photocopy machines enable us to produce high-quality, inexpensive copies of printed matter.

Xerography depends on electricity and photoconductivity. A photoconductive material is one that maintains an electric charge in the darkness but loses it when exposed to light.

The first step in today's photocopy machine is to give an electric charge to a selenium roll.

Next, a bright light is shined on the page to be copied. Using lenses, it is projected on the selenium roll. Where the image is bright, the electric charge disappears. Where it is dark, the charge is maintained. Thus, the electric charge on the selenium roll corresponds to the page.

Then, the selenium roll receives a coating of a special dust that sticks only where there is an electric charge. This is then transferred to a piece of paper, which is the photocopy.

3. Rewrite one of the process descriptions in Exercise 2.

4. Write a set of instructions for one of the following activities. Include appropriate graphic aids. In a brief note preceding the instructions, indicate your audience.

- **a.** how to fill out a tax return
- **b.** how to change a bicycle tire
- **c.** how to parallel-park a car
- **d.** how to study a chapter in a text
- **e.** how to light a fire in a fireplace
- **f.** how to make a cassette-tape copy of a record
- **g.** how to tune up a car
- **h.** how to read a corporate annual report
- **i.** how to tune a guitar
- **j.** how to take notes in a lecture class

5. In an essay of 250 to 300 words, evaluate the effectiveness of the following set of installation instructions for a set of outdoor shutters.

OUTDOOR SHUTTER INSTALLATIONS

1. Draw a 4″ to 6″ light vertical line from the top and bottom, 1″ from the sides of the shutter. Locate screw locations by positioning shutter next to window and marking screw holes. If six or eight screws are to be used, make sure they are equidistant.
2. Drill 3/16″ holes at marked locations. DO NOT DRILL INTO HOUSE. Drilling should be done on the ground or a workbench.
3. Mount the shutter screws.

   Notes: 1. For drilling into cement, use a cement bit.
   2. Always wear eye protection when drilling.
4. Make sure shutters are mounted with the top side up.
5. If you are going to paint shutters, do so before mounting them.

6. In an essay of 250 to 300 words, evaluate the effectiveness of the following set of installation instructions for a sliding door for a shower.

INSTALLATION INSTRUCTIONS

CAUTION: SEE BOX NO. 1 BEFORE CUTTING ALUMINUM HEADER OR SILL

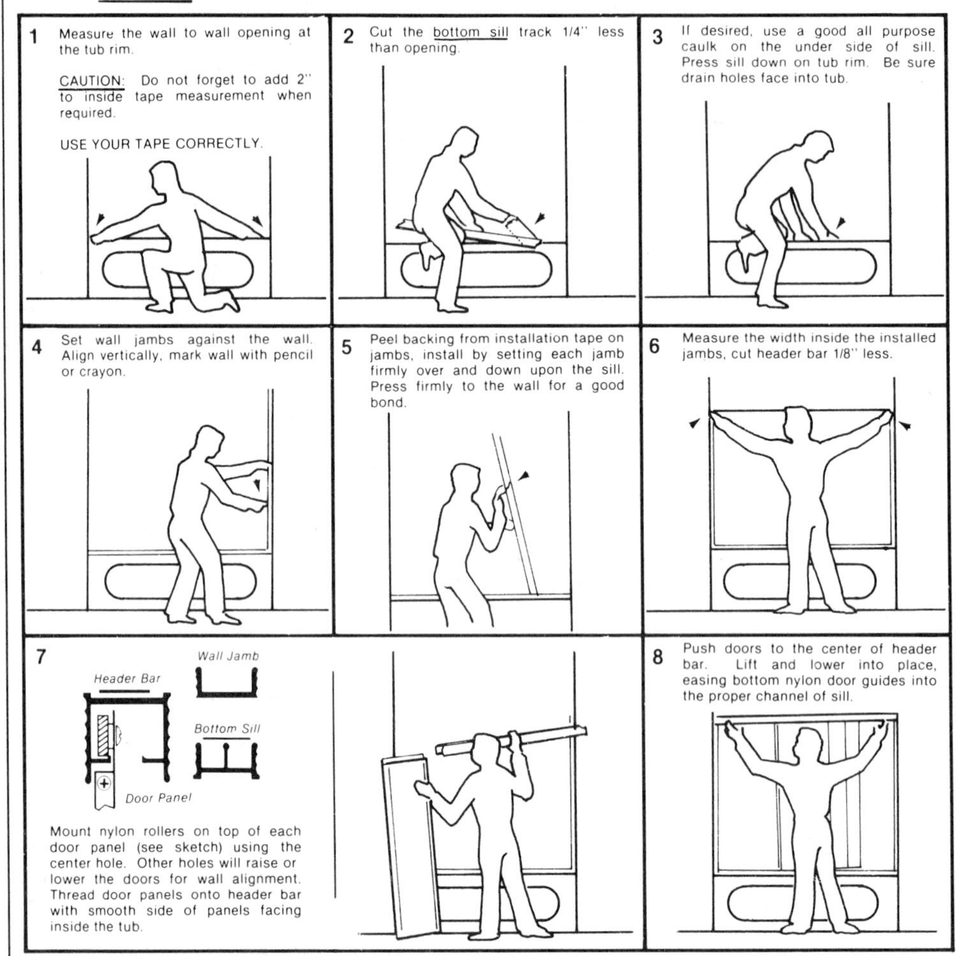

1 Measure the wall to wall opening at the tub rim.
CAUTION: Do not forget to add 2" to inside tape measurement when required.
USE YOUR TAPE CORRECTLY.
2 Cut the bottom sill track 1/4" less than opening.
3 If desired, use a good all purpose caulk on the under side of sill. Press sill down on tub rim. Be sure drain holes face into tub.
4 Set wall jambs against the wall. Align vertically, mark wall with pencil or crayon.
5 Peel backing from installation tape on jambs, install by setting each jamb firmly over and down upon the sill. Press firmly to the wall for a good bond.
6 Measure the width inside the installed jambs, cut header bar 1/8" less.
7
Wall Jamb
Header Bar
Bottom Sill
Door Panel
Mount nylon rollers on top of each door panel (see sketch) using the center hole. Other holes will raise or lower the doors for wall alignment. Thread door panels onto header bar with smooth side of panels facing inside the tub.
8 Push doors to the center of header bar. Lift and lower into place, easing bottom nylon door guides into the proper channel of sill.

TRIDOR MODEL ONLY:

To reverse direction of panels, raise panels out of bottom track and slide catches past each other thereby reversing direction so that shower head does not throw water between the panels.

HARDWARE KIT CONTENTS

TUDOR MODEL

4 nylon bearings
4 ball bearing screws # 8-32 × 3/8"

TRIDOR MODEL

6 nylon bearings
6 ball bearing screws # 8-32 × 3/8"

REFERENCES

Change your own motor oil, it's easy. n.d. Kendall Refining Company.

Harris, Martin L. 1979. *Introduction to data processing*, 2nd edition. New York: Wiley.

CHAPTER TEN

GRAPHIC AIDS

Graphic aids are the "pictures" of technical writing: photographs, diagrams, charts, graphs, and tables. Few technical documents contain only text, because graphic aids offer several benefits that sentences and paragraphs alone cannot provide.

First, graphics are visually appealing. Watch people pick up a report and skim through it. They almost automatically stop at the graphic aids and begin to study them. Readers are intrigued by graphics; that in itself increases the effectiveness of your communication.

Second, effective graphic aids are easy to understand and remember. Try to convey—using only words—what a simple hammer looks like. It's not easy to describe the head and get it to fit on the handle correctly. But in 10 seconds you could draw a simple diagram that would clearly show the hammer's design.

Third, graphic aids are almost indispensable in demonstrating relationships, which form the basis of most technical writing. For example, if you wanted to show the profits of the 10 largest corporations in a given year, a paragraph would be a complicated jumble of statistics. A simple graph, however, would provide a meaningful and memorable picture. Graphic aids are also effective in showing relationships over time: the number of nuclear power plants completed each year over the last decade, for example. And, of course, graphic aids can show the relationships among several variables over time, such as the numbers of four-cylinder, six-cylinder, and eight-cylinder cars manufactured in the United States during each of the last five years.

CHARACTERISTICS OF EFFECTIVE GRAPHIC AIDS

A graphic aid is like a paragraph: it has to be clear and understandable alone on the page, and it must be meaningfully related to the larger discussion. When you are writing a paragraph, it is easy to remember these two tasks: having to provide a transition to the paragraph that follows reminds you of the need for overall coherence. When you are creating a graphic aid, however, it is easy to put the rest of the document out of your mind temporarily as you concentrate on what you "draw."

The tendency, therefore, is to forget to relate the graphic aid to the text. Although it is true that some graphics merely illustrate points made in the text, others must be explained to the readers. Graphic aids illustrate facts, but they can't explain causes, results, and implications. The text must do that.

Effective graphic aids are

1. appropriate to the writing situation
2. labeled completely
3. placed in an appropriate location
4. integrated with the text

1. *A graphic aid should be appropriate to the writing situation.* The first question you have to answer once you have decided that some body of information would be conveyed more effectively as a graphic aid than as text is "What type of graphic aid would be most appropriate?" Some kinds of graphic aids are effective in showing deviations from a numerical norm, some are effective in showing statistical trends, and so forth. Some types of graphic aids are very easy to understand. Other, more sophisticated types contain considerable information, and the reader will need help to understand them. Think about your writing situation: the audience and purpose of the document. Can your readers handle a sophisticated graphic aid? Or will your purpose in writing be better served by a simpler graphic aid? A pie chart might be perfectly appropriate in a general discussion of how the federal government

spends its money. However, so simple a graphic would probably be inappropriate in a technical article addressed to economists.

2. *A graphic aid should be labeled completely.* Every graphic aid (except a brief, informal one) should have a clear and informative title. The columns of a table, and the axes of a graph, should be labeled fully, complete with the units of measurement. The lines on a line graph should also be labeled. Your readers should not have to guess whether you are using meters or yards as your unit of measure, or whether you are including in your table statistics from last year or just this year. If the information in the graphic aid was not discovered or generated by you, the source of the information should be cited.

3. *A graphic aid should be placed in an appropriate location.* Graphic aids can be placed in any number of locations. If your readers need the information in order to understand the discussion, put the graphic aid directly after the pertinent point in the text—or as soon after that point as is feasible. If the information functions merely as support for a point that is already clear, or as elaboration of it, the appendix is probably the best location for the graphic aid.

Understanding your audience and purpose is the key to deciding where to put a graphic aid. For example, you might be trying to convince your supervisors that they should *not* put any more money into a project. You know the project is going to fail. However, you also know that your supervisors like the project: they think it will succeed, and they have supported it openly. In this case, the graphic aid conveying the crucial data should probably be placed in the body of the document, for you want to make sure your supervisors see it. If the situation were less controversial—for instance, you know your readers are already aware that the project is in trouble—the same graphic aid could probably be placed in an appendix. Only the "bottom line" data would be necessary for the discussion in the body.

4. *A graphic aid should be integrated with the text.* Integrating a graphic aid with the text involves two steps. First, introduce the graphic aid. Second, make sure the readers understand what the graphic aid means.

Whenever possible, refer to a graphic aid before it appears in the text. The ideal situation is to place the graphic aid on the same page as the reference. If the graphic is included as an appendix at the end of a document, tell your readers where to find it: "For the complete details of the operating characteristics, see Appendix B, page 24."

Writers often neglect to explain clearly the meaning of the

graphic aid. When you introduce a graphic aid, ask yourself whether a mere paraphrase of its title will be clear: "Figure 2 is a comparison of the costs of the three major types of coal gasification plants." If you want the graphic to make a point, don't just hope your readers will perceive what that point is. State it explicitly: "Figure 2 shows that a high-sulfur bituminous coal gasification plant is presently more expensive than either a low-sulfur bituminous or anthracite plant, but more than half of its cost is cleanup equipment. If these expenses could be eliminated, high-sulfur bituminous would be the least expensive of the three types of plants."

TYPES OF GRAPHIC AIDS

There are literally dozens of different types of graphic aids. Many organizations employ graphic artists, whose job is to devise informative and attractive visuals. This discussion, however, will concentrate on the basic kinds of graphic aids that can be constructed by a writer who lacks special training or equipment.

The graphic aids used in technical documents can be classified into two basic categories: tables and figures. Tables are lists of data—usually numbers—arranged in columns. Figures are everything else: graphs, charts, diagrams, photographs, and the like. Generally, tables and figures have their own set of numbers: the first table in a document is Table 1; the first figure is Figure 1. In documents of more than one chapter (as in this book, for instance), the graphic aids are usually numbered chapter by chapter. Figure 3–2, for example, would be the second figure in Chapter 3.

TABLES

Tables easily convey large amounts of information, especially quantitative data, and often provide the only means of showing several variables for a number of items. For example, if you want to show the numbers of people employed in six industries in ten states, a table would probably be the best graphic aid to use. Tables lack the visual appeal of figures, but they can handle much more information with complete accuracy.

The parts of a table are shown in Figure 10–1.

In addition to giving each table a number (for example, "Table 1"), identify the table with a substantive title. Create a title that encompasses the items you are comparing, as well as the criteria of comparison. For example:

FIGURE 10-1

Parts of a Table

Table 1

Title
(subtitle)

Stub Heading	Column Heading 1	Column Heading 2	Column Heading 3
Stub Category 1			
Item A . . .	data[a]	data	data
Item B . . .	data	data	data
Item C . . .	data	data[b]	data
Item D . . .	data	data	data
Stub Category 2			
Item E . . .	data	data	data[c]
Item F . . .	data	data	data
Item G . . .	data	data	data

Notes: [a]Footnote
[b]Footnote
[c]Footnote
Source:

Mallard Population in Rangeley, 1978-1982

Grain Sales by the United States to the Soviet Union in 1981

The Growth of the Robotics Industry in Japan, 1975-1980

Multiple Births in the Industrialized Nations in 1982

Note that tables are enumerated and titled above the data. The number and title are centered horizontally.

If all the data in the table are expressed in the same unit, indicate that unit under the title:

Farm Size in the Midwestern States

(in Hectares)

If the data in the different columns are expressed in different units, indicate the unit in the column heading.

	Population (in millions)	Per Capita Income (in thousands of U.S. dollars)

Provide footnotes for any information that needs to be clarified. Also at the bottom of the table, below any footnotes, add the source of your information (if you did not generate it yourself).

The stub is the left-hand column, in which you list the items being compared in the table. Arrange the items in the stub in some logical order: big to small, important to unimportant, alphabetical, chronological, and so forth. If the items fall into several categories, you can include the names of the categories in the stub:

```
Sunbelt States
 Arizona.....................................................
 California..................................................
 New Mexico .................................................

Snowbelt States
 Connecticut.................................................
 New York ...................................................
 Vermont.....................................................
```

If the items in the stub are not grouped in logical categories, skip a line every four or five items to help the reader follow the rows across the table. Leader dots, a row of dots that links the stub and the next column, are also useful.

The columns are the heart of the table. Within the columns, arrange the data as clearly and logically as you can. Line up the numbers consistently.

```
 3,147
   365
46,803
```

In general, don't change units. If you use meters for one of your quantities, don't use feet for another. If, however, the quantities are so dissimilar that your readers would have a difficult time understanding them as expressed in the same units, inconsistent units might be more effective.

```
 3 hr.
12 min.
 4 sec.
```

This listing would probably be easier for most readers to understand than a listing in which all quantities were expressed in hours, minutes, or seconds.

Use leader dots if a column contains a "blank" spot: a place where there is no appropriate data:

```
 3,147
  . . .
46,803
```

Make sure, however, that you don't substitute leader dots for a quantity of zero.

```
 3,147
     0
46,803
```

Figure 10–2 is an example of an effective table. Notice in this table that the writer has used decked headings in his two columns. Rather than having to write "Valve #1 Initial" and "Valve #1 Final," he simply stacked the "Valve #1" over the words "Initial" and "Final."

FIGURES Every graphic aid that is not a table is a figure. The discussion that follows covers the principal types of figures: bar graphs, line graphs, charts, diagrams, and photographs.

BAR GRAPHS Bar graphs provide a simple and effective means of representing different quantities so that they can be compared at a

FIGURE 10–2

Table

Table 6. *Test Results for Valves #1 and 2*

	VALVE #1		VALVE #2	
	INITIAL	FINAL	INITIAL	FINAL
Maximum Bypass Cv	**43.1**	**43.1**	**48.1**	**48.1**
Minimum Recirc. Flow (GPM)	**955**	**955**	**930**	**950**
Number of Pilot Threads Exposed	**+3**	**+3**	**+3**	**+2**
Main △P at rated flow (psid)	**4.5**	. . .	**4.5**	. . .

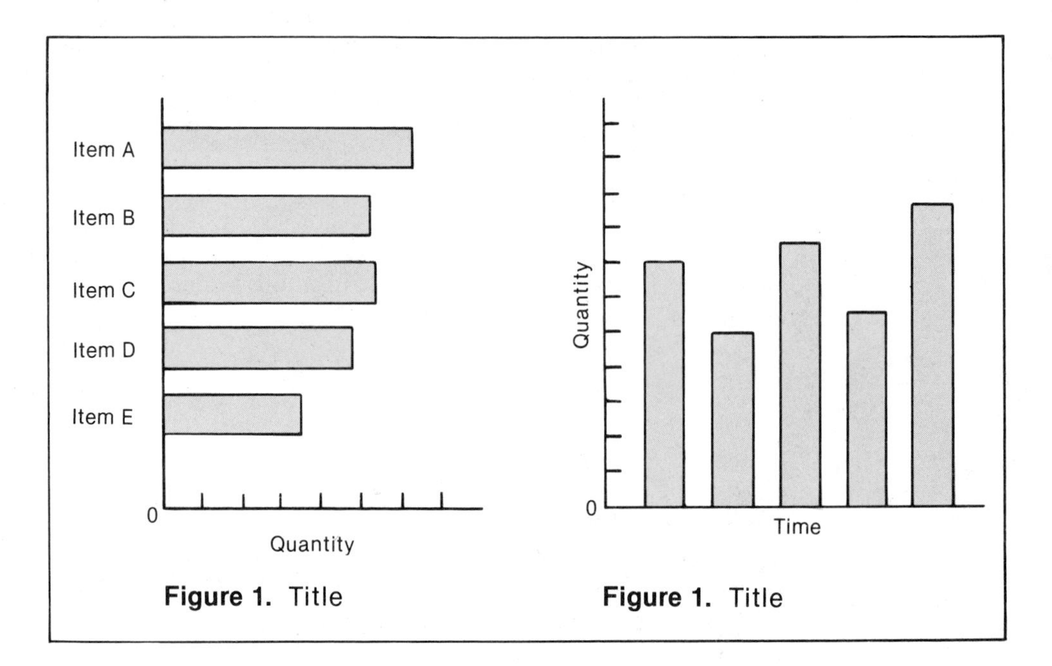

FIGURE 10–3

Structure of a Horizontal and a Vertical Bar Graph

glance. The principle behind bar graphs is that the length of the bar represents the magnitude of the quantity. The bars can be drawn horizontally or vertically. Horizontal bars are generally preferred for showing different items at any given moment (such as quantities of different products sold during a single year), whereas vertical bars are used for showing how the same item varies over time (such as month-by-month sales of a single product). These distinctions are not ironclad, however; as long as the axes are labeled carefully, your readers should have no trouble understanding you.

Figure 10–3 shows the structure of basic horizontal and vertical bar graphs. When you construct bar graphs, follow five basic guidelines.

1. *Number the axes at regular intervals.* Use any convenient scale (such as one inch on the paper equals $1,000), and then draw the bars. Graph paper, of course, makes your job easier.

2. *If at all possible, begin the quantity scale at zero.* This will ensure that the bars accurately represent the quantities. Notice in Figure 10–4 how misleading a graph is if the scale doesn't begin at zero. Version *a* is certainly more dramatic than version *b*. In version *a*, the difference in the lengths of the bars suggests that Item A is much greater than Item B, and that Item B is much greater than Item C.

If it is not practical to start the quantity scale at zero, break the quantity axis clearly, as in Figure 10-5.

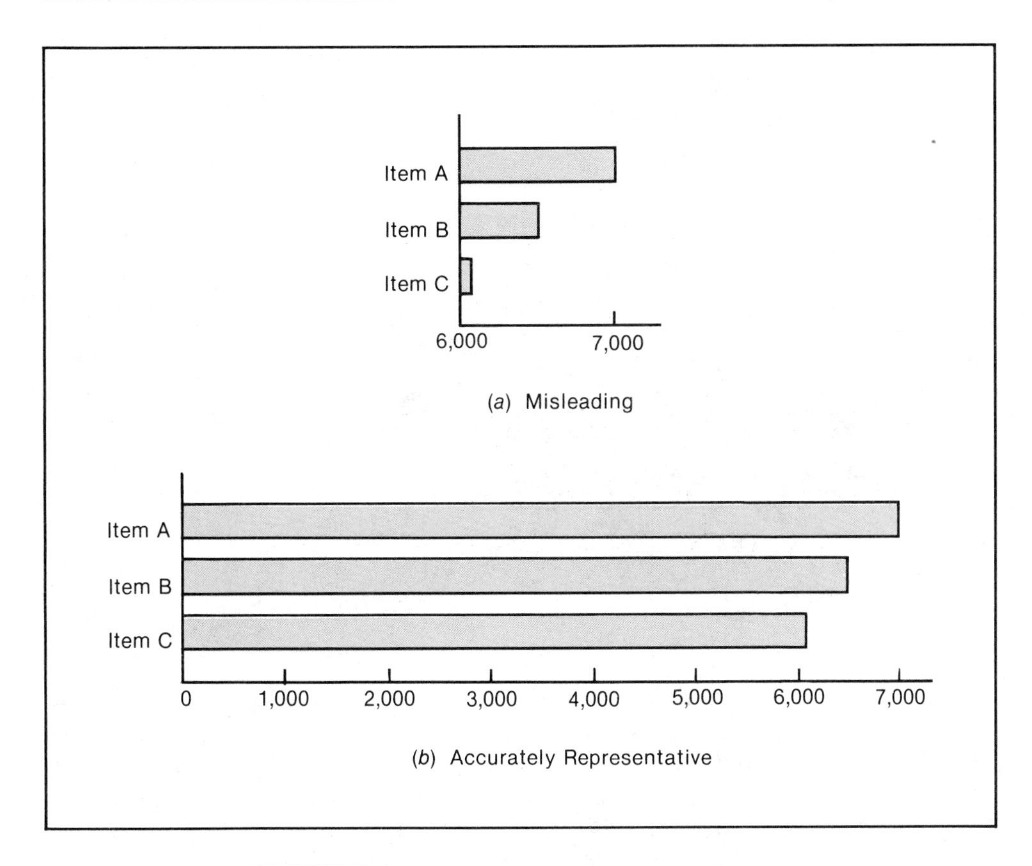

FIGURE 10–4
Misleading and Accurately Representative Bar Graphs

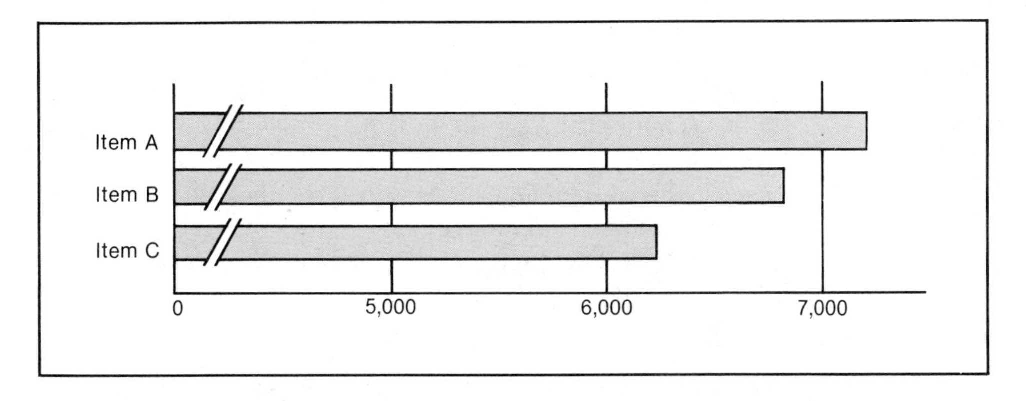

FIGURE 10–5
A Bar Graph with the Quantity Axis Clearly Broken

3. *Use tick marks or grid lines to signal the amounts.* Ticks are the little marks drawn to the axis:

Grid lines are ticks extended through the bars:

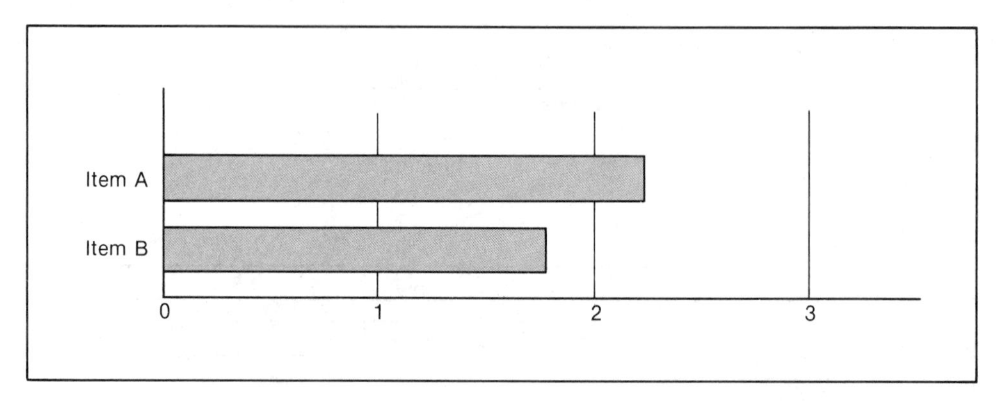

Grid lines are usually necessary only if you have several bars, some of which would be too far away from tick marks for readers to gauge the quantity easily.

4. *Arrange the bars in a logical sequence.* In a vertical bar graph, chronology usually dictates the sequence. For a horizontal bar graph, arrange the bars in descending-size order beginning at the top of the graph, unless some other logical sequence seems more appropriate.

5. *Make all bars equally wide, and put the same amount of space between them.* The bars should be wide enough to be seen easily. The space between the bars should be a little narrower than the bars themselves.

Figure 10–6 shows an effective bar graph.

Notice that figures are titled underneath. Unlike tables, which are generally read from top to bottom, figures are usually read from the bottom up. If the graph displays information that you have gathered from an outside source, cite that source in a brief note near the bottom of the graph.

The basic bar graph can be varied easily to accommodate many different communication needs. Here are a few common variations.

The *grouped bar graph*, such as that in Figure 10–7, lets you show two or three quantities for each item you are representing. Grouped bar graphs are useful for showing information such as

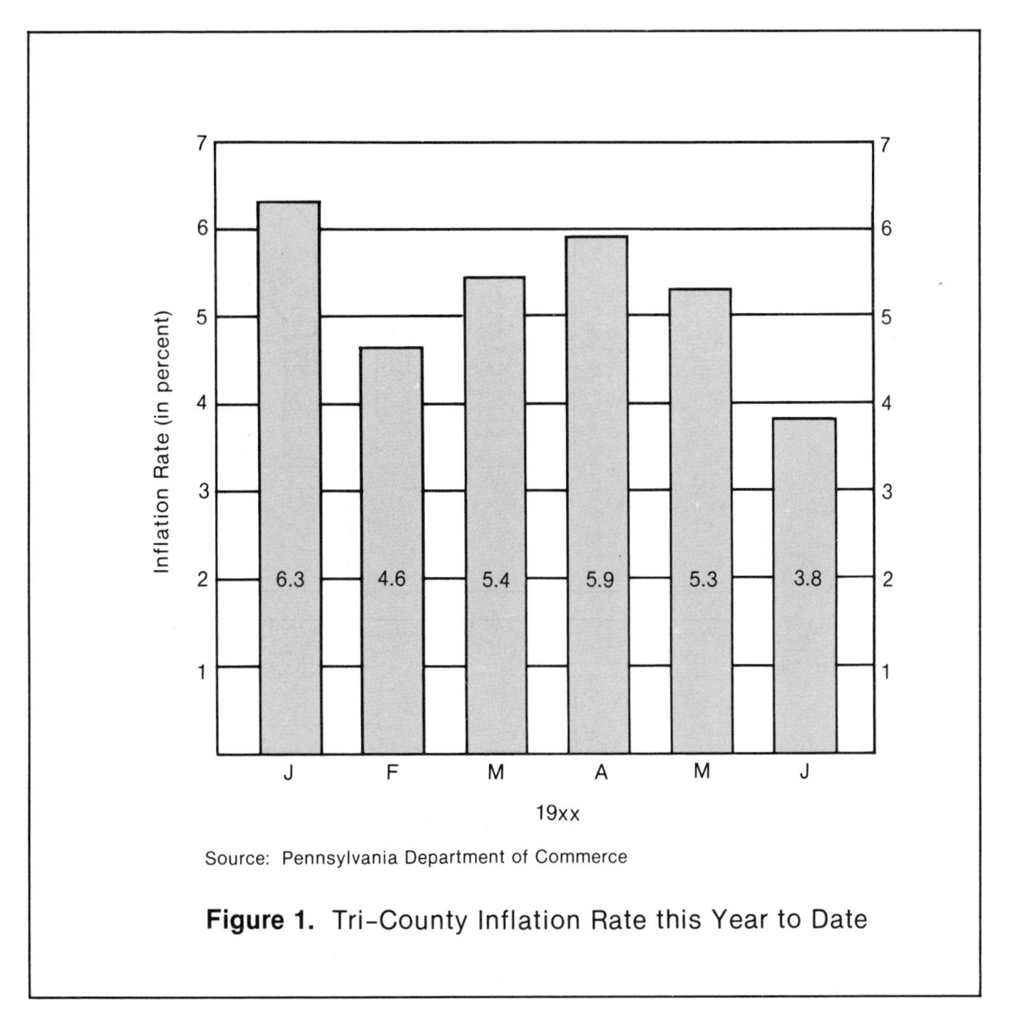

Figure 1. Tri-County Inflation Rate this Year to Date

FIGURE 10–6
Bar Graph

the numbers of full-time and part-time students at several universities. One kind of bar represents the full-time students; the other, the part-time. To distinguish the bars from each other, use hatching (striping) or shading, and label one set of bars or provide a key.

Another way to show this kind of information is through the *subdivided bar graph*, shown in Figure 10–8. A subdivided bar graph adds Aspect I to Aspect II, just as wooden blocks are placed on one another. Although the totals are easy to compare in a subdivided bar graph, the individual quantities (except those that begin on the horizontal axis) are not.

Related to the subdivided bar graph is the *100% bar graph*,

FIGURE 10–7
Grouped Bar Graph

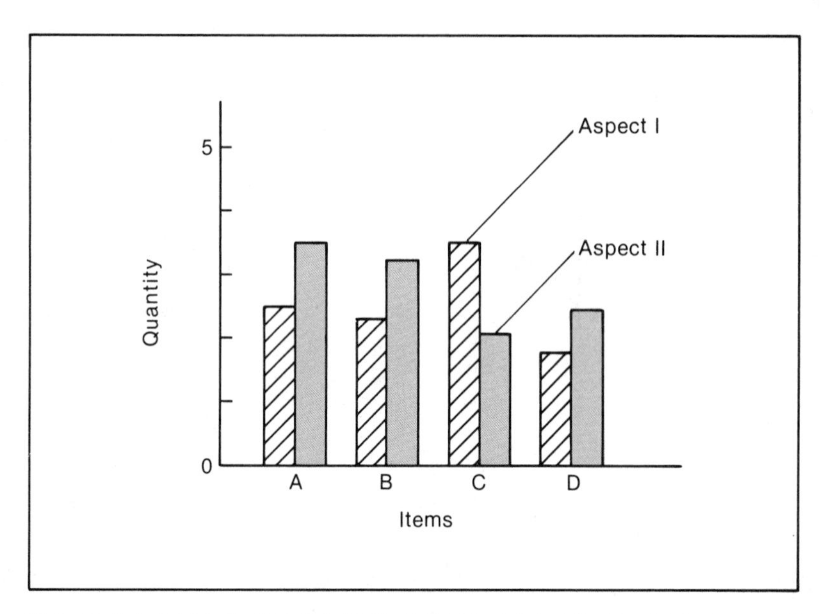

FIGURE 10–8
Subdivided Bar Graph

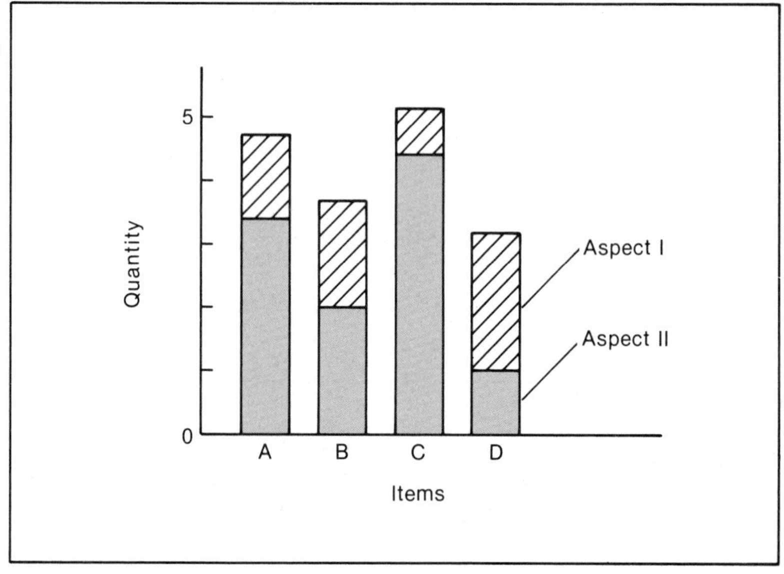

which enables you to show the relative proportions of the elements that make up several items. Figure 10–9 shows a 100% bar graph. This kind of graph is useful in portraying, for example, the proportion of full-scholarship, partial-scholarship, and no-scholarship students at a number of colleges.

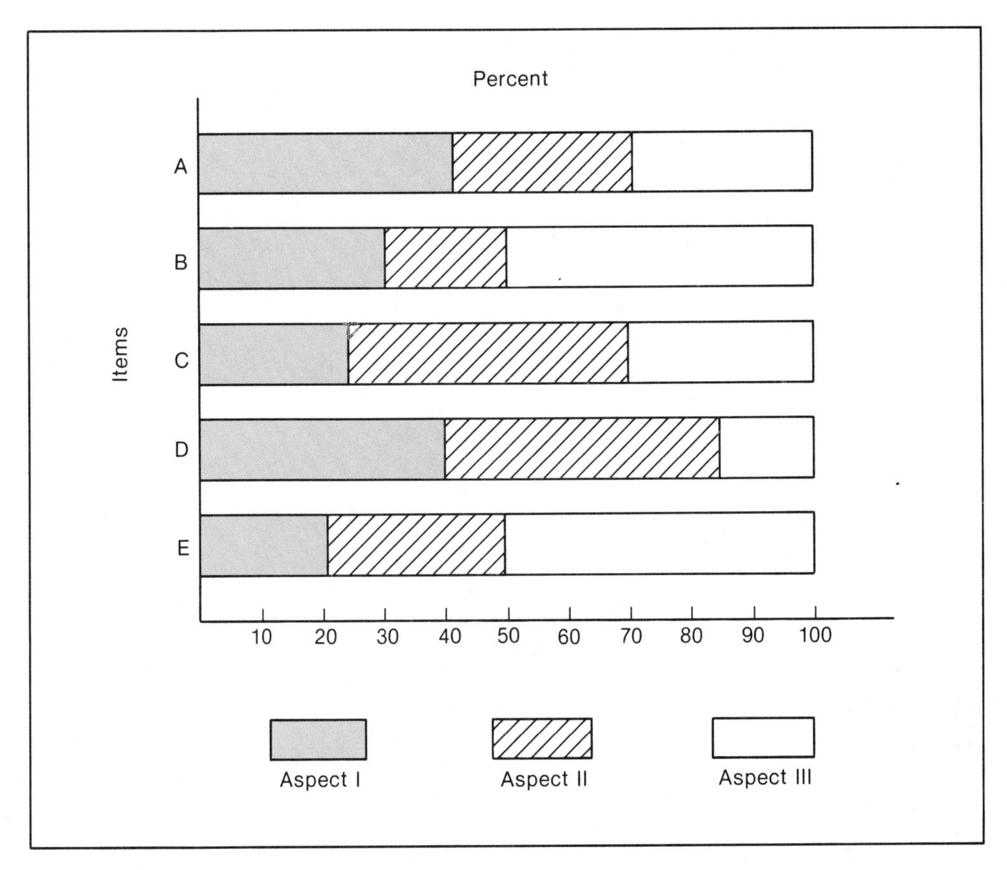

FIGURE 10–9
100% Bar Graph

The *deviation bar graph*, shown in Figure 10–10, lets you show how various quantities deviate from a norm. Deviation bar graphs are often used when the information contains both positive and negative values, as with profits and losses. Bars on the positive side of the norm line represent profits; on the negative side, losses.

Pictographs are simple graphs in which the bars are replaced by series of symbols that represent the items (see Figure 10–11). Pictographs are generally used only to enliven statistical information for the general reader. The quantity scale is usually replaced by a statement that indicates the numerical value of each symbol. This kind of pictograph almost always arranges the symbols horizontally.

LINE GRAPHS Line graphs are like vertical bar graphs, except that in line graphs the quantities are represented not by bars but by points linked by a line. This line traces a pattern that in a bar

FIGURE 10-10
Deviation Bar Graph

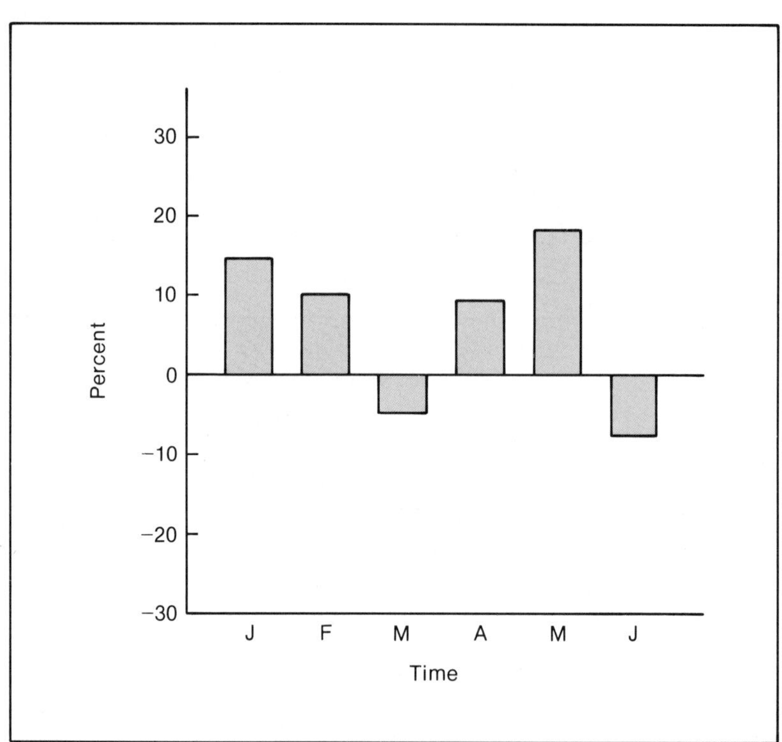

FIGURE 10-11
Pictograph

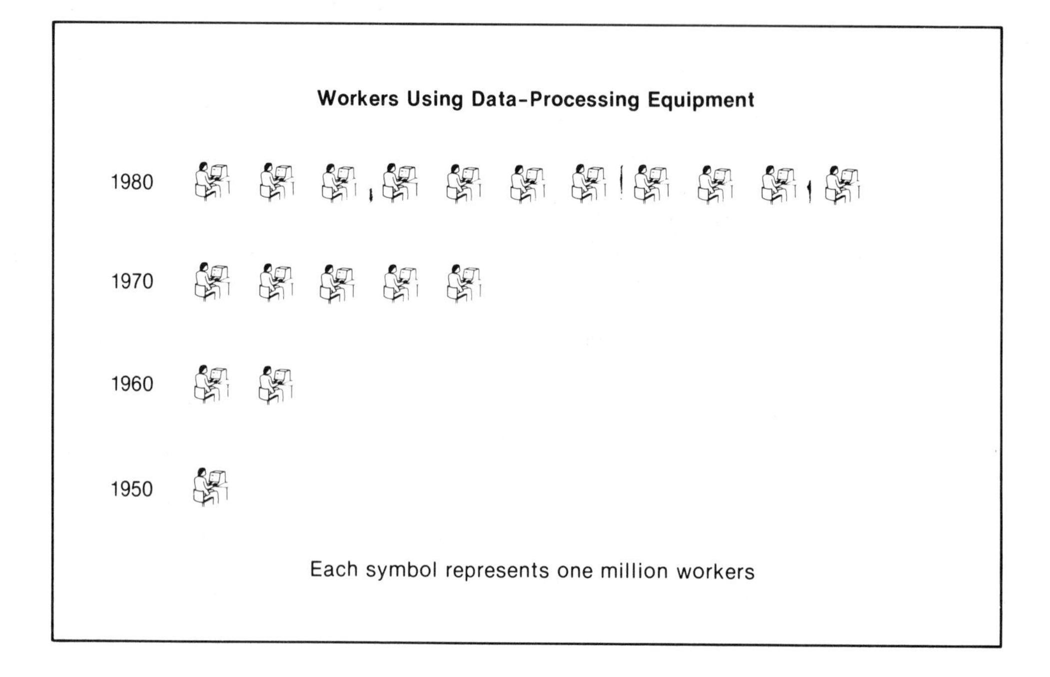

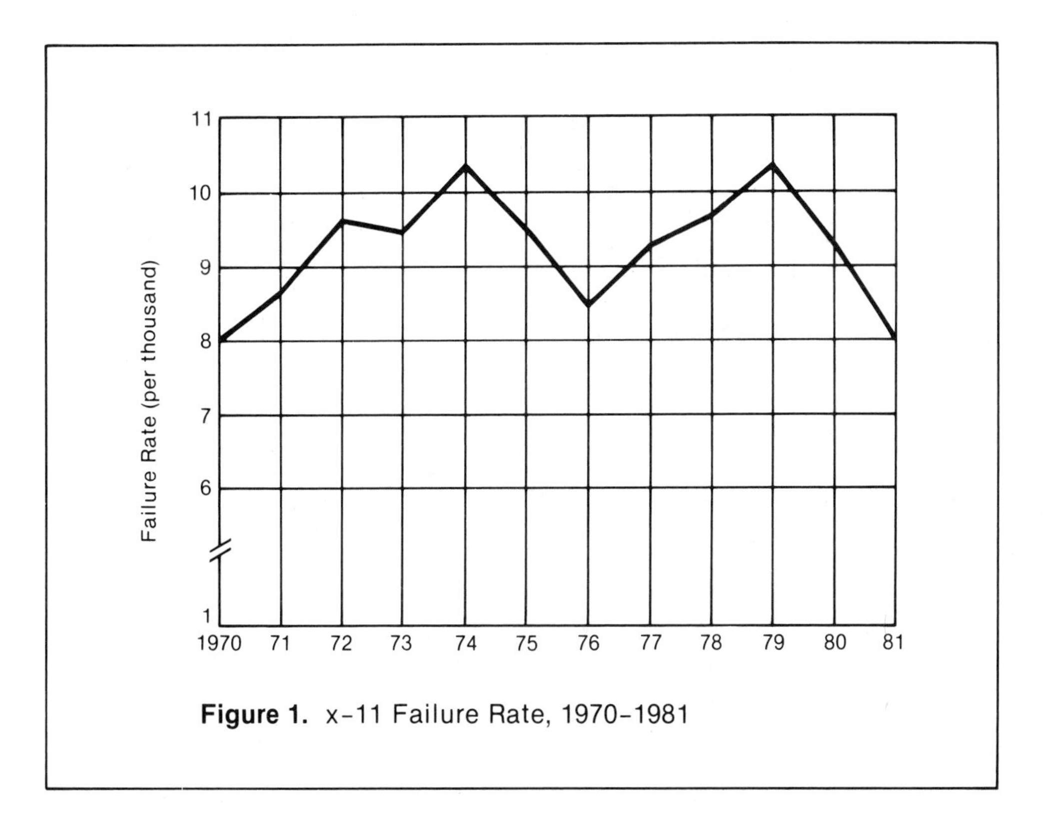

Figure 1. x-11 Failure Rate, 1970-1981

FIGURE 10-12

Line Graph with a Truncated Axis

graph would be formed by the highest point of each bar. Line graphs are used almost exclusively to show how the quantity of an item changes over time. Some typical applications of a line graph would be to portray the month-by-month sales figures or production figures for a product or the annual rainfall for a region over a given number of years. A line graph focuses the reader's attention on the change in quantity, whereas a bar graph emphasizes the actual quantities themselves. Figure 10–12 shows a typical line graph.

An additional advantage of the line graph for demonstrating change is that it can accommodate much more data. Because three or four lines can be plotted on the same graph, you can compare trends conveniently. Figure 10–13 shows a multiple-line graph. However, if the lines intersect each other often, the graph will be unclear. If this is the case, draw separate graphs.

The principles of constructing a line graph are similar to those used for a vertical bar graph. The vertical axis, which charts the quantity, should begin at zero; if it is impractical to begin at zero because of space restrictions, clearly indicate a break in the axis, as

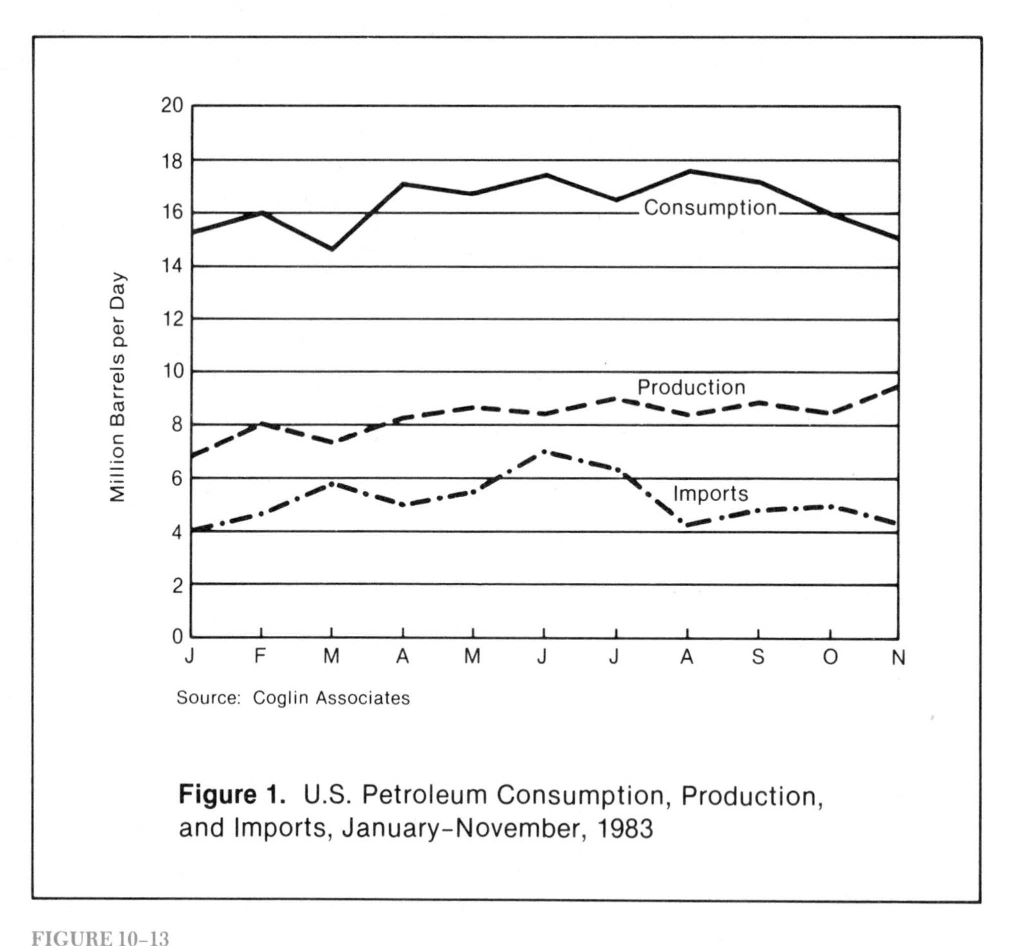

Figure 1. U.S. Petroleum Consumption, Production, and Imports, January-November, 1983

FIGURE 10–13
Multiple-line Graph

Figure 10–12 does. Where precision is required, use grid lines—horizontal, vertical, or both—rather than tick marks.

Two common variations on the line graph—the *stratum graph* and the *ratio graph*—deserve mention. A stratum graph shows an overall change, and then breaks down the total change into its constituent parts. Figure 10–14 contains two related stratum graphs that are easy to read—with a little practice. The top graph shows that total energy consumption in the United States in 1973 was approximately 74 quadrillion Btu. Of this total, approximately 18 quadrillion Btu was used for transportation, approximately 32 quadrillion Btu was used by industry (50 minus 18), and approximately 24 quadrillion was used by the residential and commercial sector (74 minus 50).

A *ratio graph* (sometimes called a semilogarithmic graph or chart) is a means of portraying the rate of change of two or more

Consumption

Consumption of Energy by End-Use Sector

Yearly

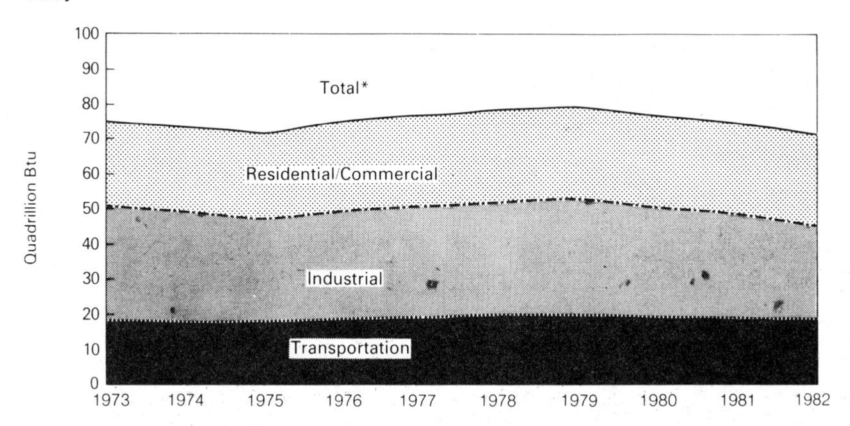

Monthly

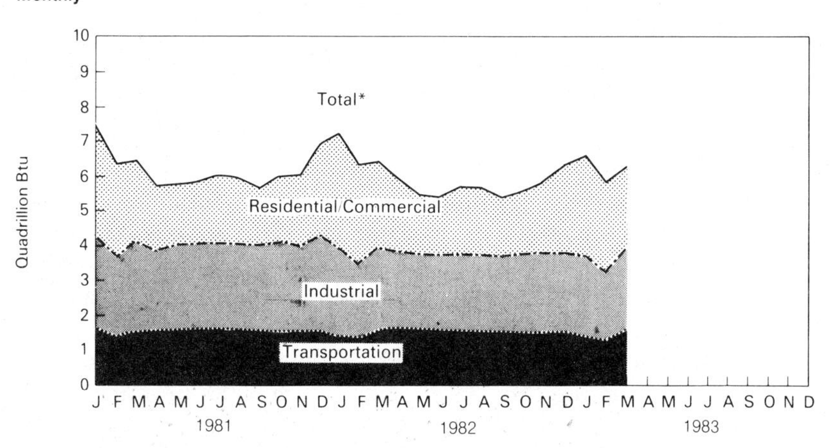

*Btu consumption for all sectors were cumulated to create total.

FIGURE 10–14

Stratum Graphs

FIGURE 10–15

Ratio Graph

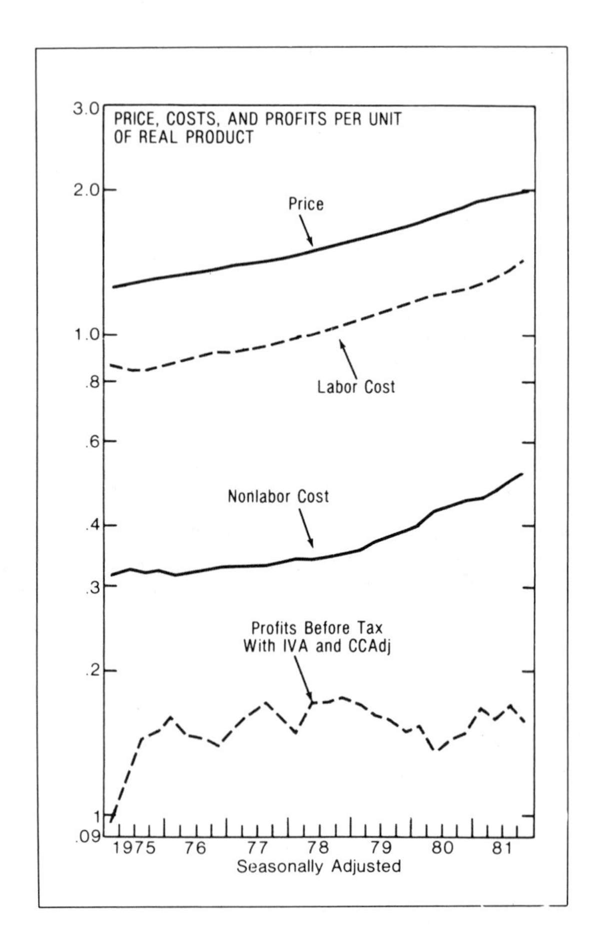

items of vastly different magnitudes. For example, you might wish to compare the month-by-month sales of a large corporation with those of a small one. You would have great trouble making a vertical axis that would accommodate a company with sales of $20,000 a month and one with sales of $20,000,000. Even if you had a giant piece of paper, the graph could not reflect a true relation between the companies. If both companies increased their sales at the same rate (such as 2% per month), the small company's line would appear relatively flat, whereas the big company's line would shoot upward, just because of the large quantities involved.

To solve this problem, the ratio graph compresses the scale of the vertical axis more and more as the quantity increases. Specially prepared logarithmic paper is ruled so that the vertical axis increases by a factor of ten instead of a factor of one: that is, on log paper an increase of one inch in the vertical scale, for example, signifies a jump from 10 to 100, not 10 to 20. Figure 10–15 shows how a ratio graph works. In this figure, the vertical axis is staggered to accommodate the smaller quantities. For instance, the distance on the vertical axis between 0.1 and 0.2 is greater than the distance between 2.0 and 3.0.

CHARTS Whereas tables and graphs present statistical information, most charts convey more-abstract relationships, such as causality or hierarchy. (The pie chart, which is really just a circular rendition of the 100% bar graph, is the major exception.) Many forms of tables and graphs are well known and fairly standard. By contrast, only a few kinds of charts—such as the organization chart and flow chart—follow established patterns. Most charts reflect original concepts and are created to meet specific communication needs.

The *pie chart* is a simple but limited design used for showing the relative size of the parts of a whole. Pie charts can be instantly recognized and understood by the untrained reader: everyone remembers the perennial "where-your-tax-dollar-goes" pie chart. The circular design is effective in showing the relative size of as many as five or six parts of the whole, but it cannot easily handle more parts because, as the "slices" get smaller, judging their sizes becomes more difficult. (Very small quantities that would make a

FIGURE 10-16
Pie Chart

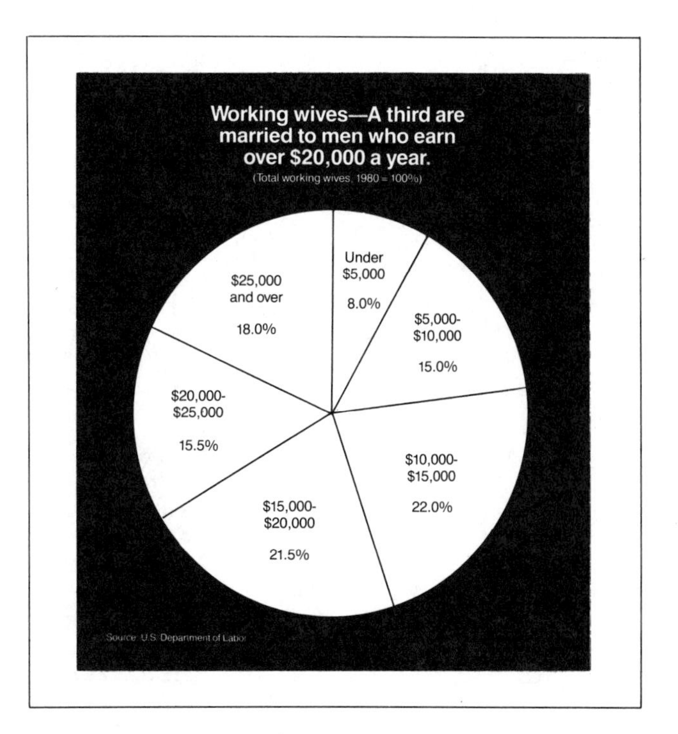

pie chart unclear can be grouped under the heading "Miscellaneous" and explained in a footnote.)

To create a pie chart, begin with the largest slice at the top of the pie and work clockwise in decreasing-size order, unless you have a good reason for arranging the slices in a different order. Label the slices (horizontally, not radially) inside the slice, if space permits. It is customary to include the percentage that each slice represents. Sometimes, the absolute quantity is added. To emphasize one of the slices—for example, to introduce a discussion of the item represented by that slice—separate it from the pie. Make sure your math is accurate as you convert percentages into degrees in dividing the circle. A percentage circle guide—a template with the circle already converted into percentages—is a useful tool. Figure 10–16 shows a basic pie chart. Notice that because of the point the writer wishes to make, the slices in Figure 10–16 are *not* arranged in size order.

A *flow chart*, as its name suggests, traces the stages of a procedure or a process. A flow chart might be used, for example, to show the steps involved in transforming lumber into paper or in synthesizing an antibody. Flow charts are useful, too, for summarizing instructions that a reader is to carry out. The basic flow

FIGURE 10–17

Open-system Flow Chart

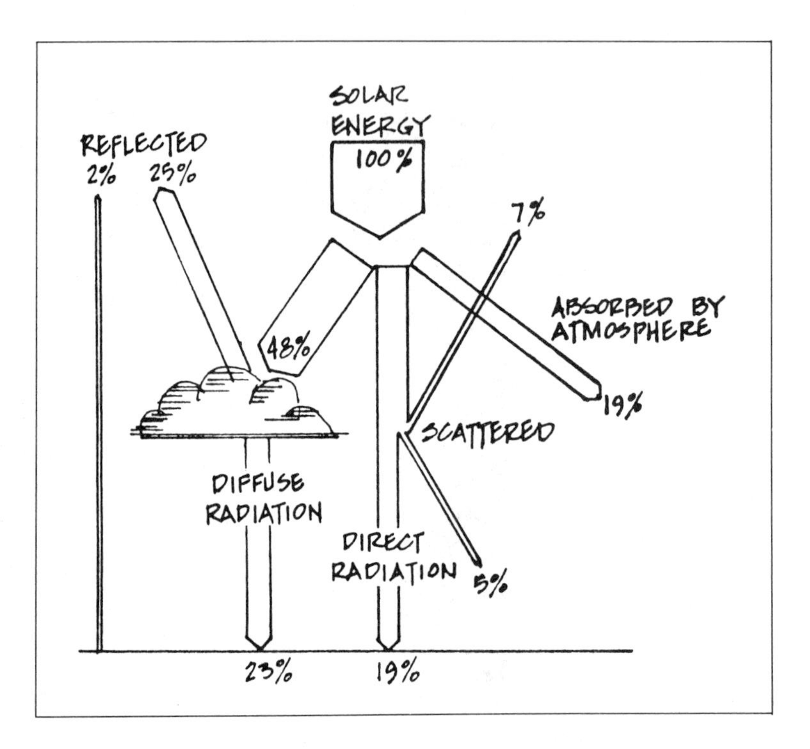

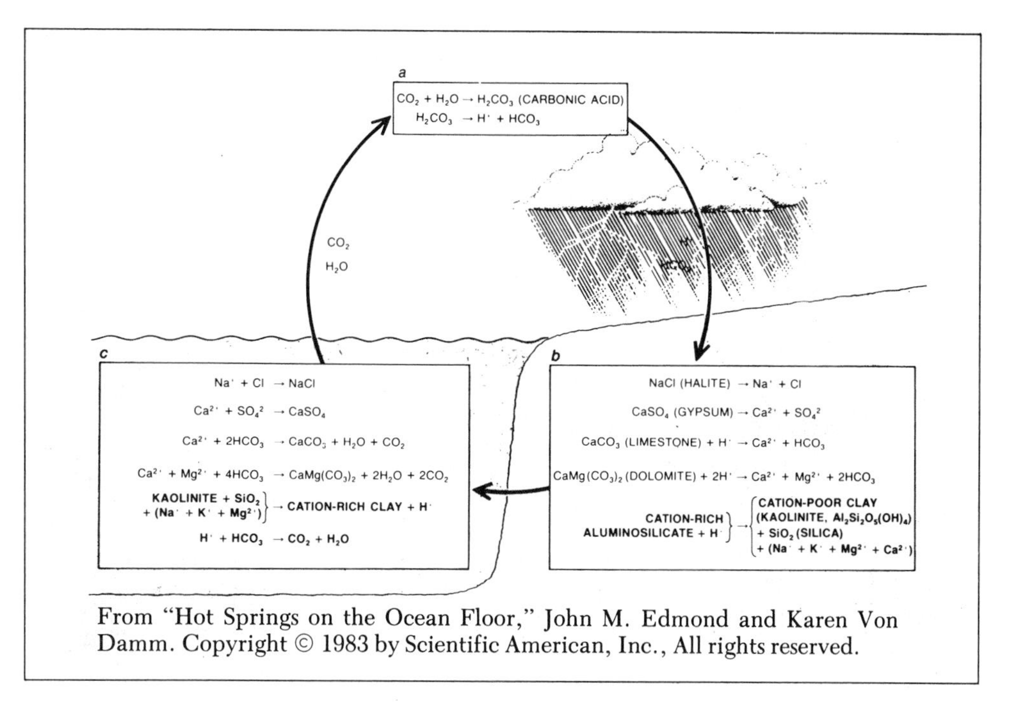

From "Hot Springs on the Ocean Floor," John M. Edmond and Karen Von Damm.

FIGURE 10–18
Closed-system Flow Chart

chart portrays stages with labeled rectangles or circles. To make it visually more interesting, use pictorial symbols instead of geometric shapes. If the process involves quantities (for example, the process of paper manufacturing might "waste" 30% of the lumber), they can be listed or merely suggested by the size of the line used to connect the stages. Flow charts can portray open systems (those that have a "start" and a "finish") or closed systems (those that end where they began). A special kind of flow chart, called a decision chart (in which the flow follows different routes depending on yes/no answers to questions), is used frequently in computer science.

Figure 10–17 shows an open-system flow chart whose bars correspond in width to the magnitude of the quantity they represent. The subject of the flow chart is the percentage of solar energy that reaches the Earth's surface. Figure 10–18 shows a closed-system flow chart.

An *organization chart* is a type of flow chart that portrays the flow of authority and responsibility in a structured organization. In most cases, the positions are represented by rectangles. The more important positions can be emphasized through the size of the boxes, the width of the lines that form the boxes, the typeface, or the use of color. If space permits, the boxes themselves can include brief descriptions of the positions, duties, or responsibilities.

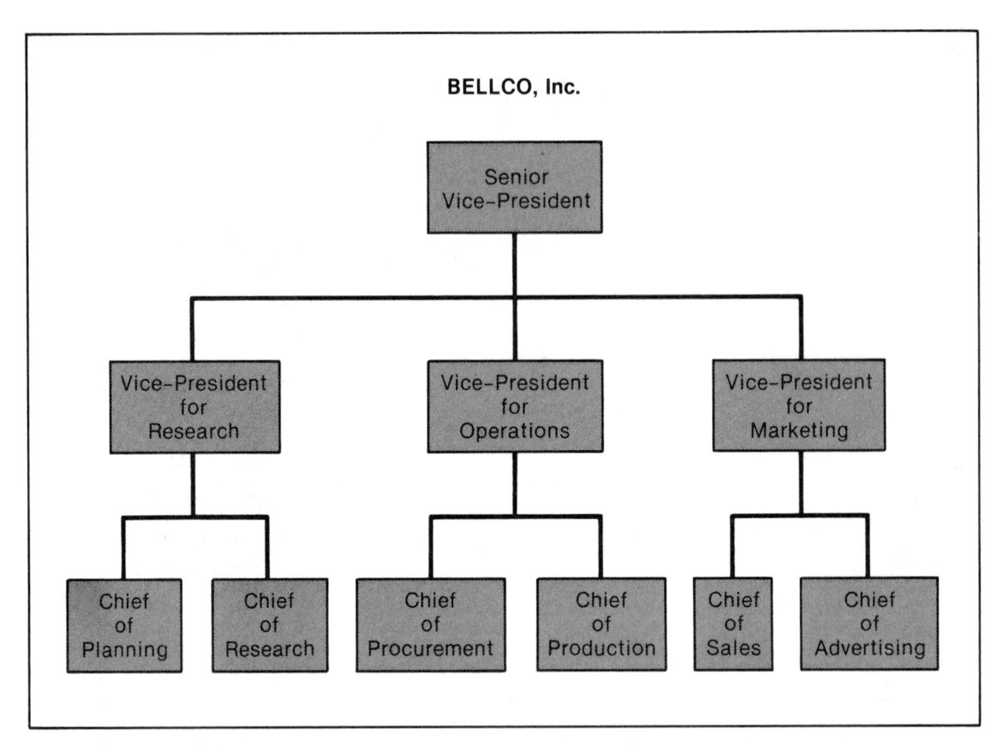

FIGURE 10–19

Organization Chart

FIGURE 10–20

Cutaway Diagram

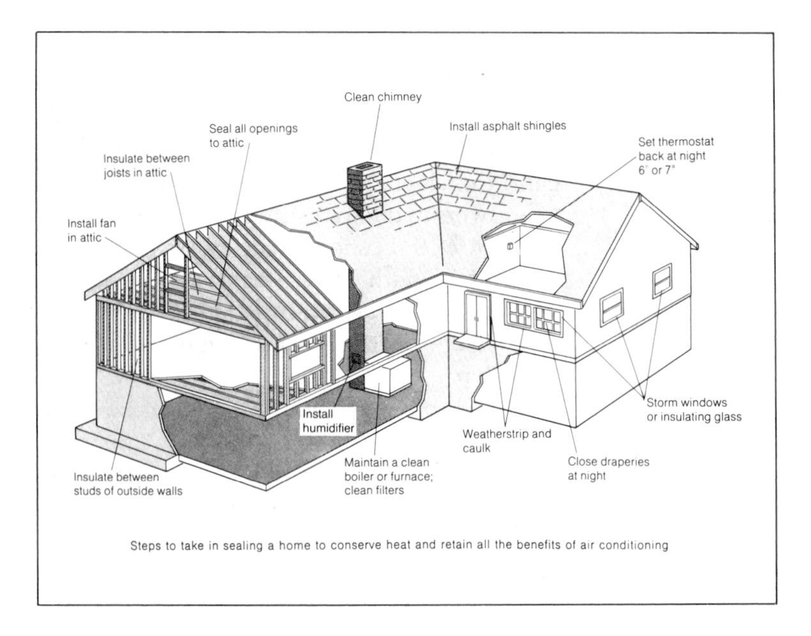

Figure 10–19 is a typical organization chart. Unlike other figures, organization charts are generally titled *above* the chart.

DIAGRAMS AND PHOTOGRAPHS In portraying physical relationships, such as those in pieces of equipment or machinery, diagrams and photographs are often the most effective types of graphic aids. Photographs are unmatched, of course, for reproducing realistic images. Recent advances in specialized kinds of photography—especially in internal medicine and biology—are expanding the possibilities of the art. Diagrams drawn by hand are used to portray perspectives that cannot be photographed. Cutaways, for example, let you "remove" a part of the surface to expose what is underneath. "Exploded" diagrams separate components from each other while maintaining their physical relationship. Figure 10–20 shows a cutaway diagram; Figure 10–21, an exploded diagram. Notice how effectively the diagrams have been labeled.

FIGURE 10–21
Exploded Diagram

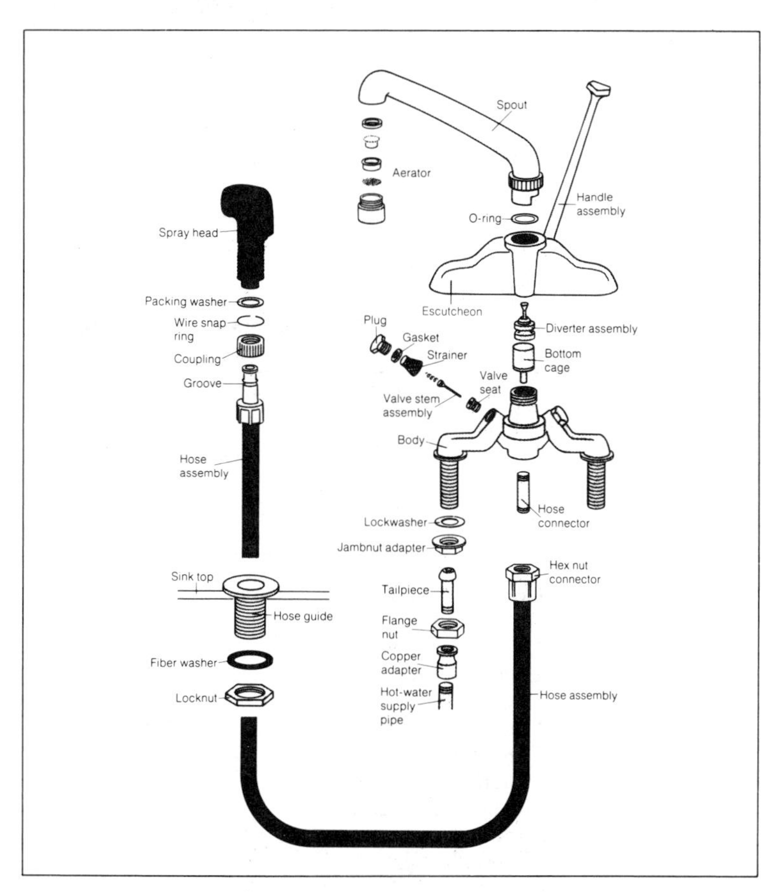

EXERCISES

1. For each of the following graphic aids, write a paragraph evaluating its effectiveness and describing how you would revise it.

a.

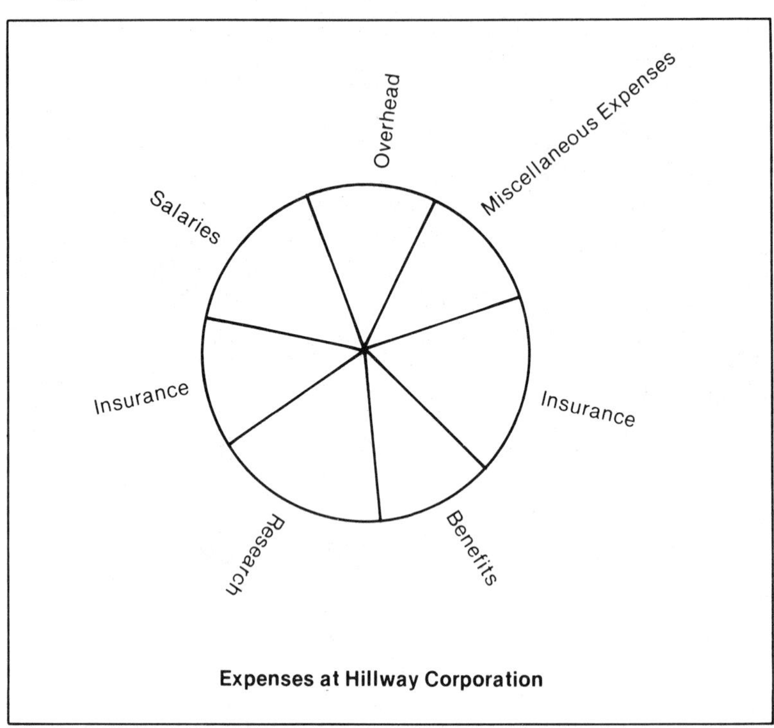

b.

Engineering and Liberal Arts Graduate Enrollment

	1979	1980	1981
Civil Engineering	236	231	253
Chemical Engineering	126	134	142
Comparative Literature	97	86	74
Electrical Engineering	317	326	401
English	714	623	592
Fine Arts	112	96	72
Foreign Languages	608	584	566
Materials Engineering	213	227	241
Mechanical Engineering	196	203	201
Other	46	42	51
Philosophy	211	142	151
Religion	86	91	72

c.

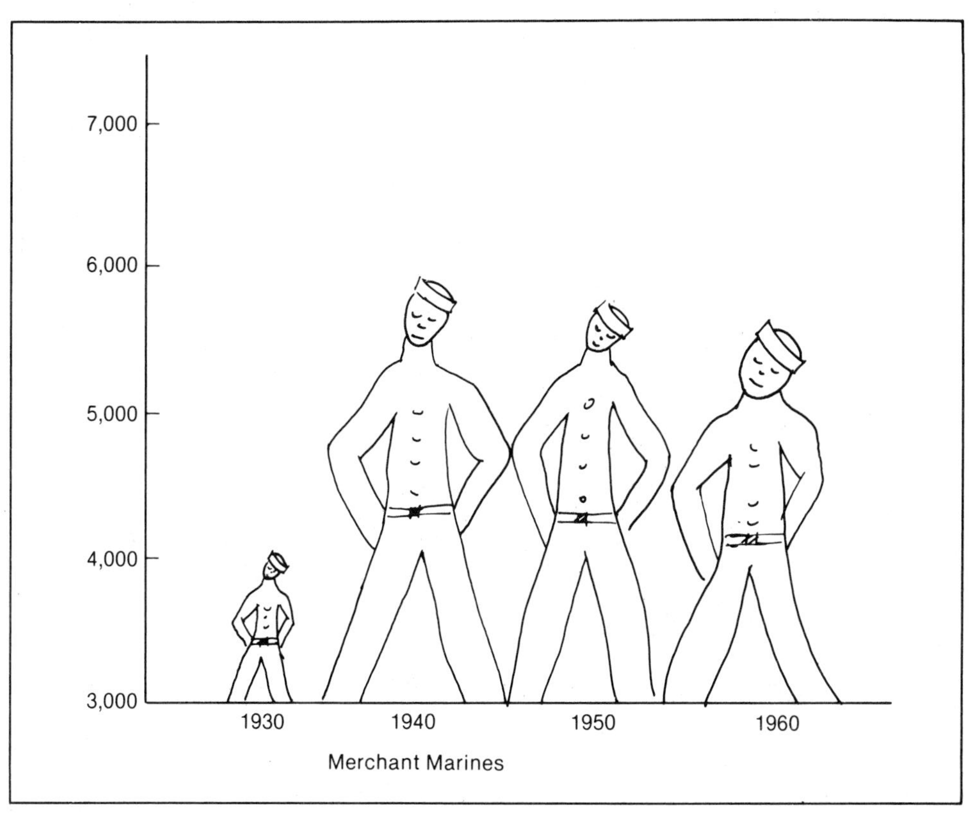
7,000
6,000
5,000
4,000
3,000
1930
1940
1950
1960
Merchant Marines

d.

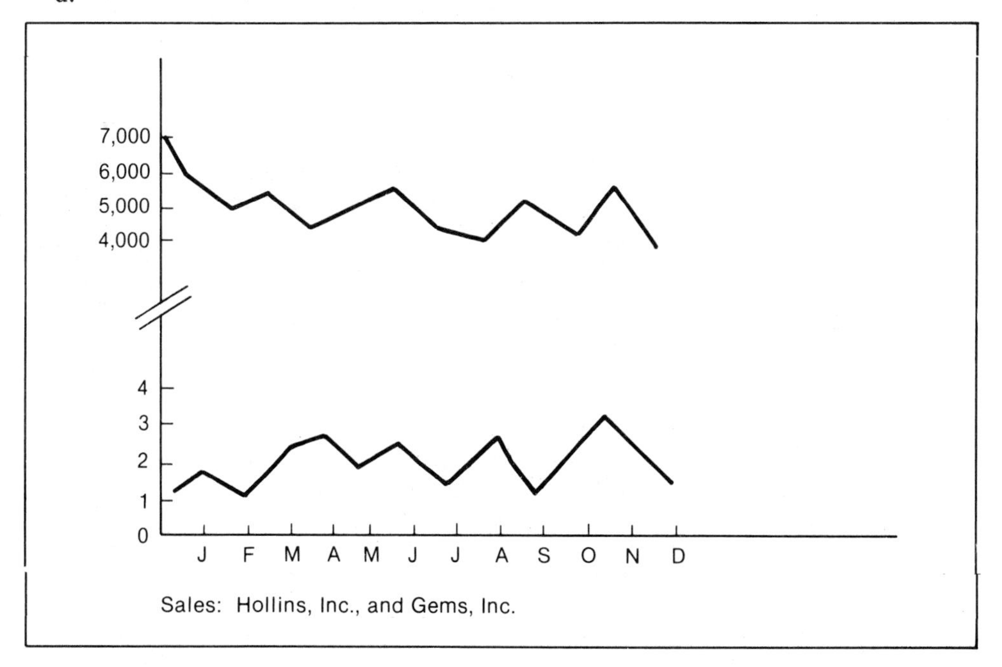
7,000
6,000
5,000
4,000
4
3
2
1
0
J
F
M
A
M
J
J
A
S
O
N
D
Sales: Hollins, Inc., and Gems, Inc.

e.

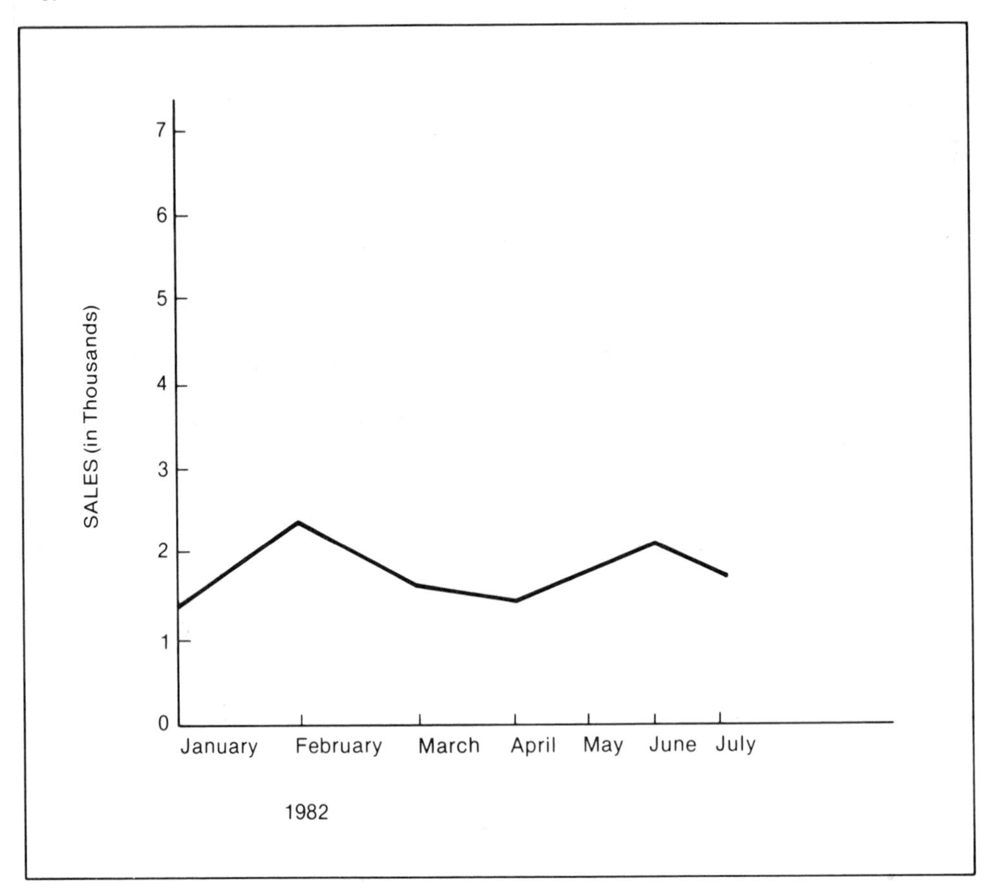

2. In each of the following exercises, translate the written information to a graphical format. Be prepared to explain why you chose the particular type of graphic aid.

a. Following are the profit and loss figures for Pauley, Inc., in early 1982: January, a profit of 6.3%; February, a profit of 4.7%; March, a loss of 0.3%; April, a loss of 2.3%; May, a profit of 0.6%.

b. The prime interest rate had a major effect on our sales. In January, the rate was 11.5%. It went up a full point in February, and another half point in March. In April, it leveled off, and it dropped two full points each in May and June. Our sales figures were as follows for the Crusader 1: January, 5,700; February, 4,900; March, 4,650; April, 4,720; May, 6,200; June, 8,425.

c. Following is a list of our new products, showing for each the profit on the suggested retail price, the factory where produced, the date of introduction, and the suggested retail price.

THE TIMBERLINE	THE FOUR SEASON
Profit 28%	Profit 32%
Milwaukee	Milwaukee
March 1984	October 1983
$235.00	$185.00

THE FAMILY EXCURSION	THE DAY TRIPPER
Profit 19%	Profit 17%
Brooklyn	Brooklyn
October 1983	May 1983
$165.00	$135.00

d. This year, our student body can be broken down as follows: 45% from the tristate area; 15% from foreign countries; 30% from the other Middle Atlantic states; and 10% from the other states.
e. In January of this year we sold 50,000 units of the BG-1, of which 20,000 were purchased by the army. In February, the army purchased 15,000 of our 60,000 units sold. In March, it purchased 12,000 of the 65,000 we sold.
f. The normal rainfall figures for this region are as follows: January, 1.5 in.; February, 1.7 in.; March, 1.9 in.; April, 2.1 in.; May, 1.8 in.; June, 1.2 in.; July, 0.9 in.; August, 0.7 in.; September, 1.3 in.; October, 1.1 in.; November, 1.0 in.; December, 1.2 in. The following rainfall was recorded in this region: January, 2.3 in.; February, 2.6 in.; March, 2.9 in.; April, 2.0 in.; May, 1.6 in.; June, 0.7 in.; July, 0.1 in.; August, 0.4 in.; September, 1.3 in.; October, 1.2 in.; November, 1.4 in.; December, 1.8 in.

PART THREE

TECHNICAL REPORTS

CHAPTER ELEVEN

ELEMENTS OF A REPORT

This chapter discusses the elements of a report—those components that are usually included in a formal report in business and industry. Few reports will have all of the elements discussed here, in the order in which they are covered; most organizations have their own format preferences. You should therefore study the style guide used in your organization. If there is no style guide, study a few of the reports in your organization's files. Successful reports are the best teaching guides; ask a colleague to suggest some samples.

The following elements will be discussed here:

1. letter of transmittal
2. title page
3. abstract
4. table of contents
5. list of illustrations
6. executive summary
7. glossary and list of symbols
8. appendix

Two crucial elements of a report are not discussed in this chapter: the body, which varies according to the type of report, and the documentation, which can also vary. For discussions of the body, refer to the individual chapters on the various types of reports: proposals (Chapter 13), progress reports (Chapter 14), and completion reports (Chapter 15). Documentation—the system of citing sources of information used in a report—is discussed in Chapter 3.

THE LETTER OF TRANSMITTAL

The letter of transmittal introduces the purpose and content of the report to the principal reader of the report. The letter is attached to the report or simply placed on top of it. Even though it might contain no information that is not included elsewhere in the report, the letter is important, because it is the first thing the reader sees. It establishes a courteous and graceful tone for the report. Letters of transmittal are customary even when the writer and the reader both work for the same organization and ordinarily communicate by memo.

The letter of transmittal gives you an opportunity to emphasize whatever you feel your reader will find particularly important or interesting in the attached materials. It also enables you to point out any errors or omissions in the materials. For example, you might want to include some information that was gathered after the report was typed or printed.

Transmittal letters generally contain most of the following elements:

1. a statement of the title, and, if necessary, the purpose of the report
2. a statement of who authorized or commissioned the project, and when
3. a statement of the methods used in the project (if they are noteworthy) or of the principal results, conclusions, and recommendations
4. an acknowledgement of any assistance you received in preparing the materials
5. a gracious offer to assist in interpreting the materials or in carrying out further projects.

Figure 11–1 provides an example of a transmittal letter. (For a discussion of letter format, see Chapter 18.)

FIGURE 11-1

Letter of Transmittal

ALTERNATIVE ENERGY, INC.
Bar Harbor, ME 00314

April 3, 19--

Rivers Power Company
15740 Green Tree Road
Gaithersburg, MD 20760

Attention: Mr. J. R. Hanson
Project Engineering Manager

Subject: Project #619-103-823

Gentlemen:

We are pleased to submit "A Proposal for the Riverfront Energy Project" in response to your request for a proposal dated February 6, 19--.

The windmill design described in the attached proposal uses the most advanced design and materials. Of particular note is the state-of-the-art storage facility described on pp. 14-17. As you know, storage limitations are a crucial factor in the performance of a generator such as this.

In preparing this proposal, we inadvertently omitted one paragraph on p. 26 of the bound proposal. That paragraph is now on the page labeled 26A, attached to the front cover of the proposal. We regret this inconvenience.

If you have any questions, please do not hesitate to call us. We will gladly be of any assistance that we can.

Yours very truly,

Ruth Jeffries
Project Manager

RJ/fj
Enclosures 2

THE TITLE PAGE

The only difficult task in creating the title page is to think of a good title. The other usual elements—the date of submission and the names and positions of the writer and the principal reader—are simply identifying information.

A good title is sufficiently informative without being unwieldy. It answers two basic questions: What is the subject of the report? and What type of report is it? Several examples of effective titles follow:

FIGURE 11-2
Simple Title Page

PETROLEUM PRICES FOR THE EIGHTIES:

A FORECAST

Prepared for: Harold Breen, President
Reliance Trucking Co.

by: Adelle Byner, Manager
Purchasing Department
Reliance Trucking Co.

April 19, 1979

Choosing a Microcomputer: A Recommendation

An Analysis of the Kelly 1013 Packager

Open Sea Pollution-control Devices: A Summary

Note that a convenient way to define the type of report is to use a generic term—such as *analysis, recommendation, summary, review, guide,* or *instructions*—in a phrase following a colon.

If you are creating a simple title page, center the title (typed in full capital letters) about a third of the way down the page. Then

FIGURE 11–3
Complex Title Page

RESEARCH REPORT COVER PAGE

(Company Confidential)

Standard Technical Report No.________

Date Issued________________________

Security (Check One)______ RC______ C

Originating R & D Department________________________________

Location (Facility, City, State)____________________________

Group or Division____

WILSON CHEMICALS, INC.

____

____
(Title)

____
(Indicate Progress or Final Report)

Work Done By:
Report Written By:
Supervisor:
R & D Director:
Previous Related Reports:
Department Overhead Number:
Project Number:
Period Covered:
Notebook Number(s)

PROPRIETARY INFORMATION FOR AUTHORIZED COMPANY USE ONLY

add the reader's and the writer's positions and the date. Figure 11–2 provides a sample of a simple title page.

Most organizations have their own formats for title pages. Often, an organization will have different formats for different kinds of reports. Figure 11–3 shows the complex title page used for research reports at one company.

This company wants so much information that two pages are required. Notice that on the first page, the title of the report would be followed by a statement of whether the report is a progress report or final report. On the second page, the word *distribu-*

FIGURE 11-3
(Continued)

AUTHOR

____

SUPERVISOR

____

DIRECTOR

____

DISTRIBUTION

Complete Report

Summary Report

tion would be followed by the listing of those who are to receive the report. Like many companies, this one in fact sends the complete report to relatively few people (approximately four or five) and to the files. The summary report (the executive summary) goes to many more readers (approximately twenty or thirty).

THE ABSTRACT

An abstract is a brief technical summary—usually no more than 200 words—of the report. Like the abstract that accompanies published articles (see Chapter 16), the abstract of a report is di-

FIGURE 11–4

Descriptive Abstract

"Commercial Electroplating of Plastics:
A Feasibility Study" by Stig Washburn

Zinc and brass prices have been climbing at a rapid rate. This report studies the advantages and disadvantages of replacing metals with plastics as a substrate. Plastics are compared with zinc and brass in terms of finished-part costing, lifetime costing, social costing, weight, and corrosion resistance. The nature of chemical pretreatments for plastics is also discussed.

rected primarily to readers who are familiar with the technical subject and need to know whether to read the full report. Therefore, in writing an abstract you can use technical terminology freely and refer to advanced concepts in your field. (For managers who want a summary focusing on the managerial implications of a project, many reports that contain abstracts also contain executive summaries. See the discussion on pp. 246–249.)

Because abstracts can be useful before and after a report is read—and can even be read in place of the report—they are duplicated and kept on file in several locations within an organization: the division in which the report originated and many or all of the higher-level units of the organization. And, of course, a copy of the abstract is attached to or placed within the report. It is not unusual to find six or eight copies of an abstract somewhere in an organization. To facilitate this wide distribution, some organizations have special forms on which abstracts are typed.

The two basic types of abstracts are generally called descriptive and informative abstracts. The descriptive abstract is rapidly losing popularity, whereas the informative abstract is becoming the accepted standard.

THE DESCRIPTIVE ABSTRACT

The descriptive abstract, sometimes called the topical or table of contents abstract, does only what its name implies: it describes what the report is about. It does not provide the important results or conclusions or recommendations. It simply lists the topics covered, giving equal coverage to each. Thus, the descriptive abstract simply duplicates the information included in the table of contents. Figure 11–4 provides an example of a descriptive abstract.

THE INFORMATIVE ABSTRACT

The informative abstract presents the major information that the report conveys. Rather than merely listing topics, it states the problem, the scope and methods (if appropriate), and the major

FIGURE 11-5

Informative Abstract

"Commercial Electroplating of Plastics:
A Feasibility Study" by Stig Washburn

Zinc and brass prices have been climbing at a rapid rate. This report studies the advantages and disadvantages of replacing metals with plastics as a substrate. Plastics have a distinct economic advantage over metals in terms of reduced finished-part costing, lifetime costing, social costing, weight, and increased corrosion resistance. Metals are superior only in the areas of thermal cycling and recovery of scrap material. To convert the electroplating process from metal to plastic, a preplate chemical conditioning system must be added.

results, conclusions, or recommendations, as shown in Figure 11-5.

As these two examples clearly demonstrate, the informative abstract presents more useful information than the descriptive abstract does.

The basic structure of the informative abstract includes three elements:

1. *The identifying information.* The name of the report, the writer, and perhaps the writer's department.

2. *The problem statement.* One or two sentences that define the problem or need that led to the project. Many writers mistakenly omit the problem statement, assuming that the reader knows what the problem is. The *writer* knows, being intimately involved with the project, but the readers are likely to be totally unfamiliar with it. Without an adequate problem statement to guide them, many readers will be unable to understand the abstract.

3. *The important findings.* The final three or four sentences—the biggest portion of the abstract—state the crucial information the report contains. Generally, this means some combination of results, conclusions, recommendations, and implications for further projects. Sometimes, however, the abstract presents other information. For instance, many technical projects focus on new or unusual methods for achieving results that have already been obtained through other means. In such a case, the abstract will focus on the methods, not the results.

The structure of the descriptive abstract resembles that of the informative abstract in that both contain identifying information and a problem statement. Rather than focusing on the important findings, however, the descriptive abstract gives equal emphasis to all or most of the topics listed in the table of contents. In a descriptive report, then, the research methods, for example, will be mentioned even if they are not the focus of the project.

THE TABLE OF CONTENTS

Far too often, good reports are ruined because the writer fails to create a useful table of contents. This element is crucial to the report, because it enables different readers to turn to specific pages to find the information they want. No matter how well organized the report itself may be, a table of contents that does not make the structure clear will be ineffective.

Most people will have read the abstract when they turn to the table of contents to find one or two items in the body of the report. Because a report usually has no index, the table of contents will provide the only guide to the report's structure, coverage, and pagination. The headings listed in the table of contents are the headings that appear in the report itself. To create an effective table of contents, therefore, you first must make sure the report has effective headings—and that it has enough of them. If the table of contents shows no entry for five or six pages, the report could probably be divided into additional subunits. In fact, some tables of contents have a listing of every page in the report.

Insufficiently specific tables of contents generally result from the exclusive use of generic headings (those that describe entire classes of items) in the report. Figure 11–6 shows how inadequate a table of contents can become if it simply lists generic headings.

FIGURE 11–6

Ineffective Table of Contents

Table of Contents

Introduction	1
Materials	3
Methods	4
Results	19
Recommendations	23
References	26
Appendixes	28

To make the headings more informative, combine generic and specific items, as in the following examples:

Recommendations: Five Ways to Improve Information Retrieval

Materials Used in the Calcification Study

Results of the Commuting-Time Analysis

Then build more subheadings into the report. For example, in the "Recommendations" example, make a separate subheading for each of the five recommendations.

If you use a clear system of headings within the report, you should be able to use the same system on your contents page. In other words, you can create an informative table of contents by listing your headings in the same style in which they appear in the text, using capitalization and underlining, indentation, or an outline system (see the discussion of headings in Chapter 5):

1. *Capitalization and underlining.* Capital letters are more emphatic than small letters, and underlined words are more emphatic than nonunderlined words. You can create up to six hierarchical levels of headings using capitalization and underlining:

CATHODE RAY TUBES

CATHODE RAY TUBES

Cathode Ray Tubes

Cathode Ray Tubes

cathode ray tubes

cathode ray tubes

2. *Indentation.* Less-important headings can be indented from the left-hand margin of the page:

```
First-level heading
  Second-level heading
    Third-level heading
  Second-level heading
    Third-level heading
      Fourth-level heading
      Fourth-level heading
```

3. *Outline-style headings.* Arabic and Roman numbers and capital and small letters can be combined to identify various hierarchical levels, as in

a traditional outline. Outline-style headings generally employ indentation as well.

```
I.  First-level heading
    A.  Second-level heading
        1.  Third-level heading
        2.  Third-level heading
    B.  Second-level heading
        1.  Third-level heading
            a.  Fourth-level heading
            b.  Fourth-level heading
```

Figure 11–7 shows how you can combine generic and specific headings and use the resources of the typewriter to structure a report effectively. The report is titled "Methods of Computing the Effects of Inflation in Corporate Financial Statements: A Recommendation." This table of contents combines all three of the techniques discussed.

The table of contents in Figure 11–7 is effective for several reasons. First, managers can find the executive summary quickly and easily. Second, all of the other readers can find the information they are looking for, because each substantive section is listed separately. An additional advantage of a specific table of contents is that it gives the readers a clear idea of the scope and structure of the report before they start to read it.

A note about pagination is in order. The abstract pages of a report are generally not numbered, but other preliminary elements (for example, a preface or acknowledgments page) are numbered with lowercase roman numerals (i, ii, and so forth) centered at the bottom of the pages. The report itself is generally numbered with Arabic numerals (1, 2, and so forth) in the upper right-hand corner of the pages. Some organizations include on each page the total number of pages (1 of 17, 2 of 17, and so forth) so the readers will be sure they have the complete document.

After creating the table of contents, review the report to make sure that the headings in the table of contents are written just as they appear in the report. The wording and the format must be the same: if the heading is typed in full capital letters and preceded by a roman numeral in the report, it must appear that way in the table of contents. Such correspondence between the table of contents and the body of the report helps the readers understand and remember the structure of the report.

FIGURE 11-7

Effective Table of Contents

CONTENTS

THE LIST OF ILLUSTRATIONS

A list of illustrations is a table of contents for the figures and tables of a report. (See Chapter 10 for a discussion of figures and tables.) If the report contains figures but not tables, the list is called a *list of figures*. If the report contains tables but not figures, the list is called a *list of tables*. If the report contains both figures and tables, figures are listed separately, before the list of tables, and the two lists together are called a *list of illustrations*.

Some writers will begin the list of illustrations on the same page as the table of contents; others prefer a separate page for the list of illustrations. If the list of illustrations begins on a separate page, it is listed in the table of contents. Figure 11–8 provides an example of a list of illustrations.

THE EXECUTIVE SUMMARY

The executive summary (sometimes called the *epitome*, the *executive overview*, the *management summary*, or the *management overview*) is a one-page condensation of the report. Its audience is made up, of course, of managers, who rely on executive summaries to cope with the tremendous amount of paper crossing their desks every day. The reason for the executive summary is that managers do not need or want a detailed and deep understanding of the various projects undertaken in their organizations; this kind of understanding would in fact be impossible for them, because of limitations in time and specialization. What managers *do* need is a broad understanding of the projects and how they fit together into a coherent whole. Consequently, a one-page (double spaced) maximum for the executive summary has become almost an unwritten standard.

The special needs of managers dictate a two-part structure for the executive summary:

1. *Background*. Because managers are not necessarily technically competent in the writer's field, the background of the project is discussed clearly and completely. The specific problem—what was not working or not working effectively or efficiently, or what potential modification of a procedure or product had to be analyzed—is stated explicitly.

2. *Major findings and implications*. Managers are not interested in the details of the project, so the methods—the largest portion of the report—rarely receive more than one or two sentences. The conclusions and recommendations, however, are discussed in a full paragraph.

FIGURE 11-8
List of Illustrations

LIST OF ILLUSTRATIONS

For instance, if the research and development division at an automobile manufacturer has created a composite material that can replace steel in car axles, the technical details of the report might deal with the following kinds of questions:

1. How was the composite devised?
2. What are its chemical and mechanical structures?
3. What are its properties?

The managerial implications, on the other hand, involve other kinds of questions:

1. Why is this composite better than steel?
2. How much do the raw materials cost? Are they readily available?
3. How difficult is it to make the composite?
4. Are there physical limitations to the amount we can make?
5. Is the composite sufficiently different from similar materials to prevent any legal problems?
6. Does the composite have other possible uses in cars?

The executives don't care about chemistry; they want to know how this project can help them make a better automobile for less money.

For several reasons, the executive summary poses a great challenge to writers. First, the brevity of the executive summary requires an almost unnatural restraint. Having spent some weeks or even months collecting data on a complex subject, writers find it difficult to reduce all of that information to one page.

Second, the executive summary usually is written specifically for a nontechnical audience. Most writers are trained to write to specialists in their field. Beginning with their training in high school science courses, writers were asked to address an audience (their teachers) that knew at least as much about the subject as they did. Learning and using the technical vocabulary and concepts was an important part of this training. In communicating with nonspecialists, however, writers must avoid or downplay specialized vocabulary and explain the subject without referring to advanced concepts.

Third, and most important, writers are so used to thinking in terms of supporting their claims with hard evidence that the notion of simply making a claim—and then not substantiating it—seems almost sacrilegious to them. Although some managers would probably like to read the full report, they simply don't have the time. They must trust the writer's technical accuracy.

Unfortunately, it is easier to define why executive summaries are a challenge to write than it is to suggest ways to write them easily. Implicit in this discussion is an obvious point: you must try to ignore most of the technical details of the project and think, instead, of the manager's needs. The expression *the bottom line* is useful to keep in mind when you must focus on the managerial implications of the project. Somewhat less obvious is another suggestion: give the draft of the executive summary to someone who has had nothing to do with the project—preferably someone outside the field. That person should be able to read the page and understand the basics of what the project means to the future of the organization.

Placement of the executive summary can be important. The current practice in business and industry is to place the executive summary before the detailed discussion. To highlight the executive summary further, writers commonly make it equal in importance to the entire detailed discussion. This strategy is signaled in the table of contents, in which the report as a whole is divided into two units. The executive summary becomes unit one of the report, and the detailed discussion unit two. In the traditional organiza-

FIGURE 11–9

Executive Summary

SUMMARY

Zinc and brass prices have risen more than 14% per year over the last three years. Because zinc and brass purchases account for more than half of our electroplating costs, a project was undertaken to determine if plastic could substitute for zinc and brass in our electroplating operation and to determine the extent of the modifications necessary to convert to plastic.

Conclusions

1. Plastics have a distinct economic advantage over metals in terms of finished-part costing, lifetime costing, social costing, weight, and corrosion resistance.

2. Metals are superior in terms of physical properties and recovery of scrap products.

Recommendation

Because current estimates are that plastic will remain cheaper than metals in electroplating, and because new plastics with better physical properties are being formulated, we recommend that our electroplating operation be converted from metals to plastics. The capital expenditure will be approximately $200,000, a sum that will be paid back within three years.

tion of a report, the executive summary would be unit one and the detailed discussion units two through, say, seven. Figure 11–9 provides an example of an effective executive summary.

Notice the differences between this executive summary and the informative abstract (Figure 11–5). The abstract focuses on the technical subject itself: whether plastic can replace zinc and brass in electroplating operations. The executive summary concentrates on whether plastics can replace zinc and brass in electroplating operations *at this one company*. Consequently, the problem is stated in terms of its economic implications—its symptoms. The problem with the metals is not that they aren't working well; rather, they are becoming too expensive. Notice in particular how at the start of the second sentence the writer fills in the executives on a technical point they might not know: ". . . zinc and brass purchases account for more than half of our electroplating costs. . . ."

The third sentence also focuses on a managerial concern: that raw-materials costs make up only a part of the total cost. In comparing two raw materials, managers will want to know how much it would cost to switch from one to another. Especially effective in this executive summary is the recommendations section, which clearly spells out the initial costs and the pay-back period. *Those* are the bottom-line figures.

THE GLOSSARY AND LIST OF SYMBOLS

A glossary is an alphabetical list of definitions. A glossary is particularly useful if you are addressing a multiple audience that includes readers who will not be familiar with the technical vocabulary used in your report.

Instead of slowing down the detailed discussion by defining technical terms as they appear, you can use an asterisk or some other form of notation to inform your readers that the term is defined in the glossary. A footnote at the bottom of the page on which the first asterisk appears serves to clarify this system for the readers. For example, the first use of a term defined in the glossary might occur in the following sentence in the detailed discussion of the report: "Thus the positron* acts as the. . . ." At the bottom of the page, add

*This and all subsequent terms marked by an asterisk are defined in the Glossary, page 26.

Although the glossary is generally placed near the end of the report, right before the appendixes, it can also be placed right after the table of contents. This placement is appropriate if the glossary is brief (less than a page) and defines terms that are essential for managers likely to read the body of the report. Figure 11–10 provides an example of a glossary.

A list of symbols is structured like a glossary, but rather than defining words and phrases, it defines the symbols and abbreviations used in the report. Don't hesitate to include a list of symbols if you suspect that some of your readers will not know what your symbols and abbreviations mean or might misinterpret them. Like the glossary, the list of symbols may be placed before the appendixes or after the table of contents. Figure 11–11 provides an example of a list of symbols (in this case, abbreviations).

THE APPENDIX

An appendix is any section that follows the body of the report (and the list of references or bibliography, glossary, or list of symbols). Appendixes provide a convenient way to convey information that is too bulky to be presented in the body or that will be of interest to only a small number of the report's readers. For the sake of conciseness in the report proper, this information is separated from the body. Examples of the kinds of materials that are usually found in appendixes include maps, large technical diagrams or

FIGURE 11–10
Glossary

GLOSSARY

byte:	a binary character operated upon as a unit and, generally, shorter than a computer word.
error message:	an indication that an error has been detected by the system.
hard copy:	in computer graphics, a permanent copy of a display image that can be separated from a display device. For example, a display image recorded on paper.
parameter:	a variable that is given a constant value for a specified application and that may denote that application.
record length:	the number of words or characters forming a record.

FIGURE 11–11
List of Symbols

LIST OF SYMBOLS

CRT	cathode-ray tube
H_z	hertz
rcvr	receiver
SNR	signal-to-noise ratio
uhf	ultra high frequency
vhf	very high frequency

charts, computations, computer printouts, test data, and texts of supporting documents.

Appendixes, which are usually lettered rather than numbered (Appendix A, Appendix B, etc.), are listed in the table of contents and are referred to at the appropriate points in the body of the report. Therefore, they are accessible to any reader who wants to consult them.

Remember that an item in an appendix is titled "Appendix." The item is not called "Figure" or "Table," even if it would have been so designated had it appeared in the body of the report.

WRITER'S CHECKLIST

In the rush to compile a report, it is easy to forget important elements of the format. The following checklist is intended to help you make sure you have included the appropriate format elements and written them correctly.

1. Does the transmittal letter
   a. Clearly state the title and, if necessary, the subject and purpose of the report? _____
   b. State who authorized or commissioned the report? _____
   c. Briefly state the methods you used? _____
   d. Summarize your major results, conclusions, or recommendations? _____
   e. Acknowledge any assistance you received? _____
   f. Courteously offer further assistance? _____

2. Does the title page
   a. Include a title that both suggests the purpose of the report and identifies the type of report it is? _____
   b. List the names and positions of both you and your principal reader? _____
   c. Include the date of submission of the report and any other identifying information? _____

3. Does the abstract
   a. List your name, the report title, and any other identifying information? _____
   b. Clearly define the problem that led to the project? _____
   c. Briefly describe (when appropriate) the research methods? _____
   d. Summarize the major results, conclusions, or recommendations? _____

4. Does the table of contents
   a. Clearly identify the executive summary? _____
   b. Contain a sufficiently detailed breakdown of the major sections of the body of the report? _____
   c. Reproduce the headings as they appear in your report? _____
   d. Include page numbers? _____

5. Does the list of illustrations (or tables or figures) include all of the graphic aids included in the body of the report? _____

6. Does the executive summary
   a. Clearly state the problem that led to the project? _____
   b. Explain the major results, conclusions, recommendations, and/or managerial implications of your report? _____

c. Avoid technical vocabulary and concepts that the managerial audience is not likely to know? _____

7. Does the glossary include definitions of all of the technical terms your readers might not know? _____

8. Does your list of symbols include all of the symbols and abbreviations your readers might not know? _____

9. Do your appendixes include the supporting materials that are too bulky to present in the report body or that will interest only a small number of your readers? _____

EXERCISES

1. For each of the following letters of transmittal, write a one-paragraph evaluation. Consider each letter in terms of clarity, comprehensiveness, and tone.

a.

Dear Mr. Smith:

The enclosed report, "Robots and Machine Tools," discusses the relationship between robots and machine tools.

Although loading and unloading machine tools was one of the first uses for industrial robots, this task has only recently become commonly feasible. Discussed in this report are concepts that are crucial to remember in using robots.

If at any time you need help understanding this report, please let me know.

Sincerely yours,

b.

Dear General Smith:

Along with Milwaukee Diesel, we are pleased to submit our study on potential fuel savings and mission improvement capabilities that could result from retrofitting the Air Force's C-3 patrol aircraft with new, high-technology recuperative or intercooled-recuperative turboprop engines.

Results show that significant benefits can be achieved, but because of weight and drag installation penalties, the new recuperative or intercooled-recuperative engines offer little additional savings relative to a new conventional engine.

Sincerely,

c.

Dear Mr. Smith:

Enclosed is the report you requested on the effects consumer preferences will have on the future use of natural flavors.

The major object of this project was to assess the strength of the apparent consumer desire for naturally flavored food products stemming from the controversy surrounding artificial additives. The lesser objective was to examine the other factors affecting the development and production of natural flavors.

I have found that food purchasers seem to be more interested in the taste of their foods than they are in the source of their ingredients. Moreover, the use of natural flavors is complicated by problems such as high cost, lack of availability, and inconsistent quality. Therefore, I have recommended that Smith & Co. not enter the natural flavor market at this time.

If you require any additional information, please do not hesitate to contact me.

Sincerely yours,

2. For each of the following report titles, write a three- or four-sentence evaluation. How clear an idea does each title give you of the subject of the report? Of the type of report it is? Based on your analysis, rewrite each title.

a. "Recommended Forecasting Techniques for Haldane Company"
b. "Investigation of Computerized Tax Return Systems"
c. "Analysis of Multigroup Processing Techniques for Pennco Billing System"
d. "Site Study for Proposed Shopping Mall in Speonk, N.Y."
e. "Robotics in Japan"
f. "A Detailed Description of the Factors That Led to the Great Depression of 1929"
g. "A Study of Disc Cameras"
h. "Teleconferencing"
i. "The Effect of Low-level Radiation on Farm Animals"
j. "Synfuels—Fact or Hoax?"

3. For each of the following abstracts, write a one-paragraph evaluation. How well does each abstract define the problem, methods, and important results, conclusions, and recommendations involved?

a.

"Proposal to Implement an Amplifier Cooling System"

In professional sound reinforcement applications, such as discotheques, the sustained use of high-powered amplifiers causes these units to operate at high temperatures. These high temperatures cause a loss in efficiency and ultimately audible distortion, which is undesirable at disco volume levels. A cooling system of forced air must be implemented to maintain stability. I propose the use of separate fans to cool the power amplifiers, and of the best components to eliminate any distorting signals.

b.

"Design of a New Computer Testing Device"

The modular design of our new computer system warrants the development of a new type of testing device. The term modular design indicates that the overall computer system can be broken down into parts or modules, each of which performs a specific function. It would be both difficult and time consuming to test the complete system as a whole, for it consists of sixteen different modules. A more effective testing method would check out each module individually for design or construction errors prior to its installation into the system. This individual testing process can be accomplished by the use of our newly designed testing device.

The testing device can selectively call or "address" any of the logic modules. To test each module individually, the device can transmit data or command words to the module. Also, the device can display the status or condition of the module on a set of LED displays located on the front panel of the device. In addition, the device has been designed so that it can indicate when an error has been produced by the module being tested.

c.

"A Recommendation for a New System of Monitoring Patients with Implanted Cardiac Pacemakers"

The monitoring system used today to test pacemaker patients is inadequate at the data distribution and storage end. Often doctors receive 20 to 50 reports a week from individual tests on patients who, because of clerical and administrative difficulties, cannot be evaluated. This report recommends a digital time-sharing system for data storage and instant recall of test results at hospital locations throughout the country. With the current system and the recommendations in this report, pacemaker monitoring will reduce data storage procedures, extend the useful life of individual pacemakers, provide doctors with current and reliable information on individual heart conditions, and supply patients with personal support and professional care.

4. For each of the following tables of contents, write a one-paragraph evaluation. How effective is each table of contents in highlighting the executive summary, defining the overall structure of the report, and providing a detailed guide to the location of particular items?

a. from "An Analysis of Corporations v. Sole Proprietorships":

TABLE OF CONTENTS

b. from "Recommendation for a New Incentive Pay Plan: the Scanlon Plan":

CONTENTS

c. from "Initial Design of a Microprocessor-Controlled FM Generator":

TABLE OF CONTENTS

5. For each of the following executive summaries, write a one-paragraph evaluation. How well does each executive summary present concise and useful information to the managerial audience?

a. from "Analysis of Large-Scale Municipal Sludge Composting as an Alternative to Ocean Sludge-Dumping":

Coastal municipalities currently involved with ocean sludge-dumping face a complex and growing sludge management problem. Estimates suggest that treatment plants will have to handle 65% more sludge in 1985 than in 1975, or approximately seven thousand additional tons of sludge per day. As the volume of sludge is increasing, traditional disposal methods are encountering severe economic and environmental restrictions. The EPA has banned all ocean sludge-dumping as of January 1, 1981. For these reasons, we are considering sludge composting as a cost-effective sludge management alternative.

Sludge composting is a 21-day biological process in which waste-water sludge is converted into organic fertilizer that is aesthetically acceptable, essentially pathogen-free, and easy to handle. Composted sludge can be used to improve soil structure, increase water retention, and provide nutrients for plant growth. At $150 per dry ton, composting is currently almost three times as expensive as ocean dumping, but effective marketing of the resulting fertilizer could dramatically reduce the difference.

b. from "Recommendation for a New Incentive Pay Plan: the Scanlon Plan":

Summary

The incentive pay plan, Payment by Results, in operation at Cargo Corporation of America's Trenton plant, is not working effectively. Since the implementation of this plan, union and management have experienced increasing conflicts. The employee turnover ratio and absenteeism both have risen markedly, and productivity output has increased only slightly.

An extensive research project was conducted with the objective of providing a solution to the plant's problem. Originally, four possible solutions were considered: (1) a revised Payment by Results, (2) Fixed Hourly Wage (no incentive plan at all), (3) Measured Daywork, (4) Scanlon Plan. After careful consideration and investigation, the best alternative was found to be the Scanlon Plan.

The Scanlon Plan is a companywide bonus system to reward increases in productivity and efficiency. The plan has two major parts: (1) a base productivity norm, and (2) a system of work committees. The base productivity norm is based on performance over a period compared to a base period. Any improvement over the base period is called a bonus pool to be shared by <u>all</u> employees. Many different formulas have been designed to calculate the norm, but the most common is total payroll divided by the sales value of production (sales-value of production method).

Most companies implement the Scanlon Plan with departmental production committees, all reporting to one central screening committee. The individual committees meet regularly and discuss suggestions for improvements. The production committee meets formally only once a month for an hour or so. At this meeting they review the suggestions that deal with operating improvements. Any issues that they do not have the authority to deal with--for example union matters, wages, bonuses, etc.--are forwarded to the central screening committee. This committee is composed of employees from a representative cross-section of the company, including union officials. There is no voting by the committee on suggestions, but there is thorough discussion of all points of view when there is a disagreement. After careful consideration, management makes the final decision.

Managers who have implemented the plan have found that employees are much more compromising toward change, particularly technological change. Experience has shown that thorough discussion of planned changes enables employees to realize that the new technology is not going to bring with it all sorts of ills that cannot be solved satisfactorily.

Companies that have used the plan have not experienced any resentment from the presiding union, and actually relations have improved. In many union-management relationships, the most difficult problem for the union leader is to get top management to sit down and discuss their various problems. A basic prerequisite of the Scanlon approach is that management be "willing to listen."

The Scanlon Plan is not for companies that are seeking a gimmick that will solve their problems. It is not a substitute for good management. The message of the plan is simple: operations improvement is an area in which management, the union, and employees can work together without strife.

c. from "Applying Multigroup Processing to the Gangloff Billing Account":

Gangloff Accounting is divided into seven geographical areas. Previously, end-of-the-month billing was processed for each group separately. Multigroup processing allows for processing all the groups simultaneously.

Multigroup processing is beneficial to both the data processing department and the accounting department in many ways. Running all of the groups together will cut the number of job executions from 108 to 14, an 87% difference. This difference accounts for a 60% saving in computer time and a 30% saving in paper. With multigroup processing, all sense switches would be eliminated. With the sense switches done away with and the tremendous decrease in computer operator responsibilities, the chance of human error in the execution will diminish significantly.

CHAPTER TWELVE

MEMO REPORTS

The memorandum is the workhorse of technical writing. It is the basic means of written communication between two persons—or between individuals and groups—in an organization. Each day, the average employee is likely to receive as many as a half-dozen memos and send out another half-dozen. Whereas most memos convey routine news addressed to several readers, many organizations are turning to the memo as an efficient format for brief technical reports.

To facilitate the transfer of information, most organizations have distribution routes—lists of employees who automatically receive copies of memos that pertain to their area of expertise or responsibility. A technician, for example, might be on several different distribution routes: one that includes all employees, one that includes the technical staff, and one that includes all employees involved with a particular project.

This chapter concentrates not on the "FYI" memo—the simple communication addressed "For Your Information"—but on the more substantive kinds of brief technical reports that generally are written in memo form, such as directives, responses to inquiries, trip reports, and field reports.

Like all technical writing, the memo should be clear, accurate, concise, accessible, and correct. But unlike letters, memos need not be ceremonious. Memo writing is technical writing with its sleeves rolled up.

Clarity, conciseness, and accessibility are encouraged by a mechanical device: headings. The printed forms on which memos are written begin with a subject heading—a place for the writer to define the subject. If you use this space efficiently, you can define not only the subject but also the purpose of the memo, and thus begin to communicate immediately.

In the body of the memo, headings help your readers understand the message you are conveying. For example, the simple word *Results* at the start of a paragraph tells your readers what you are describing and thus enables them to decide whether or not to read the discussion. (See Chapter 5 for a further discussion of headings.)

Conciseness is also encouraged by the size of the forms. Most organizations have at least two sizes of memo forms: a full page and a half page. Often, writers are reluctant to put down only two or three lines on a full page, so they start to pad the message. The smaller page reduces this temptation.

The following discussion describes in detail how to write effective memos.

THE STRUCTURE OF THE MEMO

The memo is made up of two components: the identifying information and the body.

THE IDENTIFYING INFORMATION

Basically, a memo looks like a streamlined letter. The salutation ("Dear Mr. Smith") and the complimentary close ("Sincerely yours") are eliminated. Some writers put their initials or signature next to their typed names or at the end of their memos; others don't bother. The inside address—the mailing address of the reader—is replaced by a department name or an office number, generally listed after the person's name. Sometimes no "address" at all is given.

Almost all memos have five elements at the top: the logo or a brief letterhead of the organization and the "To," "From," "Subject," and "Date" headings. Some organizations have a "copies" or "cc" (carbon copy) heading as well. "Memo," "Memorandum," or "Interoffice" might be printed on the forms.

Organizations sometimes have preferences about the ways in which the headings are filled out. Some organizations prefer that the full names of the writer and reader be listed. Others want only the first initials and the last names. Job titles are sometimes used. If the organization for which you work does not object, include your own job title and that of your reader. In this way, the memo will be complete and informative for a reader who refers to it after you or your reader has moved on to a new position as well as for readers elsewhere in the organization who might not know you and your reader. The names of those persons receiving copies of the memo (generally photocopies, not carbons) are listed either alphabetically or in descending order of organizational rank. In listing the date, write out the month (March 4, 19-- or 4 March 19--); do not use the all-numeral format (3/4/--). Foreign-born people could be confused by the numerals, because in many countries the first numeral identifies the day, not the month.

The subject heading deserves special mention. Don't be *too* concise. Avoid naming only the subject, such as "Tower Load Test"; rather, add a clue identifying what it is about the test you wish to address. For instance, "Tower Load Test Results" or "Results of Tower Load Test" would be much more informative than "Tower Load Test," which does not tell the reader whether the memo is about the date, the location, the methods, the results, or any number of other factors related to the test.

The top of a memo is designed to identify the writing situation as efficiently as possible. The writer names himself or herself, the audience, and the subject, ideally with some indication of purpose.

The following examples show several common styles of filling in the identifying information of a memo.

AMRO MEMO

To: B. Pabst
From: J. Alonso
Subject: MIXER RECOMMENDATION FOR PHILLIPS
Date: 11 June 19--

NORTHERN PETROLEUM COMPANY INTERNAL CORRESPONDENCE

Date: January 3, 19--
To: William Weeks, Director of Operations
From: Helen Cho, Chemical Engineering Dept.
Subject: Trip Report--Conference on Improved Procedures for Chemical Analysis Laboratory

HARSON ELECTRONICS	MEMORANDUM		
To:	John Rosser, Accounting	From:	Andrew Miller, Technical Services
		Subject:	Budget Revision for FY 19--
cc:	Dr. William De Leon, President John Grimes, Comptroller	Date:	March 11, 19--

INTEROFFICE

To:	C. Cleveland	cc:	B. Aaron
From:	H. Rainbow		K. Lau
Subject:	Shipment Date of Blueprints to Collier		J. Manuputra W. Williams
Date:	2 October 19--		

The second and all subsequent pages of memos are typed on plain paper. The following information is typed in the upper left-hand corner of each page:

1. the name of the recipient
2. the date of the memo
3. the page number

Often, writers will define the communication as a memo and repeat the primary names as well. A typical second page of such a memo begins like this:

Memo to:	J. Alders	April 6, 19--
from:	R. Rossini	Page 2

THE BODY OF THE MEMO

The average memo has a very brief "body." A message about the office closing early during heavy snow, for example, requires only one or two sentences. However, memos that convey complex technical information, and those that approach one page or longer (memos are always single-spaced on the typewriter), are most effective when they follow a basic structure. This structure gives you the same sense of direction that a full-scale outline does as you plan a formal report. A useful structure for most substantive memos includes four parts:

1. purpose statement
2. summary
3. discussion
4. action

This structure—including any subsections—should be incorporated explicitly into the memo. Use headings and lists (see Chapter 5) to clarify the structure.

Headings make the memo easier to read, easier *not* to read, and easier to refer to later. Headings are not subtle; they define clearly what the discussion that follows is about and thus improve the reader's comprehension. In addition, headings represent a courtesy to executive readers who are not interested in the details of the memo. These readers can read the purpose and summary and stop if they need no further information. Finally, headings enable readers to isolate quickly the information they need after the memo has been filed away for a while. Rather than having to reread a three-page discussion, they can turn directly to the summary, for instance, or to a subsection of the discussion.

Lists help your reader understand the memo. If, for instance, you are making three points, you can list and enumerate them consistently in the different sections of the memo. Point number one under the "Summary" heading will correspond to point number one under the "Recommendations" heading, and so forth.

PURPOSE STATEMENT Memos are reproduced and distributed very freely. The majority of those that you receive any given day might be only marginally relevant to you. Many readers, after starting to read their incoming memos, ask, "Why is he telling me this?" The first sentence of the body should answer that question. Following are a few examples of purpose statements.

```
I want to tell you about a problem we're having with the pressure on
the main pump, because I think you might be able to help us.
```

```
The purpose of this memo is to request authorization to travel to the
Brownsville plant Monday to meet with the other quality inspectors.
```

```
This memo presents the results of the internal audit of the Phoenix
branch that you authorized March 13, 19--.
```

I want to congratulate you on the quarterly record of your division.

This memo confirms our phone call of Tuesday, June 10, 19--.

Notice that a purpose statement need not be long and complicated. In fact, the best purpose statements are concise and direct. Make sure your purpose statement has a verb that clearly communicates what you want the memo to accomplish, such as *to request, to explain,* or *to authorize.* (See Chapter 2 for a more detailed discussion of purpose.) Some students of the techniques of logical argumentation and persuasion object to a direct statement of purpose—especially when the writer is asking for something, as in the example about requesting travel authorization. Rather than beginning with a direct statement of purpose, the standard argument structure for such a request would open with the reasons that the trip is necessary, the trip's potential benefits, and so forth. Then the writer would conclude the memo with the actual request: "For these reasons, I am requesting authorization to. . . ." Although some readers would rather have the reasons presented first, far more would prefer to know immediately why you have written. There are two basic problems with the standard argument structure: (1) some readers will toss the memo aside if you seem to be rambling on about the Brownsville plant without getting to the point; and (2) some readers will feel as if you are trying to manipulate them, to talk them into doing something they don't want to do. The purpose statement sacrifices subtlety in favor of directness.

SUMMARY Along with the purpose statement, the summary forms the core of the memo. It helps all the readers follow the subsequent discussion, enables executive readers to skip the rest of the memo, and serves as a convenient reminder of the main points. Following are some examples of summaries:

The proposed revision of our bookkeeping system would reduce its errors by 80% and increase its speed by 20%. The revision would take two months and cost approximately $4,000. The payback period would be less than one year.

The conference was of great value. The lectures on new coolants suggested techniques that might be useful in our Omega line, and I met three potential customers who have since written inquiry letters.

The analysis shows that lateral stress was the cause of the failure. We are now trying to determine why the beam could not sustain a lateral stress weaker than that it was rated for.

In March, we completed Phase II (Design) on schedule. At this point, we anticipate no delays that will jeopardize our projected completion date.

The summary should reflect the length and complexity of the memo. It might range in length from one simple sentence to a long and technical paragraph. If possible, the summary should reflect the structure of the memo. For example, the discussion following the first sample summary should explain, first, the proposed revision of the bookkeeping system and, second, its two advantages: fewer errors and increased speed. Next should come the discussion of the costs in terms of time and money. Finally, the memo should discuss the payback period.

DISCUSSION The discussion is the elaboration of the summary. It is the most detailed and technical portion of the memo. Generally, the discussion begins with a background paragraph. Even if you think the reader will be familiar with the background, it is a good idea to include a brief recap, just to be safe. Also, the background will be valuable to a reader who refers to the memo later.

The background discussion is, of course, as individual as the memo of which it is a part; however, some basic guidelines are useful. If the memo defines a problem—for example, a flaw detected in a product line—the background might discuss how the problem was discovered or present the basic facts about the product line: what the product is, how long it has been produced, and in what quantities. If the memo reports the results of a field trip, the background might discuss why the trip was undertaken, what its objectives were, who participated, and so forth.

Following is a background paragraph from a memo requesting authorization to have a piece of equipment retooled.

Background

The stamping machine, a Curtiss Model 6143, is used in the sheet-metal shop. We bought it in 1976 and it is in excellent condition. However, since we switched the size of the tin rolls last year, the stamping machine no longer performs up to specifications.

After the background comes the detailed discussion of the message itself. Here you present whatever details are necessary to give your readers a clear and complete idea of what you have to say. The detailed discussion might be divided into the subsections of a more formal report: materials, equipment, methods, results, conclusions, and recommendations. Or it might be made up of headings that pertain specifically to the subject you are discussing.

Small tables or figures might also be included; larger ones should be attached as appendixes to the memo.

The discussion section of the memo can be developed according to any of the basic patterns for structuring technical documents:

1. chronological
2. spatial
3. general to specific
4. more important to less important
5. problem-methods-solutions

These patterns are discussed in detail in Chapter 4.

Following is the detailed discussion section from a memo written by a salesman working for the "XYZ Company," which makes electronic typewriters. The XYZ salesman met an IBM salesman by chance one day, and they talked about the XYZ typewriters. The XYZ salesman is writing to his supervisor, telling her what he learned from the IBM salesman and also what XYZ's research and development (R&D) department told him in response to the comments of the IBM salesman.

DISCUSSION

Salesman's Comments:

In our conversation, he talked about the strengths of our machines and then mentioned two problems: excessive ribbon consumption and excessive training time.

In general, he had high praise for the XYZ machines. In particular, he liked the idea of the rotary and linear stepping motor. Also, he liked having all the options within the confines of the typewriter. He said that although he knows we have some reliability problems, the machines worked well while he was training on them.

The major problem with the XYZ machines, he said, is excessive ribbon consumption. According to his customers who have XYZ machines, the $5 cartridge lasts only about two days. This adds up to about $650 a year, about a third the cost of our basic Model A machine.

The minor problem with the machines, he said, is that most customers are used to the IBM programming language. Since our language is very different, customers are spending more time learning our system than they had anticipated. He didn't offer any specifics on training-time differences.

R&D's Response:

I relayed these comments to Steve Brown in R&D. Here is what he told me.

Ribbon Consumption: A recent 20% price reduction in the 4.1" cartridge should help. In addition, in a few days our 4.9" correctable cartridge--with a 40% increase in character output--should be ready for shipment. R&D is fully aware of the ribbon-consumption problem, and will work on further improvements, such as thinner ribbon, increased diameter, and new cartridges.

Training Time: New software is being developed that should reduce the training time.

If I can answer any questions about the IBM salesman's comments, please call me at x1234.

The basic pattern of this discussion is chronological: the writer describes first his discussion with the IBM salesman and then the response from R&D. Within each of these two subsections the basic pattern is more-important-to-less-important: the ribbon-consumption problem is more serious than the training-time problem, so ribbon consumption is discussed first.

ACTION Most memo reports present information that will eventually be used in formulating or modifying major projects or policies. These memo reports will become parts of the files on these projects or policies. Some reports, however, require follow-up action more immediately, by either the writer or the readers. For example, a writer addressing a group of supervisors might define what he or she is going to do about a problem discussed in the memo. A supervisor might use the action component to delegate tasks for other employees to accomplish. In writing the action component of a memo, be sure to define clearly *who* is to do *what* and *when* it is due. Following are two examples of action components.

Action:

I would appreciate it if you would work on the following tasks and have your results ready for the meeting on Monday, June 9.

1. Henderson to recalculate the flow rate.
2. Smith to set up meeting with the regional EPA representative for some time during the week of February 13.
3. Falvey to ask Armitra in Houston for his advice.

Action:

To follow up these leads, I will do the following this week:

1. Send the promotional package to the three companies.
2. Ask Customer Relations to work up a sample design to show the three companies.

3. Request interviews with the appropriate personnel at the three companies.

Notice in the first example that although the writer is the supervisor of his readers, he uses a polite tone in this introductory sentence.

TYPES OF MEMOS

Each memo is written by a specific writer to a specific audience for a specific purpose. Every memo is unique. Nevertheless, it is possible to define broad categories of memos according to the functions they fulfill. This section of the chapter discusses four basic types of memos: the directive, the response to an inquiry, the trip report, and the field/lab report.

Notice as you read about each type of memo how the purpose-summary-discussion-action strategy is tailored to the occasion. Pay particular attention to the headings, lists, and indentation used to highlight the structure of the examples.

THE DIRECTIVE In a directive memo, you define a policy or procedure you want your readers to follow. If possible, explain the reason for the directive; otherwise, it might seem like an irrational order rather than a thoughtful request. For short memos, to prevent the appearance of bluntness, the explanation should precede the directive. For longer memos, the actual directive might precede the detailed explanation. This strategy will ensure that the readers will not overlook the directive. Of course, the body of the memo should begin with a polite explanatory note, such as the following:

The purpose of this memo is to establish a uniform policy for dealing with customers who fall more than sixty days behind in their accounts. The policy is defined below under the heading "Policy Statement." Following the statement is an explanation of our rationale.

Figure 12–1 provides an example of a directive.

Notice in this example that the directive is stated as a request, not an order. Unless your readers have ignored previous requests, a polite tone is the most effective.

Also note that in spite of its brevity and simplicity, this example follows, without headings, the purpose-summary-discussion-action structure. Purpose is identified on the subject line, the first paragraph is a combination of summary and discussion (extensive discussion is hardly necessary in this situation), and the second paragraph dictates the specific action to be taken.

FIGURE 12–1

Directive

Quimby Interoffice

Date: March 19, 19--
To: All supervisors and sales personnel
From: D. Bartown, Engineering
Subject: Avoiding customer exposure to sensitive information outside Conference Room B.

It has come to our attention that customers meeting in Conference Room B have been allowed to use the secretary's phone directly outside the room. This practice presents a problem: the proposals that the secretary is typing are in full view of the customers. Proprietary information such as pricing can be jeopardized unintentionally.

In the future, would you please escort any customers or non-Quimby personnel needing to use a phone to the one outside the Estimating Department? Your cooperation in this matter will be greatly appreciated.

THE RESPONSE TO AN INQUIRY Often you might be asked by a colleague to provide information that cannot be communicated on the phone because of its complexity or importance. In responding to such an inquiry, the purpose-summary-discussion-action strategy is particularly useful. The purpose of the memo is simple: to provide the reader with the information he or she requested. The summary states the major points of the subsequent discussion and calls the reader's attention to any parts of it that might be of special importance. The action section (if it is necessary) defines any relevant steps that you or some other personnel are taking or will take. Figure 12–2 provides an example of a response to an inquiry.

Notice in this sample memo how the numbered items in the summary section correspond to the numbered items in the discussion section. This parallelism enables the reader to find quickly the discussion he wants.

THE TRIP REPORT A trip report is a record of a business trip written after the employee returns to the office. Most often, a trip report takes the form of a memo. The key to writing a good trip report is to remember that your reader is less interested in an hour-by-hour description of what happened than in a carefully structured discussion of what was important. If, for instance, you attended a professional conference, don't list all the presentations—simply attach the agenda or program if you think your reader will be inter-

FIGURE 12–2

Response to an Inquiry

NATIONAL INSURANCE COMPANY MEMO

To: J. M. Sosry, Vice President
From: G. Lancasey, Accounting
Subject: National's Compliance with the Federal Pay Standards
Date: February 2, 19--

Purpose: This memo responds to your request for an assessment of our compliance with the Federal Non-Inflationary Pay and Price Behavior Standards.

Summary:

1. We are in compliance except for a few minor violations.
2. Legal Affairs feels we are exercising "good faith," a measure of compliance with the Standards.
3. Data Processing is currently studying the costs and benefits of implementing data processing of the computations.

The following discussion elaborates on these three points.

Discussion:

1. We are in compliance with the Standards except for the following details related to fringe benefits.

    a. The fringe benefits of terminated individuals have not yet been eliminated from our calculations. The salaries have been eliminated.

    b. The fringe benefits associated with promotional increases have not yet been eliminated from our calculations.

    c. The fringe benefits of employees paid $7,800 or less have not yet been eliminated from our calculations.

2. I met with Joe Brighton of Legal Affairs last Thursday to discuss the question of compliance. Joe is aware of our minor violations. His research, including several calls to Washington, suggests that the Standards define "good faith" efforts to comply for various-size corporations, and that we are well within these guidelines.

3. I talked with Ted Ashton of Data Processing last Friday. They have been studying the costs/benefits of implementing data processing for the calculations. Their results won't be complete until next week, but Ted predicts that it will take up to two months to write the program internally. He is talking to representatives of computer service companies this week, but he doubts if they can provide an economical solution.

   As things stand now--doing the calculations manually--we will need three months to catch up, and even then we will always be about two weeks behind.

Action: I have asked Ted Ashton to send you the results of the cost/benefits study when they are in. I have also asked Joe Brighton to keep you informed of any new developments with the Standards.

If I can be of any further assistance, please let me know.

FIGURE 12-3

Trip Report

Dynacol Corporation

INTEROFFICE COMMUNICATION

To: G. Granby, R&D
From: P. Rabin, Technical Services
Subject: Trip Report--Computer Dynamics, Inc.
Date: September 20, 19--

Purpose:
This memo presents my impressions of the Computer Dynamics technical seminar of September 18. The purpose of the seminar was to introduce their new HP-500 line of computers.

Summary:
In general, I was not impressed with the new line. Most of the functions Computer Dynamics touted could be performed by relatively inexpensive peripherals. The only hardware that might be of interest to us is their graphics terminal, which I'd like to talk to you about.

Discussion:
Computer Dynamics offers several models in its 500 series, ranging in price from $11,000 to $45,000. The top model has a 32K memory. Although it's very fast at matrix operations, this feature would be of little value to us. The other models are described as "new," able to do data-base management and time-share applications. However, we could upgrade our current system with FELIX to do data-base management and with PC-88 for time-share applications.

I was disturbed by some of the answers offered by the Computer Dynamics representatives. Their insistence that "on line" systems generation takes only a few minutes of down time convinced no one. Whether this misinformation was an error or a lie I don't know, but it did not give me confidence in their other claims.

The most interesting item, for our purposes, was the graphics terminal. Because of its programmable keys, it would be convenient for the inexperienced user. Integrating their terminal with

ested. Communicate the important information you learned—or describe the important questions that didn't get answered. If you traveled to meet with a client (or a potential client), don't describe everything that happened. Focus on what your reader is interested in: how to follow up on the trip and maintain a good business relationship with the client.

In most cases, the standard purpose-summary-discussion-action structure is appropriate for this type of memo. It is a good idea to mention briefly the purpose of the trip—even if your reader might already know its purpose. By doing this, you will be our system could cost $4,000 and some 4-5 person-months before we could get a graphics output. But I think that we want to go in the direction of graphics terminals, and this one looks very good.

Recommendation:
I'd like to talk to you, when you get a chance, about our plans for the addition of graphics terminals. I think we should have McKinley and Rossiter take a look at what's available. Give me a call (x3442) and we'll talk.

providing a complete record for future reference. In the action section, list either the pertinent actions you have taken since the trip or what you recommend that your reader do. Figure 12–3 provides an example of a typical trip report.

Notice in this example that the writer and reader appear to be relatively equal in rank: the informal tone of the "Recommendation" section suggests that they have worked together before. Despite this familiarity, however, the memo is clearly organized to make it easy for the reader to read and refer to later, or to pass on to another employee who might follow up on it.

FIGURE 12–4

Lab Report

Lobate Construction

MEMO

To: C. Amalli
From: W. Kabor
Subject: Inspection of Chemopump after Run #9
Date: 6 January 19--

cc: A. Beren
S. Dworkin
N. Mancini

Purpose:
This memo presents the findings of my visual inspection of the Chemopump after it was run for 30 days on Kentucky #10 coal and requests authorization to carry out follow-up procedures.

Problem:
The inspection was designed to determine if the new Chemopump were compatible with Kentucky #10, our lowest-grade coal. In preparation for the 30-day test run, the following three modifications were made:

1. New front bearing housing buffer plates of tungsten carbide were installed.
2. The pump casting volute liner was coated with tungsten carbide.
3. New bearings were installed.

Summary:
A number of small problems with the pump were observed, but nothing serious and nothing surprising. Normal break-in accounts for the wear. The pump accepted the Kentucky #10 well.

Findings:
The following problems were observed:

1. The outer lip of the front-end bell was chipped along two thirds of its circumference.

FIELD AND LAB REPORTS Many organizations use a memo format for reports on inspection and maintenance procedures carried out on systems and equipment. These memos, known as field or lab reports, include the same information that high-school lab reports do—the problem, methods, results, and conclusions—but they de-emphasize the methods and add a recommendations section (if appropriate).

A typical field or lab report, therefore, has the following structure:

1. purpose of the memo
2. problem leading to the decision to perform the procedure

2. Opposite the pump discharge, the volute liner received a slight wear groove along one third of its circumference.
3. The impeller was not free-rotating.
4. The holes in the front-end bell were filled with insulating mud.

The following components showed no wear:

1. The 5½" impeller.
2. The suction neck liner.
3. The discharge neck liner.

Conclusions:
The problems can be attributed to normal break-in for a new Chemopump. The Kentucky #10 coal does not appear to have caused any extraordinary problems. The new Chemopump seems to be operating well.

Recommendations:
I would like authorization to modify the pump as follows:

1. Replace the front-end bell with a tungsten carbide-coated front-end bell.
2. Replace the bearings on the impeller.
3. Install insulation plugs in the holes in the front-end bell.

I recommend that the pump be reinspected after another 30-day run on Kentucky #10.

If you have any questions, please call me at x241.

3. summary
4. results
5. conclusions
6. recommendations
7. methods

Sometimes several of these sections can be combined. Purpose and problem often are discussed together, as are results and conclusions.

The lab report shown in Figure 12–4 illustrates some of the possible variations on this standard report structure.

Notice the following points about this example:

1. A single word—"visual"—constitutes the discussion of the inspection procedure in the purpose section. Nothing else needs to be said.

2. In the "findings" section, the writer lists the "bad news"—the problems—before the "good news." This is a logical order, because the bad news is of more significance to the readers.

3. By including the last sentence, the writer makes it easy for the reader to get in touch with her to ask questions or authorize the recommended modifications.

WRITER'S CHECKLIST

The following checklist covers the basic formal elements included in most memo reports.

1. Does the identifying information
   a. Include the names and (if appropriate) the job positions of both you and your readers? _____
   b. Include a sufficiently informative subject heading? _____
   c. Include the date? _____

2. Does the purpose statement clearly tell the readers why you are asking them to read the memo? _____

3. Does the summary
   a. Briefly state the major points developed in the body of the memo? _____
   b. Reflect the structure of the memo? _____

4. Does the discussion section
   a. Include a background paragraph? _____
   b. Include headings to clarify the structure and content? _____

5. Does the action section clearly and politely identify tasks that you or your readers will carry out? _____

EXERCISES

1. As the manager of Lewis Auto Parts Store, you have noticed that some of your salespeople are smoking in the showroom. You have received several complaints from customers. Write a memo defining a new policy: salespeople may smoke in the stockroom but not in the showroom.

2. There are 20 secretaries in the six departments at your office. Although they are free to take their lunch hours whenever they wish, some-

times several departments have no secretarial coverage between 1:00 and 1:30 P.M. Write a memo to the secretaries, explaining why this lack of coverage is undesirable and asking for their cooperation in staggering their lunch hours.

3. You are a senior with an important position in a school organization, such as a technical society or the campus newspaper. The faculty advisor to the organization has asked you to explain, for your successors, how to carry out the responsibilities of the position. Write a memo in response to the request.

4. The boss at the company where you last worked has phoned you, asking for your opinions on how to improve the working conditions and productivity. Using your own experiences, write a memo responding to the boss's inquiry.

5. If you have attended a lecture or presentation in your area of academic concentration, write up a trip report memo to an appropriate instructor assessing its quality.

6. Write up a recent lab or field project in the form of a memo to the instructor of the course.

7. The following memos could be improved in tone, substance, and structure. Revise them to increase their effectiveness, adding any reasonable details.

a.

KLINE MEDICAL PRODUCTS

Date: 1 September 19--
To: Mike Framson
From: Fran Sturdiven
Subject: Device Master Records

The safety and efficiency of a medical device depends on the adequacy of its design and the entire manufacturing process. To ensure that safety and effectiveness are manufactured into a device, all design and manufacturing requirements must be properly defined and documented. This documentation package is called by the FDA a "Device Master Record."

The FDA's specific definition of a "Device Master Record" has already been distributed.

Paragraph 3.2 of the definition requires that a company define the "compilation of records" that makes up a "Device Master Record." But we have no such index or reference for our records.

Paragraph 6.4 says that any changes in the DMR must be authorized in writing by the signature of a designated individual. We have no such procedure.

These problems are to be solved by 15 September 19--.

b.

Southwestern Gas Interoffice

To: John Harlan
From: Robert Kalinowski
Subject: Official vehicle
Date: July 31, 19--

Someone has told us that our vehicles have been seen at recreational sites, such as beaches. I don't have to remind you that the vehicles are for <u>official</u> <u>use</u> <u>only</u>. This practice must stop.

c.

Viking National Bank Memorandum

To: George Delmore, Expense Recording
From: David Derahl, Internal Audit
Subject: Escheat Procedures
Date: Dec. 6, 19--

Undeliverable checks received by the originating department should be voided immediately. The originating department should obtain a copy of the voided check and forward the original to Expense Recording. Efforts to determine a valid mailing address should be initiated by the originating department. If the check is less than seven months old and remains undeliverable, the liability should be recorded in the originating department's operating expense account.

This memo should define our recommended policy on escheat procedures. I hope it answers your question.

d.

CAPITAL ENERGY, INC. MEMO

To: L. Abrams, Technical Staff
From: M. Cornish, R&D
Subject: Separation by Gravity
Date: May 3, 19--

Here is the information you requested last week.

At 120°F the difference in densities is only about 0.01 g/ml. At higher temperatures, up to 160°, the difference decreases. The best temperature was 80°, but still I got some oil mixed with the water. The separation at 120° was very slight.

Based on this limited data, I don't think gravity will work at Reynoldsville.

e.

Diversified Chemicals, Inc.
Memo

Date: August 27, 19--
To: R. Martins
From: J. Speletz
Subject: Charles Research Conference on Corrosion

The subject of the conference was high-temperature dry corrosion. Some of the topics discussed were

1 - thin film formation and growth on metal surfaces. The lectures focused on the study of oxidation and corrosion by spectroscopy.

2 - the use of microscopy to study the microstructure of thick film formation on metals and alloys. The speakers were from the University of Colorado and MIT.

3 - one of the most interesting topics was hot corrosion and erosion. The speakers were from Penn State and Westinghouse.

4 - future research directions for high-temperature dry corrosion were discussed from five viewpoints.

   1 - university research
   2 - government research
   3 - industry research
   4 - European industry research
   5 - European government research

5 - corrosion of ceramics, especially the oxidation of Si_3N_4. One paper dealt with the formation of Si ALON, which could be an inexpensive substitute for Si_3N_4. This topic should be pursued.

f.

Korvon Laboratories--Memo

To: Ralph Eric
From: Walt Kavon
Subject: "Computers in the Laboratory"
Date: May 1, 19--

The seminar on "Computers in the Laboratory" was held in New York on April 22, 19--. Approximately 30 managers of labs of various sizes attended.

The leader of the seminar, Mr. Daniel Moore, presented a program that included the following topics:

Modern analytical instrumentation
Maintaining quality in quality assurance
Harnessing the power of computers
Capital investments: justifying costs to management

The subjects of minicomputers versus terminals and how to increase reliability and reduce costs were discussed by several computer manufacturers' representatives.

My major criticism of the session was the ineffective leadership of Dr. Moore. Frequently, he read long passages from published articles. Often, he was very disorganized. I did enjoy, however, meeting the other lab managers.

g.

Technical Maintenance, Inc. Memo

TO: Rich Abelson
FROM: Tom Donovan
SUBJECT: Dialysis Equipment
DATE: 10/24/--

The Clinic that sent us the dialysis equipment (two MC-311's) reported that it could not regulate the temperature precisely enough.

I found that in both 311's, the heater element did not turn off. The temperature control circuit has an internal trim potentiometer that required adjustment. It is working correctly now.

I checked out the temperature control system's independent backup alarm system that will alarm and shut down the system if the temperature reaches 40°C. It is working properly.

The equipment has been returned to the client. After phoning them, I learned that they have had no more problems with it.

h.

FREEMAN, INC. INTEROFFICE

To: C. F. Grant
From: R. C. Nedden
Subject: Testing of Boiler Water for MAS and Beron 855
cc: J. A. Jones
M. H. Miller

The research boiler group submitted 5 samples of boiler water containing 0-20 ppm MAS based upon boiler cycles. These samples were filtered and analyzed for MAS turbidimetrically. Two of the 5 samples gave a measurable response.

After this test, all 5 samples were spiked with 10 ppm of MAS. The spike recovery was theoretical in the 3 samples that did not contain measurable MAS. The spike recovery was lower than theoretical in the samples containing detectable MAS. This implies a possible matrix effect.

A similar set of tests was run on 7 boiler water samples to determine the Beron 855 content. It was found to be well below the predicted level. Three of the samples gave a measurable response, while the other 4 gave responses of less than 0.1 ppm. The spike recovery in all seven samples averaged 92% of theoretical.

The above data suggest that while the test works reasonably well on spiked samples, the levels of boiler polymers present are not predictable from cycles and feedrates.

Please let me know if you have any questions on this data.

CHAPTER THIRTEEN

PROPOSALS

Most projects undertaken by organizations, as well as most changes made within organizations, begin with proposals. The document that persuades Fairlawn Rehabilitation Center, for example, to have Hawkins Medical Supplies, Inc., write a handbook on the role of thermoplastics in orthotics and limb prosthetics is an *external proposal* prepared by Hawkins. Similarly, when a local police department wishes to purchase a new fleet of police cars, automobile manufacturers and customizers interested in securing the contract submit external proposals that detail the cost, specifications, and delivery schedule of their products. In short, when one organization purchases goods or services from another, the decision to purchase is usually based on the strength of the supplier's external proposal.

When an employee suggests to his or her supervisor that the organization purchase a new word processor or restructure a department, a similar but generally less elaborate document—the *internal proposal*—is used. This chapter will discuss both types of proposals. Because the external proposal is the more formal and detailed of the two, it will be discussed at greater length and will serve as the model for the internal proposal, which borrows formal aspects of the external proposal according to the demands of the internal situation.

THE EXTERNAL PROPOSAL

No organization produces all of the products or provides all of the services it needs. Paper clips and company cars have to be purchased. Offices have to be cleaned and maintained. Sometimes, projects that require special expertise—such as sophisticated market analyses or feasibility studies—have to be carried out. With few exceptions, any number of manufacturers would love to provide the paper clips or the cars, and a few dozen consulting organizations would happily conduct the studies.

For those seeking the product or service, it is a buyer's market. In order to get the best deal, most organizations require that their potential suppliers compete for the business. By submitting a proposal, the supplier attempts to make the case that it deserves the contract.

A vast network of contracts spans the working world. The United States government, the world's biggest customer, spent about $150 billion in 1980 on work farmed out to firms that submitted proposals. The defense and aerospace industries, for example, are almost totally dependent on government contracts. But proposal writing is by no means limited to government contractors. One auto manufacturer buys engines from another, and a company that makes spark plugs buys its steel from another company. In fact, most products and services are purchased on a contractual basis.

External proposals are generally classified as either solicited or unsolicited. A solicited proposal is written in response to a request from a potential customer. An unsolicited proposal has not been requested; rather, it originates with the potential supplier.

When an organization wants to purchase a product or service, it publishes one of two basic kinds of statements. An IFB—"information for bid"—is used for standard products. When the federal government needs pencils, for instance, it lets suppliers know that it wants to purchase, say, one million no. 2 pencils with attached erasers. The supplier that offers the lowest bid wins the contract. The other kind of published statement is an RFP—"request for proposal." An RFP is issued when the product or service is customized rather than standard. The automobiles that a police department buys are likely to differ from the standard consumer model: they might have different engines, cooling systems, suspensions, and upholstery. The RFP that the police department offers might be a long and detailed document, a set of technical specifications. The supplier that can provide the automobile that most closely resembles the specifications—at a reasonable price—will probably win the contract. Sometimes, the RFP is a more general statement of goals. The customer is in effect asking the suppliers to create their own designs or describe how they will achieve the specified goals. The supplier that offers the most persuasive proposal will probably be successful.

Most suppliers respond to RFPs and IFBs published in newspapers or received in the mail. Government RFPs and IFBs are published in the journal *Commerce Business Daily* (see Figure 13–1). The vast majority of proposals are solicited through these channels.

An unsolicited proposal looks essentially like a solicited proposal, except of course that it does not refer to an RFP. Even though the potential customer never formally requested the proposal, in almost all cases the supplier was invited to submit the proposal after the two organizations met and discussed the project informally. Because proposals are expensive to write, suppliers are very reluctant to submit them without any assurances that the potential customer will study them carefully. Thus, the term *unsolicited* is only partially accurate in this context.

External proposals—both solicited and unsolicited—can culminate in contracts of several different types: a flat fee for a product or a one-time service; a leasing agreement; or a "cost-plus" contract, under which the supplier is reimbursed for the actual costs plus a profit set at a fixed percentage of the costs.

THE ELEMENT OF PERSUASION

An electronics company that wants a government contract to build a sophisticated radar device for a new jet aircraft might submit a three-volume, 2,000-page proposal. A stationery supplier offering an unusual bargain, on the other hand, might submit to a medium-sized company a simple statement indicating the price for which it would deliver a large quantity of 20-pound, 8½- by-11-inch, 25% cotton fiber bond paper, along with an explanation of why the deal ought to be irresistible. One factor links these two very different kinds of proposals: both will be analyzed carefully and skeptically. The government officials will worry about whether the supplier will be able to live up to its promise: to build, on schedule, the best radar device at the best price. With perhaps a dozen suppliers competing for the contract, the officials will know only that a number of companies want the work; they can never be sure—not even after the contract has been awarded—that they have made the best choice. The office manager would be reluctant to spend more than has been budgeted for the short term

Issue No. PSA-8365; June 28, 1983 **COMMERCE BUSINESS DAILY**

U.S. Army Engineer District, New Orleans Corps of Engineers, Foot of Prytania St, PO Box 60267, New Orleans LA 70160

● **H -- NAVIGATION ECONOMIC STUDIES** within the boundaries of the New Orleans district. This will be an indefinite quantities option contract (guaranteed amount of $5,000) for 2 years; est $400,000 to $650,000. Evaluation criteria and their relative order of importance are contained in the request for proposals. Call 504/838-2890 for info (no collect calls accepted). RFP DACW29-83-R-0114. Issue date is 11 Jul 83. Closing date is 8 Aug 83. See note 13. (174)

Defense Supply Service-Washington, Room 1D245, The Pentagon, Washington DC 20310 (Attn: Kenneth J Farr)

★ **H -- MASTER RESTATIONING PLAN: ISSUES AND ANSWERS**—Negotiations conducted on a sole source basis with Harold Rosenbaum Associates, 111 South Bedford Street, Suite 101, Burlington MA 01803. See note 46. (174)

USDA, Farmers Home Administration, Directorate and Administrative Services Division, Room 6110, South Building, Washington DC 20250, Attn: Jacqueline V Wilson

H -- FURNISH FACTUAL DATA CREDIT REPORTS ON INDIVIDUAL FARMERS HOME ADMINISTRATION (FMHA) LOAN APPLICANTS to cover the 2 year period prior to the date of the order ticket. In addition to the basic reports, antecedent and supplemental credit reference reports may be furnished when needed to provide full credit report info for the 2 year reporting period. Bids are requested for the following areas: 1. Continental U.S. 2. Alaska 3. Hawaii 4. Puerto Rico 5. Virgin Islands 6. American Samoa 7. Guam and 8. Western Pacific Trust Territories. IFB-FmHA-83-56 to be issued by the latter part of July. Written requests for copies of this should include two self addressed mailing labels. (174)

EPA, Contracts Management Division (MD-33), Office of Administration, Attn: NCCM-W, Research Triangle Park, NC 27711

H -- REVIEW AND ANALYSIS OF BIOLOGICAL-SCIENTIFIC DATA. EPA is seeking a contractor that has the capability of providing completely unbiased biological science review, evaluation and advice that also has national credibility in the scientific community. The contractor must have no commercial ties with pesticide manufacturers, distributors, or major users, yet be familiar with pesticide regulatory decision support system issues. Additionally, the contractor will be required to have access to biological scientists in the many disciplines that can be called upon to provide advice, review and evaluations of info or products. RFP DU-83-A204. (174)

U.S. Department of Labor, Office of Procurement Operations, Room S-1521, 200 Constitution Avenue, NW, Washington DC 20210

H -- PERFORM STUDIES DEALING WITH THE ECONOMIC ASPECTS OF THE FOLLOWING AREAS: (1) Youth Labor Market Issues, (2) Job Transitions, (3) Compensating Wage Differentials, (4) Training and Human Capital Formation, (5) Economics of Collective Bargaining, (6) Private Pension Regulation and Policy. Prospective offerors may request a copy of RFP L/A 83-32 within ten days of this notice. Only written requests will be honored. (174)

U.S. Dept of Labor, Employment and Training Administration, 1371 Peachtree Street, NE, Atlanta GA 30367

H -- CORRECTION: RFP 83-RIV-JC-0011—Recruiting and Screening for Job Corps enrollees during the period Oct 1, 1983 to Sept 30, 1984, is amended to change pre-proposal conference from June 24, 1983, 2:00 pm local time-1371 Peachtree Street, NE, Room 403, Atlanta GA to July 1, 1983, 2:00 pm-1375 Peachtree Street, NE-Room 276-Seminar "A", Atlanta GA. Closing date for receipt of proposals has been changed from July 22, 1983 to July 29, 1983.

U.S. Department of Labor, Office of Procurement Operations, Room S1521, 200 Constitution Avenue, NW, Washington DC 20210

H -- COMPUTER SYSTEMS ANALYSIS, DESIGN AND IMPLEMENTATION OF ENHANCEMENTS TO THE BUREAU OF LABOR STATISTICS MANAGEMENT INFO SYSTEM, U.S. Department of Labor. Prospective offerors may request a copy of RFP L/A 83-28 within ten days of this notice. Only written requests will be honored. (174)

Naval Sea System Command, Washington DC 20362, NAVSEA 0251B, Brownell, 202/692-8000

★ **H -- ENGINEERING AND TECHNICAL SERVICES SUPPORT—ANALYSIS AND EVALUATION OF TRAINING METHODS IN SUPPORT OF AEGIS COMBAT TRAINING PROGRAM.** Negotiations will be conducted with Data-Desgin Laboratories, 1755 South Jefferson Davis Highway, Arlington VA 22202—RFP N00024-83-R-5153. See notes 41 & 46.

★ **H -- PROFESSIONAL, TECHNICAL AND ENGINEERING MANAGEMENT SERVICES TO SUPPORT THE SSBN UNIQUE SONARS PROGRAM.** RFP N00024-84-R-6015 to Tracor Inc, Applied Sciences Group, 1601 Research Blvd, Rockville MD 20850. See notes 41 & 46.

★ **H -- ENGINEERING AND TECHNICAL SUPPORT FOR THE UK TRIDENT DETAIL DESIGN EFFORT**—RFP N00024-83-R-2226 will be issued to Booz, Allen and Hamilton Inc, 4330 East West Highway, Bethesda MD 20814. FMS directed procurement, see note 46.

★ **H -- ENGINEERING AND TECHNICAL SUPPORT FOR THE AEGIS SHIPBUILDING PROJECT.** Negotiations will be conducted with SYSCON Corporation, 1000 Thomas Jefferson Street, NW, Washington DC 20007—RFP N00024-83-R-5152. See notes 41 & 46. (174)

Naval Sea Systems Command, Code 0266D, Washington DC 20362, Attn: Lt Brian J Cowan, 202/692-0951

H -- EXPERIMENTAL, DEVELOPMENTAL OR RESEARCH SUPPORT SERVICES WORK TO PERFORM COST ANALYSES AND TRADE-OFF STUDIES TO DETERMINE COST EFFECTIVENESS AND RISH RAMIFICATIONS OF VARIOUS MULTI-SOURCE AND MULTIYEAR PROCUREMENT ALTERNATIVES. 30,000 man hours. The work is required to be performed in support of the MK 48/ADCAP Torpedo Program, for a period of Nov 1, 1983 through Oct 1, 1984, with an option for an additional 60,000 man hours for the period Nov 1, 1983 through Oct 31, 1986. See notes 57, 59, 65, 68, 73 & 80. All interested parties are invited to attend the bidder's conference at Naval Sea Systems Command, Crystal City, Building National Center Three (NC3), Room 3S11 at 0900 on 28 Jun 83 in Washington DC. (174)

Contracts Branch, Harry Diamond Laboratories, Adelphia MD 20783, Attn: L Scruggs, DELHD-PR-CA, 202/394-1200

★ **H -- SERVICES FOR TECHNICAL AND MANAGEMENT SUPPORT FOR THE DARCOM TARGET SIGNATURES PROGRAM**—1 job. Negotiations will be conducted with System Support Associates Inc, 6201 Leesburg Pike-Suite 215, Falls Church, VA 22044, under Sol DAAK21-83-R-9149. See note 46. (174)

Procurement Division, SSC, Fort Benjamin Harrison IN 46216, Attn: Mrs C Maguire, Contract Specialist, 317/549-5761

H -- EFFECTIVE WRITING EVALUATORS at Fort Benjamin Harrison IN—For the period 1 Oct 83 through 30 Sept 84—IFB DABT15-83-B-0064—Est opening date 8 Aug 83.

H -- EFFECTIVE WRITING EVALUATORS at Fort Benjamin Harrison IN—For the period 1 Oct 83 through 30 Sept 84—IFB DABT15-83-B-0065—Est opening date 10 Aug 83. (174)

FIGURE 13-1

An Extract from Commerce Business Daily

in order to save money over the long term and would no doubt be suspicious about the paper's quality.

The key to proposal writing, then, is persuasion. The writers must convince the readers that the *future benefits* will outweigh the *immediate and projected costs.* Basically, external proposal writers must clearly demonstrate that they

1. understand the reader's needs
2. are able to fulfill their own promises
3. are willing to fulfill their own promises

UNDERSTANDING THE READER'S NEEDS The most crucial element of the proposal is the definition of the problem the project is intended to solve. This would seem to be mere common sense: How can you expect to write a successful proposal if you do not demonstrate that you understand the problem? Yet the people who evaluate proposals—whether they be government readers, private foundation officials, or managers in small corporations—agree that an inadequate or inaccurate understanding of the problem is the largest single weakness of the proposals they see.

Sometimes, bad definitions of the problem originate with the client: the writer of the RFP fails to convey the problem. More often, however, the writer of the proposal is at fault. The supplier might not read the RFP carefully and simply assume that he understands his client's needs. Or perhaps the supplier, in response to a request, knows that he cannot satisfy a client's needs but nonetheless prepares a proposal detailing a project that he can complete, hoping either that the reader won't notice or that no other supplier will come closer to responding to the real problem. It is easy for suppliers to concentrate on the details of what *they* want to do rather than on what is really required. But most readers will toss the proposal aside as soon as they realize that it does not respond to their situation. If you are responding to an RFP, study it thoroughly. If there is something in it you don't understand, get in touch with the organization that issued it; the organization will be happy to try to clarify it, for a bad proposal wastes everybody's time. Your first job as a proposal writer is to demonstrate your grasp of the problem.

If you are writing an unsolicited proposal, analyze your audience carefully. How can you define the problem so that your reader (or readers) will understand it? Keep the reader's needs (even if the reader is oblivious to them) and, if possible, background in mind. Concentrate on how the problem has decreased productivity or product or service quality. When you submit an unsolicited proposal, your task in many cases is to convince your readers that a problem exists. Even when you have reached an understanding with some of your customer's representatives, your proposal will still have to be approved by other officials.

DESCRIBING WHAT YOU PLAN TO DO Once you have shown that you understand why something needs to be done, describe what you plan to do. Convince your readers that you can solve the problem you have just described. Discuss your approach to the subject: indicate the procedures and equipment you would use. Create a complete picture of how you would get from the first day of the project to the last. Many inexperienced writers of proposals underestimate the importance of this description. They believe that they need only convince the reader of their enthusiasm and good faith. Unfortunately, few readers are satisfied with assurances—no matter how well intentioned. Most look for a detailed, comprehensive plan that shows that the writer has actually started to do the work.

Writing a proposal is a gamble. You might spend days or months putting it together, only to have it rejected. What can you do with an unsuccessful proposal? If the rejection was accompanied by an explanation, you might be able to learn something from it. In most cases, however, all you can do is file it away and absorb the loss. In a sense, the writer takes the first risk in working on a proposal, which, statistically, is likely to be rejected. The reader who accepts a proposal also takes a risk in authorizing the work, for he or she does not know whether the work will be satisfactory.

No proposal can anticipate all of your readers' questions about what you plan to do, of course. This situation would require that the project already be completed. But the more work you have done in planning the project before you submit the proposal, the greater the chances are that you will be able to do the project successfully if you are given the go-ahead. A full discussion of your plan is effective, too, for reasons of psychology: it suggests to your readers that you are interested in the project itself, not just in winning the contract or in receiving authorization.

DEMONSTRATING YOUR PROFESSIONALISM After showing that you understand the problem and have a well-conceived plan of attack, demonstrate that you are the kind of person—or yours is the kind of organization—that *will* deliver what is promised. Many other persons or organizations could probably carry out the project. You want to convince your readers that you have the pride, ingenuity, and perseverance to solve the problems that inevitably occur in any big undertaking. In short, you want to show that you are a professional.

One major element of this professionalism is a work schedule, sometimes called a task schedule. This schedule—which usually takes the form of a graph or chart—shows when the various phases of the project will be carried out. In one sense, the work schedule is a straightforward piece of information that enables your readers to see how you would apportion your time. But in another sense, it reveals more about your attitudes toward your work than about what you will actually be doing on any given day. Anyone with even the slightest experience with projects knows that things rarely proceed according to plan: some tasks take more time than anticipated, some take less. A careful and detailed work schedule is therefore really another way of showing that you have done your homework, that you have attempted to foresee any problems that might jeopardize the success of the project.

Related to the task schedule is generally some system of quality control. Your readers will want to see that you have established procedures to evaluate the effectiveness and efficiency of your work on the project. Sometimes, quality-control procedures consist of technical evaluations carried out periodically by the project staff. Sometimes, the writer will build into the proposal provisions for on-site evaluation by recognized authorities in the field or by representatives of the potential client.

Most proposals conclude with a budget—a formal statement of how much the project will cost.

THE STRUCTURE OF THE PROPOSAL

Most proposals follow a basic structural pattern. If the authorizing agency provides an IFB, an RFP, or a set of guidelines, follow it to the letter. If guidelines have not been supplied, or you are writing an unsolicited proposal, use the following structure to write a clear and persuasive proposal:

1. summary
2. introduction
3. proposed program
4. qualifications and experience
5. appendixes
6. budget

SUMMARY For any proposal of more than a few pages, provide a brief summary. Many organizations impose a length limit—for example, 250 words—and ask the writer to type the summary, sin-

FIGURE 13–2

Summary of a Proposal

Summary: Hawkins Medical Supplies, Inc., proposes to provide to Fairlawn Rehabilitation Center the first chapter of a handbook on the role of polymers in orthotics and limb prosthetics. This chapter is intended to enable Fairlawn's physicians and technicians, who do not currently have the technical expertise in thermoplastics, to play a greater role in the manufacture of orthotics and prosthetics. The proposed first chapter of the handbook will provide a general background on the polymers used in orthotics and prosthetics and will complement the laboratory-procedures chapters that Fairlawn Rehabilitation Center plans to have written by another organization. The proposed first chapter, fully illustrated and presented in a loose-leaf binder, will be presented within nine weeks of the acceptance of this proposal. The total budget for this project is $11,199.

gle-spaced, on the title page. The summary is crucial, because in many cases it will be the only item the readers study in their initial review of the proposal.

The summary covers the major elements of the proposal but devotes only a few sentences to each. To write an effective summary, first define the problem in a sentence or two. Next, describe the proposed program. Then provide a brief statement of your qualifications and experience. Some organizations wish to see the completion date and the final budget figure in the summary; others prefer that these data be displayed in a separate location on the title page along with other identifying information about the supplier and the proposed project.

Figure 13–2 shows an effective summary taken from a proposal written in response to a telephone inquiry from a local rehabilitation center that was seeking to commission a technical manual to be used as a handbook and reference test (Buckley 1979). A sample internal proposal is included at the end of this chapter.

INTRODUCTION The body of the proposal begins with an introduction. Its function is to define the background and the problem.

In describing the background, you probably will not be telling your readers anything they don't already know (except, perhaps, if your proposal is unsolicited). Your goal here is to show them that you understand the context of the problem: the circumstances that led to the discovery, the relationships or events that will affect the problem and its solution, and so forth.

In discussing the problem, be as specific as possible. Whenever you can, *quantify* the problem. Describe it in monetary terms, because the proposal itself will include a budget of some sort and you want to be able to convince your readers that spending money on what you propose represents a wise investment. Don't say that a design problem is slowing down production; say that it is costing $4,500 a day in lost productivity.

Figure 13–3 is the introduction to the plastics handbook proposal.

PROPOSED PROGRAM Once you have defined the problem, you have to say what you are going to do about it. As noted earlier, the proposed program demonstrates clearly how much work you have already done. The goal here is, as usual, to be specific. You won't fool anyone by saying that you plan to "gather the data and analyze it." How will you gather it? What techniques will you use to analyze it? Every word you say—or don't say—will give your readers evidence on which to base their decision. If you know your business, the proposed program will show it. If you don't, you'll inevitably slip into meaningless generalities or include erroneous information that undermines the whole proposal.

If your project concerns a subject that has been written about in the professional literature, show your readers that you are familiar with the scholarship by referring to the pertinent studies. Don't, however, just toss a bunch of references onto the page. For example, it is ineffective to write, "Carruthers (1981), Harding (1982), and Vega (1982) have all researched the relationship between acid rain levels and groundwater contamination." Rather, use the recent research to sketch in the necessary background and provide the justification for your proposed program. For instance:

> Carruthers (1981), Harding (1982), and Vega (1982) have demonstrated the relationship between acid rain levels and groundwater contamination. None of these studies, however, included an analysis of the long-term contamination of the aquifer. The current study will consist of . . .

A proposed program might include just one reference to recent research. If the proposal concerns a topic that has been researched

FIGURE 13-3

Introduction to a Proposal

Introduction: Hawkins Medical Supplies, Inc., proposes to provide, for the Fairlawn Rehabilitation Center, the first chapter of a handbook on the role of thermoplastics in orthotics and limb prosthetics.

Prior to 1960, most orthotics [supports and braces] and prosthetics [artificial limbs] were made of metal. The 1960s saw the development and refinement of new plastics that showed great potential for these medical applications. Today, technology has produced several kinds of plastics that are both lighter and less conspicuous than metals for orthotics and prosthetics.

The problem that this proposal addresses is that the physicians and lab technicians who design, create, and test these plastic orthotics and prosthetics do not have sufficient technical expertise in plastics technology. As a result, the manufacture of these plastic orthotics is more time consuming and more expensive than necessary. Last year, for example, the Fairlawn Rehabilitation Center paid outside firms over $27,000 for plastic orthotics and prosthetics. Because Fairlawn already has the facilities and equipment necessary to manufacture these items, it believes that it could reduce these costs substantially by doing the work itself. Thus it wishes to commission a handbook on the properties of polymers used in this field to enable Fairlawn physicians and technicians to assume a greater role in the manufacture of plastic orthotics and prosthetics. Hawkins Medical Suppliers, Inc., has been asked to propose an outline for the first chapter--an overview of the plastics used in orthotics and prosthetics.

FIGURE 13–4

Proposed Program of a Proposal

The Proposed Chapter: The chapter will be addressed to physicians and medical technicians and will assume no knowledge of chemistry beyond an introductory college-level course. The chapter will be structured so that it can be used as a reference guide. The discussion will be fully illustrated, with the text and illustrations appearing on facing pages so that the reader will not have to flip pages back and forth to find pertinent information. The chapter will be bound in a ring binder to facilitate the addition of the other handbook chapters.

Following is a detailed outline of the proposed chapter.

Chapter 1. Introduction to the Use of Polymers in Orthotics and Limb Prosthetics

A. Polymer and Polymer Terminology
   1. Is the material a thermoplastic or a thermoset?
   2. Is the material amorphous or crystalline?
   3. What are the mechanical properties of the material?
   4. What are the physical properties of the material?
      a. useful temperature range
      b. density
      c. moisture resistance

thoroughly, the proposed program might devote several paragraphs or even several pages to a discussion of recent scholarship. Figure 13–4 shows the proposed program for the plastics chapter proposal. In this example, the description of the proposed program is an outline of the final product—the chapter—that the supplier would furnish to the potential client. If the proposal described a project that required unusual activities, or activities carried out in a particular sequence or a particular way, the description of the proposed program would be a process description (see Chapter 9). For the plastics handbook, the "process" of writing the chapter is not of great interest; the potential client is much

B. Material Properties That Affect the Use of Polymers in Orthotics and Prosthetics
   1. Compatibility with skin
   2. Strength
   3. Weight
   4. Appearance
C. Description of the Polymers Currently Used in Orthotics and Prosthetics
   1. Polypropylene
   2. Polyethylene
   3. ABS (Acrylonitrile-Butadiene-Styrene)
   4. CAB (Cellulose-Acetate-Butyrate)
   5. Polycarbonate
D. Specifications That Indicate the Quality of the Polymer
   1. Melt index
   2. Shrinkage
   3. Density
   4. Orientation
   5. Additives
      a. plasticizers
      b. antioxidants

more interested in the "product"—the chapter itself. Notice that the proposed program does not discuss any research, for the project does not involve any original research.

QUALIFICATIONS AND EXPERIENCE After you have described how you would carry out the project, turn to the question of your ability to undertake it. Unless you can convince your readers that you have the expertise to turn an idea into action, your proposal will be interesting—but not persuasive.

The more elaborate the proposal, the more substantial the discussion of qualifications and experience has to be. For a small

FIGURE 13-4
(Continued)

c. flame retardants

d. stabilizers

e. colorants

f. fillers

project, a few paragraphs describing your technical credentials and those of your coworkers will usually suffice. For larger projects, the résumés of the project leader—often called the principal investigator—and the other important participants should be included.

External proposals should also include a discussion of the qualifications of the supplier's organization. Essentially similar to a discussion of personnel, this section outlines the pertinent projects the supplier has completed successfully. For example, a company bidding for a contract to build a large suspension bridge should describe other suspension bridges it has built. The discussion also

FIGURE 13-5

Qualifications-and-Experience Section of a Proposal

Hawkins Medical Supplies, Inc.: Hawkins Medical Supplies, Inc., has been a national leader in the medical supplies industry since 1938. Hawkins was one of the pioneers in the manufacture of thermoplastic orthotics and prosthetics in the early 1960s. One of the company's ongoing concerns is the quality of the supporting documentation that accompanies its products. Hawkins realizes that medical supplies can be used safely and effectively only if the physicians and other practitioners are provided with clear and authoritative documentation.

For this reason, the company employs an experienced staff of writers, editors, and illustrators, in addition to its technical staff. In the last year, Hawkins has produced twelve manuals and handbooks for clients, including major hospital and research centers. Several examples are included in Appendixes E, F, and G.

The team leader for the handbook project is William Argrave, M.D., who is also a materials engineer. After receiving his undergraduate engineering degree from the University of California (Los Angeles), Dr. Argrave attended the Yale University Medical College. For the past sixteen years, Dr. Argrave has been a leader in biomedical engineering. He holds twenty-three patents and is the author of over one hundred journal articles.

The resumes of Dr. Argrave and the other team members are included in Appendix C.

focuses on the necessary equipment and facilities the company already possesses, as well as the management structure that will assure successful completion of the project. Everyone knows that young, inexperienced persons and new firms can do excellent work. But when it comes to proposals, experience wins out almost every time. Figure 13-5 shows the qualifications-and-experience section of the plastics proposal.

APPENDIXES Many different types of appendixes might accompany a proposal. The plastics handbook proposal would include, among other items, some examples of the other manuals the supplier has already written. Another popular kind of appendix is the supporting letter—a testimonial to the supplier's skill and integrity, written by a reputable and well-known person in the field. Two other kinds of appendixes deserve special mention: the task schedule and the evaluation description.

The task schedule is almost always drawn in one of two graphical formats. The Gantt chart is a horizontal bar graph, with time displayed on the horizontal axis and tasks shown on the vertical axis. (See Chapter 10 for a discussion of bar graphs.) A milestone chart is a horizontal line that represents time; the tasks are written in under the line. Both the Gantt and milestone charts can include prose explanations, if necessary. Figure 13–6 shows that the Gantt chart is more informative than the milestone chart in that only the Gantt chart can indicate several tasks being performed simultaneously.

Much less clear-cut than the task schedule is the description of evaluation techniques. In fact, the term *evaluation* means different things to different people, but in general an evaluation technique can be defined as any procedure for determining whether the proposed program is both effective and efficient. Evaluation techniques can range from simple progress reports to sophisticated statistical analyses. Some proposals provide for evaluation by an outside agent—a consultant, a testing laboratory, or a university. Other proposals describe evaluation techniques that the supplier itself will perform, such as cost/benefit analyses.

The subject of evaluation techniques is complicated by the fact that some people think in terms of quantitative evaluations—tests to determine whether a proposed program is, for example, increasing production as much as had been hoped—whereas others think in terms of qualitative evaluations—tests of whether a proposed program is improving, say, the durability or workmanship of a product. And, of course, some people imply both qualitative and quantitative testing when they refer to evaluations. An additional complication is that projects can be tested both while they are being carried out (*formative evaluations*) and after they have been completed (*summative evaluations*).

When an RFP calls for "evaluation," experienced writers of proposals know it's a good idea to get in touch with the sponsoring agency's representatives, to determine precisely what they mean. Figure 13–7 is a description of the evaluation techniques for the plastics handbook.

BUDGET Good ideas aren't good unless they're affordable. The budget section of a proposal specifies how much the proposed program will cost.

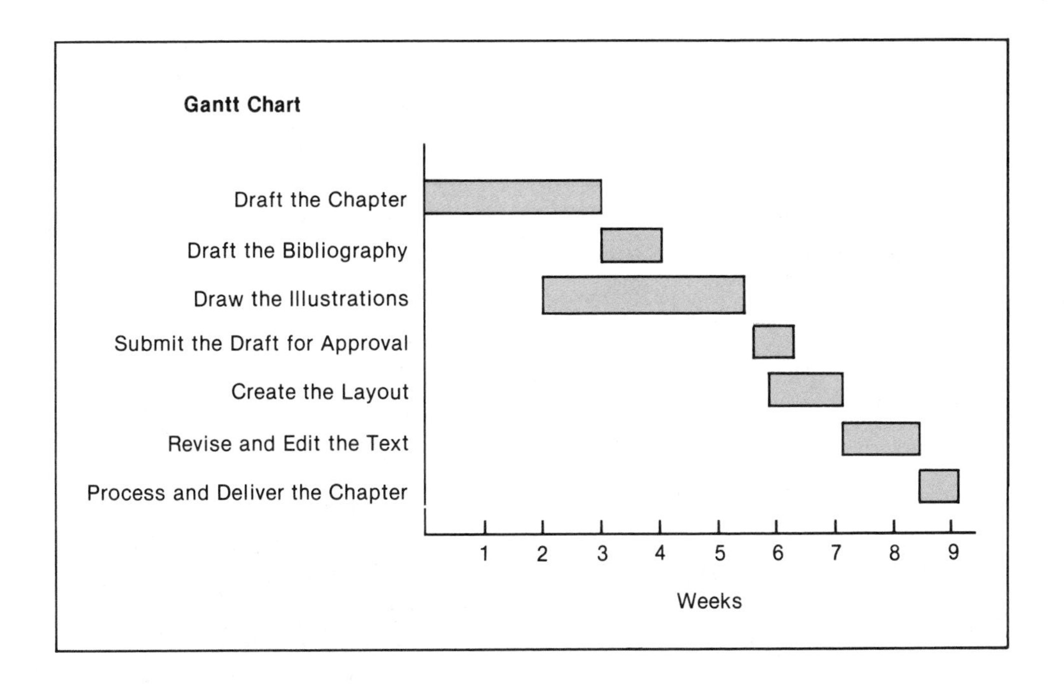

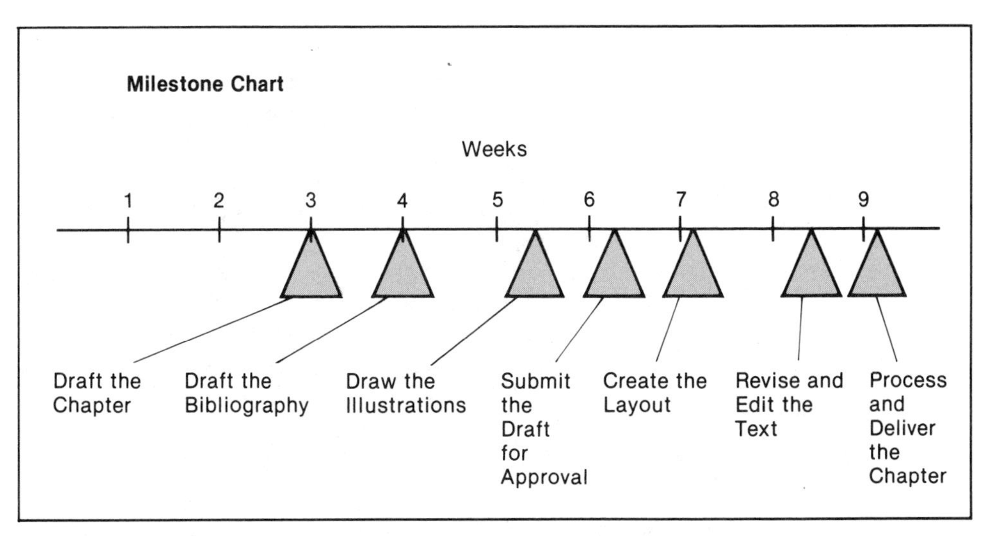

FIGURE 13–6

Gantt and Milestone Charts

FIGURE 13-7

Evaluative-techniques Section of a Proposal

```
Appendix B. Evaluation of Preliminary Draft
   Hawkins Medical Supplies, Inc., will submit to Fairlawn Reha-
bilitation Center, during week 6, the preliminary draft of the
chapter, complete with its illustrations. Within two weeks of re-
ceiving the draft, Fairlawn Rehabilitation Center will provide
Hawkins Medical Supplies with a written evaluation of the draft.
This evaluative statement will guide Hawkins Medical Supplies,
Inc., in its revisions of the draft.
```

Budgets vary greatly in scope and format. For simple internal proposals, the writer adds the budget request to the statement of the proposed program: "This study will take me about two days, at a cost—including secretarial time—of about $400" or "The variable-speed recorder currently costs $225, with a 10% discount on orders of five or more." For more complicated internal proposals and for all external proposals, a more explicit and complete budget is usually required.

Most budgets are divided into two parts: direct costs and indirect costs. Direct costs include such expenses as salaries and fringe benefits of program personnel, travel costs, and any necessary equipment, materials, and supplies. Indirect costs cover the intangible expenses that are sometimes called overhead. General secretarial and clerical expenses not devoted exclusively to the proposed program are part of overhead, as are other operating expenses such as utilities and maintenance costs. Indirect costs are usually expressed as a percentage—ranging from less than 20% to more than 100%—of the direct expenses. In many external proposals, the client imposes a limit on the percentage of indirect costs.

Figure 13-8 shows an example of a budget statement.

THE INTERNAL PROPOSAL

One day, while you're working on a project in the laboratory, you realize that if you had a new centrifuge you could do your job better and more quickly. The increased productivity would save your company the cost of the equipment in a few months. You call your supervisor and tell him about your idea. He tells you to send him a

FIGURE 13-8

Budget Section of a Proposal

Budget Itemization

for the period from 10/1/-- to 1/12/--

DIRECT COSTS

1. Salaries and Wages

Personnel	Title	Time	Amount
William Argrave	Dir. of Research	Three weeks	$ 3,800
Elaine Susman	Technical Illustrator	Three weeks	1,265
Mark Pattee	Technical Writer	One month	1,500
	Secretary	Two weeks	475
		Total Salary and Wages	$ 7,040

2. Fringe Benefits

Rate: 15% — $ 1,056

3. Supplies and Materials

Office Supplies — $ 200

Subtotal $ 8,296

INDIRECT COSTS

35% of $8,296 = — $ 2,903

Total Cost $11,199

memo describing what you want, why you want it, what you're going to do with it, and what it costs; if your request seems reasonable, he'll try to get you the money.

The memo you write is an internal proposal—a persuasive argument, submitted within an organization, for carrying out an activity that will benefit the organization, generally by saving it money. An internal proposal is simply a suggestion, made by someone within an organization, about how to improve some aspect of that organization's operations. The suggestion can be simple—to purchase an inexpensive piece of office equipment—or complicated—to hire an additional employee or even add an additional department to the organization. The nature of the suggestion determines the format it will take. A simple request might be conveyed orally, either in person or on the phone. A more ambitious request might require a brief memo. The most ambitious requests are generally conveyed in formal proposals. Often, organizations use dollar figures to determine the format of the proposal.

A proposal that would cost less than $1,000 to implement, for instance, is communicated in a brief form, whereas a proposal of more than $1,000 requires a report similar to an external proposal.

The element of persuasion is just as important in the internal proposal as it is in the external proposal. The writer must show that he or she understands the organization's needs, has worked out a rational proposed program, and is a professional who would see that the job gets done.

A careful analysis of the writing situation is the best way to start, for as usual every aspect of the document is determined by your reader's needs and your purpose.

SAMPLE INTERNAL PROPOSAL

Following is an internal proposal. The author was a librarian for an Environmental Protection Agency library that was experiencing theft problems. She was asking her supervisor for authorization to analyze the available security systems and, if appropriate, recommend one for her library (Swift 1979).

The progress report written after the project was underway is included at the end of Chapter 14. The completion report is in Chapter 15. Notes have been added to the margin of the proposal.

This proposal was written as a memo. If the proposal had been longer, it probably would have taken the form of a report. For a discussion of memos, see Chapter 12.

February 7, 1979

To: John J. Sherman, Regional Administrator, U.S. EPA Libraries

From: Catherine M. Swift, Information Specialist, EPA Region III Library

Subject: Proposal to Investigate Electronic Security Systems for EPA Region III Library

Purpose

This sentence describes the purpose of the memo and the purpose of the proposed investigation.

This memo describes a proposal to investigate the currently available electronic security systems, to determine whether any of them might be installed to solve the theft problem in the Region III library.

Summary

The problem.

The relevant scholarship.

The proposal.

Theft of library resources has reached an alarming level: over 10% of our annual acquisitions budget is currently devoted to replacement of stolen materials. Preliminary research in library journals suggests that electronic security systems can reduce from 80% to 90% of resource losses. This memo proposes that the currently available security systems be investigated according to technical, management/maintenance, and financial criteria, and that, if appropriate, a system be recommended for purchase or leasing by the Region III library.

Problem Definition

Here the problem is defined in specific terms: financial loss.

Our 1978 Resource Inventory revealed an alarming conclusion: substantial theft losses in our collection are undermining our efforts to maintain a first-class environmental library for use by EPA officials and lawyers. In 1978 alone, $2,000 in materials was stolen from our legal collection and $2,300 from our general collection. With our annual acquisitions budget of only $40,000, this theft loss cannot be tolerated.

Proposed Procedure

The writer refers to an informal discussion of the proposal with the reader and explains and justifies her proposed program.

In accordance with our telephone conversation of February 1, 1979, I would like to investigate the currently available security systems. Although research described in Library Technology Reports (1976) indicates that these systems can reduce theft losses by 80% to 90%, none of the systems is designed specifically for our special needs: most are intended for large public or university libraries. In order to determine whether any of these systems would be effective for our needs, I have devised a set of criteria by which the systems might be evaluated.

Technical Criteria:

In this discussion of the three sets of criteria, the writer shows that she has thought about the needs of the library and is able to evaluate the different antitheft systems (if she receives authorization to carry out the investigation).

1. Because an elevator is located only five feet from the checkout desk in the library, the system would have to operate without interfering with--or being interfered with by--the elevator. Some magnetic-based systems are susceptible to such interference.
2. Some systems interfere with the operation of cardiac pacemakers, hearing aids, watches, and electronic calculators. Systems that could endanger the health or property of library patrons would not be considered.
3. Because of the limited floor space in the checkout area, the system would have to be compact.

Management/Maintenance Criteria:

1. Because our staff is already working at capacity, the system would have to be easy to implement.

2. The system would have to be self-sufficient and reliable.

3. Because the library maintains a varied "forms" collection, the system would have to be adaptable to books, periodicals, microfilm, microfiche, and cassettes.

Financial Criterion:

1. Because it would have to be paid for from our limited acquisitions budget, the system would have to be as inexpensive as possible, while still fulfilling the other two sets of criteria.

Here the writer specifies the final "product" of the investigation.

After evaluating the available systems against these criteria, I will write a report that contains my results, conclusions, and recommendations. This report will provide the information necessary for you to formulate a decision on how to deal with the increasing problem of theft.

Costs

This informal budget is appropriate, because the investigation would require no unusual expenses.

This research project would require about 20 hours of my time and 4 hours of typing:

Catherine Swift	20 hours at $7.50/hour =	$150.00
Typist	4 hours at $4.00/hour =	16.00
	Total	$166.00

Credentials

The writer's credentials show that she has a good understanding of the theft problem.

As an Information Specialist in the U.S. EPA Region III Library, I have become increasingly familiar with the problem of library theft. I have participated in the last two internal circulation control and inventory checks. Last year, I instituted "Awareness Campaigns" for the library users. Posters that reminded patrons to discharge circulating materials were devised, printed, and hung throughout the library. Although these campaigns were effective in reminding "absent-minded" patrons, they did not, of course, deter patrons who were deliberately stealing our materials.

Task Schedule

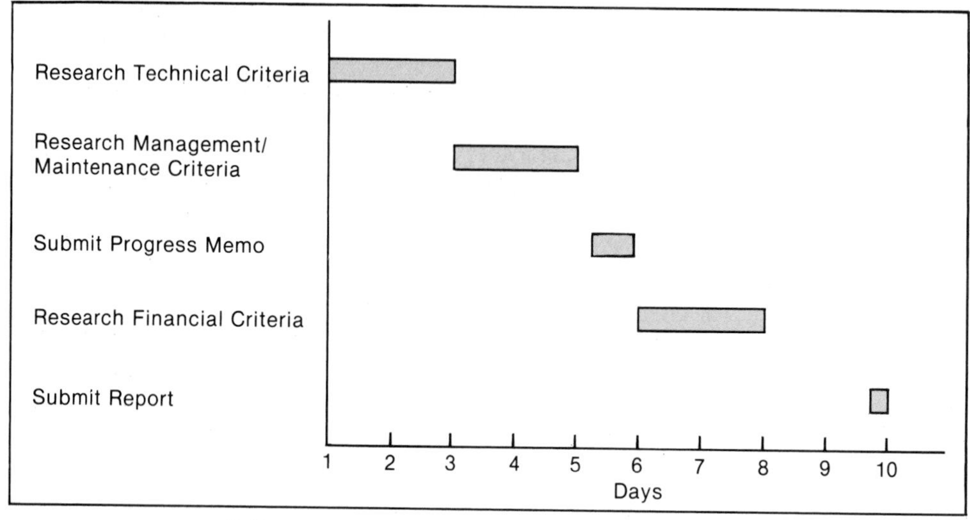

References

1. Bommer, M., and B. Ford. 1974. A cost-benefit analysis for determining the value of an electronic security system. College and Research Libraries 35, no. 4 (July): 270-79.
2. Griffith, J. W. 1978. Library thefts: A problem that won't go away. American Libraries 9, no. 4 (April): 224-27.
3. Knight, N. H. 1976. Theft detection systems--a survey. Library Technology Reports 12, no. 6 (November): 575-690.

WRITER'S CHECKLIST

The following checklist covers the basic elements of a proposal. Any guidelines established by the recipient of the proposal should of course take precedence over these general suggestions.

1. Does the summary provide an overview of
   a. The problem? ____
   b. The proposed program? ____
   c. Your qualifications and experience? ____

2. Does the introduction define
   a. The background leading up to the problem? ____
   b. The problem itself? ____

3. Does the description of the proposed program
   a. Cite the relevant professional literature? ____
   b. Provide a clear and specific plan of action? ____

4. Does the description of qualifications and experience clearly outline
   a. Your relevant skills and past work? ____
   b. Those of the other participants? ____
   c. Your department's (or organization's) relevant equipment, facilities, and experience? ____

5. Do the appendixes include the relevant supporting materials, such as a task schedule, a description of evaluation techniques, and evidence of other successful projects? ____

6. Is the budget
   a. Complete? ____
   b. Correct?

EXERCISES

1. Write a proposal for a research project that will constitute the major assignment in this course. Start by defining a technical subject that interests you. Using abstract journals and other bibliographic tools, create a bibliography of articles and books on the subject. (See Chapter 3 for a discussion of choosing a topic.) Then make up a reasonable real-world context: for example, you could pretend to be a young civil engineer whose company is considering the purchase of a new kind of earth-moving equipment. Address the proposal to your supervisor, requesting authorization to investigate the advantages and disadvantages of this new piece of equipment. Although your proposal will not include a budget, it can contain all of the other major elements of a real proposal.

2. The following proposal was written by a student who had been a summer trainee with the Navy International Logistics Control Office (NAVILCO) (Fromnic 1981). In an essay of 250–300 words, evaluate the effectiveness of the proposal. Is it clear and persuasive?

TO: Commander William Haggerty, Commanding Officer, NAVILCO
FROM: Maureen Fromnic, Finance Department Trainee, NAVILCO
DATE: February 20, 1981

SUBJECT:
Proposal to Write a Terminology and Procedures Manual for All Incoming Summer Trainees

Context:
Each summer 150 trainees are hired to aid various departments in completing backlogged work as well as to temporarily replace vacationing workers. Of all incoming trainees, 98% have never had any previous government experience. The justification for hiring inexperienced clerks is that it enables the various departments to complete projects using inexpensive labor as well as to train prospective permanent employees.

Problem:
Since the trainees work only three months, there is very little time to train them and orient them to their respective departments. Three and a half weeks are required to familiarize the new employees completely with Navy terminology and work procedures. This nonproductive period represents approximately one third of the trainee's term and $75,150 ($501 per new employee) of the taxpayers' money. Furthermore, additional time and money ($25,100) is spent as permanent employees are "borrowed" from their regular jobs and assigned to train the newcomers. By the end of the summer of 1980, $100,250 had been spent and the amount of work completed was insufficient to justify the cost of indoctrination.

Proposal:
I would like to assemble a brief but comprehensive manual defining Navy terminology and describing NAVILCO work procedures. This manual would serve as a directory for all incoming employees so that they could become oriented to our routines more quickly and efficiently than they now are.

Procedure:
The following is an outline to aid you in viewing the development of the training manual.

Description and Purpose of the Summer Employment Program:
--The origin of the Summer Employment Program
--The function of the Summer Employment Program
--The advantages of the program for the Summer Employment Trainee
--The advantages of the program for NAVILCO

Trainee Orientation into the Program:
--Navy terminology
--Navy chain of command
--Explanation of leave time and pay periods
--Responsibilities of trainees

--Type of work assignments given during the employment period
--Guidelines for correspondence preparation

Extracurricular Activities Available During Employment:
--Equal Employment Opportunity Committee
--Special Events Committee
--Compound Sports Teams
--NAVILCO social events

Conclusion:
--Attitudes of supervisors and fellow workers toward trainees
--NAVILCO expectations for the summer employment trainees

In the first section of the manual, I shall explain how this particular program originated and why it was necessary to maintain this program. In addition, I shall explain the benefits of the program for NAVILCO and for the trainee. Secondly, I shall define the various duties and responsibilities the trainee must perform. In order to prevent a language barrier, I shall discuss the various Navy terms that are ordinarily familiar only to enlisted Navy personnel. Thirdly, the manual will include the various organizations or groups available to trainees within NAVILCO. I will point out the function of each group and the role of its members. Lastly, I shall discuss how supervisors and fellow workers feel about trainees. Also, I shall indicate to the trainees the importance of fulfilling NAVILCO's expectations for them, especially if they desire future employment with the Navy.

Budget:

Maureen Fromnic	45 hours @$4.50/hr.:	$202.50
Typist	7 hours @$3.50/hr.:	24.50
	Total:	$227.00

Credentials:
I was previously a summer employment trainee for two summers. I have been a permanent employee of NAVILCO for two years. I have attended several government seminars on how to train newly employed trainees. In addition, while attending Central University, I took courses that covered government-sponsored programs and government spending for special-education programs.

Task Schedule:

research the origin and function of the program

interview supervisors, employees and past trainees

compile information and write the body of the manual

draw conclusions

proofread rough draft and make necessary adjustments

submit manual

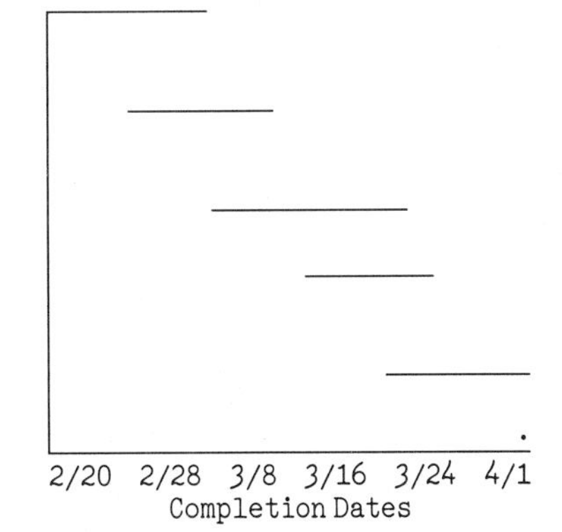

References:
1. Hills, C. A. 1980. Three ways to get a federal job. Working Woman (October): 66.
2. King, M. E. 1979. How government service can help your career. The Compound Chronicle (August): 22-24.
3. Miller, S. 1980. Career payoff in Washington, D.C. Working Woman (July): 53-56.
4. Rumaker, P. 1980. Profile of federal workers. U.S. News & World Report (16 August): 38-39.
5. U.S. Civil Service Commission. 1979. Summer Jobs. Announcement no. 414. Washington, D.C.: Government Printing Office, 1-3.

3. The following proposal was written by a student who, for purposes of the assignment, assumed the role of a consultant (Tracy 1981). The proposal is addressed to a partner in an accounting firm. The student was requesting authorization to conduct a study to determine whether the firm could save money by converting from a manual system of preparing tax returns to a computerized system. In an essay of 250 to 300 words, evaluate the effectiveness of the proposal. Is it clear and persuasive?

Proposal to Convert Manual Preparation
of Tax Returns to
Computerized Procedure

Submitted to: Mr. Saul O'Neill
Senior Partner
O'Neill and Bernstein

by: Terry Tracy, CPA

October 30, 19--

Background:

O'Neill & Bernstein is a medium-sized Public Accounting firm consisting of a 40-person audit staff and an 18-person tax department in the Philadelphia office. The volume of business has increased substantially over the past two years, and therefore you have had to hire 22 new audit personnel and 13 additional tax professionals.

Two years ago the tax department at O'Neill & Bernstein consisted of three staff members: one supervisor, one manager, and one tax partner. Presently your tax department is made up of nine staff members: three advanced staff, two supervisors, two managers, and two partners. The reason for this drastic personnel increase has been the tremendous gain in volume of tax clients. Two years ago you had 75 tax clients; today you have 225 clients.

Your professionals work an average of 12 overtime hours a week during the busy season (February 1st-April 15th, June 1st-August 15th) in order to meet the deadline dates for client tax returns.

Problem:

Over the past year you have lost three $50,000-a-year corporate tax customers and twenty individual and two partnership clients because of your failure to meet tax return deadlines. Because your workload diminishes substantially in the off season, it would not be economically feasible to hire additional tax people.

Proposal:

I believe your tax return deadlines can be met without hiring more personnel, by having all of the returns processed through a computer. I have looked into a number of companies who have established reputations in providing the computer personnel and facilities necessary to improve the efficiency, and reduce the cost of our tax department. As Avi Liveson states in his article on computer processed tax returns, "Accountants continue to make increasing use of computerized tax return preparation. Where large numbers of returns are processed, computers provide a quick, efficient method for obtaining accurate returns. Last year six million returns were prepared by computer."

Procedure:

My initial step in dealing with this proposal will be to research the implementation of computerized tax returns in other companies, and observe its effectiveness and efficiency.

The preliminary format I will use to determine whether this proposal will benefit your company is:

--Prepare an introduction discussing our problem and recommended proposal.
--Prepare a list of companies that have made the transition to computerized preparation of tax returns. This will list both the positive and negative reactions from these companies.
--Calculate the amount of time and money saved through the method discussed in my proposal.
--Prepare a list showing the alternatives of having your tax returns sent to the computer firm (through the mail) or having a computer terminal installed in your office and transmitting your information to the computer through the terminal.
--Calculate the rate of return on your original investment and how long it will take to recoup your original investment.
--Prepare a recommendation depicting the computer firm that I feel would best meet your needs, and also list the most favorable alternative companies.

Budget:

The charge for this analysis would be $3,000, payable in full within 30 days.

Credentials:

I am a graduate of Central University with a Bachelor of Science degree in accounting. I have a year-and-a-half's experience in the tax department at Carruthers and Higgins, a year's experience in the tax department with a "Big 8" CPA firm that prepared its returns through a computer, and I have worked for six months with computers that can process tax returns. I have been a freelance consultant for three years.

Task Schedule:

Discuss Problem and Proposal

Prepare List of Companies That Use Computerized Tax Preparation

Prepare List Stating Our Alternative Methods

Calculate Rate of Return on Investment

Prepare Conclusion and Recommendation

Submit Report

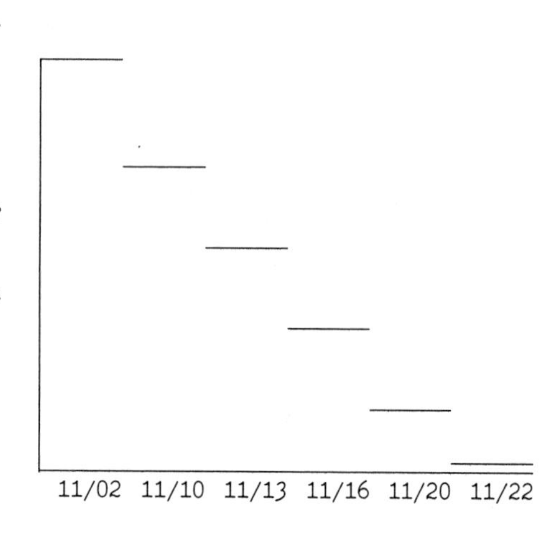

Bibliography:

1. Karter, N. A. 1976. Computer tax return preparation. Journal of Taxation (19 May): 2, 6, 8, 230.
2. Livenson, A. 1979. Computer processed tax returns. Taxation for Accountants (12 October): 144-74.
3. Meyers, B. A. 1977. How to computerize corporate tax operations. Tax Executive (6 June): 17-26.
4. Morson, D. M. 1977. Computer versus pencil for the tax professional. Tax Executive (October): 267-79.
5. Sonnichsen, G. W. 1980. Tax automation for administrative control. Tax Executive (January): 91-111.
6. Tanner, W. J. 1979. The computer: Its uses in the tax field. Journal of Taxation (19 October): 205-209.
7. Weiss, Richard. 1978. Computerized return preparation continues to be available in a wide range of services. Taxation for Accountants (September): 144-71.

REFERENCES

Buckley, L. J. 1979. Unpublished manuscript.
Fromnic, M. 1981. Unpublished manuscript.
Swift, C. M. 1979. Unpublished manuscript.
Tracy, T. 1981. Unpublished manuscript.

CHAPTER FOURTEEN

THE STRUCTURE OF THE PROGRESS REPORT

INTRODUCTION

SUMMARY

DISCUSSION

CONCLUSION

APPENDIXES

SAMPLE PROGRESS REPORTS

"Investigation of Electronic Security Systems for EPA Region III Library" / "Removing Miscible Organic Pollutants from Water Effluents"

WRITER'S CHECKLIST

EXERCISES

PROGRESS REPORTS

A progress report communicates to a supervisor or sponsor the current status of a project that has been begun but is not yet completed. As its name suggests, a progress report is an intermediate communication—between the proposal (the argument that the project be undertaken) and the completion report (the comprehensive record of the completed project).

Although progress reports sometimes describe the entire range of operations of a department or division of an organization—such as the microcellular research unit of a pharmaceutical manufacturer—they usually describe a single, discrete project, such as the construction of a bridge or an investigation of excessive pollution levels in a factory's effluents.

Progress reports allow the persons working on a project to "check in" with their supervisors or sponsors. Supervisors are vitally interested in the progress of their projects, because they have to integrate them with other present and future commitments. Sponsors (or customers) have the same interest, plus an additional one: they want the projects to be done right—and on time—because they are paying for them.

The schedule for submitting progress reports is, of course, established by the supervisor or the sponsor. A relatively short-term project—one expected to take a month to complete, for example—

might require a progress report after two weeks. A more extensive project usually will require a series of progress reports submitted on a fixed schedule, such as every month.

The format of progress reports varies widely. A small internal project might require only brief memos, or even phone calls. A small external project might be handled with letters. For a larger, more formal project—either internal or external—a formal report (see Chapter 11) generally is appropriate. Sometimes a combination of formats is used: for example, quarterly reports and memos for each of the other eight months. (See Chapter 12, "Memo Reports," and Chapter 18, "Correspondence," for discussions of these formats.) This chapter exemplifies both progress reports written as memos and those written as reports; the strategy discussed throughout applies to progress reports of every length and format.

The following example might help to clarify the role of progress reports. Suppose your car is old and not worth much. When you bought it for $200 three years ago, you knew you would be its third—and last—owner. Recently you've noticed some problems in shifting. You bring the car to your local mechanic, who calls within an hour to report that a small gasket has to be replaced, at a total cost of $25. Pleased with the mechanic's progress report, you tell him to do the work and that you'll pick up the car later in the afternoon.

If all projects in business and industry went this smoothly—with no major technical problems and no unanticipated costs—every progress report would be simple to write and a pleasure to read. The writer merely would have to photocopy the task schedule from the original proposal, check off the tasks that have been completed, and send the page on to the supervisor or sponsor. The progress report would be a happy confirmation that the project personnel have indeed accomplished what they promised they would do by a certain date. The reader's task would be simple and pleasant: to congratulate the project personnel and tell them to continue.

Unfortunately, most projects in the real world don't go so smoothly. Suppose that you had received a different phone call from the mechanic. Your transmission is ruined, he tells you; a rebuilt transmission will cost about two hundred dollars; a new one, four hundred. "What do you want me to do?" he asks. You tell him you'll call back in a few minutes. You sort through your options:

1. bring your car to another mechanic for another estimate

2. have the rebuilt transmission installed
3. have the new transmission installed
4. take the bus to the garage, retrieve the license plates, and arrange to have the car sold for scrap

This version of the story illustrates why progress reports are crucial. Managers—you, in the case of the car—need to know the progress of their projects so that they can make informed decisions when things go wrong. The problem might not be what it was thought to be when the proposal was written, or unexpected difficulties might have hampered the working methods. Perhaps equipment failed; perhaps personnel changed; perhaps prices went up. An experienced manager could list a hundred typical difficulties.

The progress report is, to a large extent, the original proposal updated in the light of recent experience. If the project is proceeding smoothly, you simply report the team's accomplishments and future tasks. If the project has encountered difficulties—if the anticipated result, the cost, or the schedule has to be revised—you need to explain clearly and fully what happened and how it will affect the overall project. Your tone should be objective, neither defensive nor casual. Unless ineptitude or negligence caused the problem, you're not to blame. Regardless of what kind of news you are delivering—good, bad, or mixed—your job is the same: to provide a clear and complete account of your team's activities and to forecast the next stage of the project.

THE STRUCTURE OF THE PROGRESS REPORT

Progress reports vary considerably in structure, because of differences in format and length. Written as a one-page letter, a progress report is likely to be a series of traditional paragraphs. As a brief memo, it might also contain section headings. As a report of more than a few pages, it might contain the elements of a formal report.

Regardless of these differences, most progress reports share a basic structural progression. The writer

1. introduces the readers to the progress report by explaining the objectives of the project and providing an overview of the whole project
2. summarizes (if appropriate) the progress report

FIGURE 14-1

Introduction to a Progress Report

INTRODUCTION

This is the first monthly progress report on the project to develop a new testing device for our computer system. The goal of the project is to create a device that can debug our current and future systems quickly and accurately.

The project is divided into four phases:

I. define the desired capabilities of the device

II. design the device

III. manufacture the device

IV. test the device

3. discusses the work already accomplished and the work that remains to be done, and speculates on the future promise and problems of the project
4. provides a conclusion that evaluates the progress of the project

If appropriate, appendixes are attached to the report.

INTRODUCTION

The introduction provides background information that orients the reader to the report. First, of course, it identifies the document as a progress report and identifies the span of time the report covers. If more than one progress report has been (or will be) submitted, the introduction places the report in the proper sequence—for example, it might be the third quarterly progress report. Second, the introduction states the objectives of the project. These objectives have already been defined in the proposal, yet they are generally repeated in each progress report for the benefit of the readers, who are likely to be following the progress of a number of projects simultaneously. And third, the introduction provides a brief statement of the phases of the project. Figure 14-1 provides an example of an introduction to a progress report.

SUMMARY

Progress reports of more than a few pages are likely to contain a summary section, which, like all other summaries, provides a brief overview of the contents of the progress report for those readers who do not need to know all of the technical details. Often, the summary enumerates the accomplishments achieved during the period covered in the report and then comments on the current

FIGURE 14-2

Summary of a Progress Report

SUMMARY

In November, Phase I of the project--to define the desired capabilities of the testing device--was completed.

The device should have two basic characteristics: versatility and simplicity of operation.

1. Versatility. The device should be able to test different kinds of logic modules. To do this, it must be able to "address" each module and ask it to perform the desired task.
2. Simplicity of Operation. The device should be able to display the response from the module being tested so that the operator can tell easily whether the module is operating properly.

   Phase II of the project--to design the device--is now underway.

We would like to meet with Research and Development to discuss a problem we are having in designing the device so that it can "address" each module. Please see "Discussion" and "Conclusion," below.

work. Notice how the summary shown in Figure 14-2 calls the reader's attention to a problem described in greater detail later on in the progress report.

DISCUSSION

The points listed in the summary are elaborated in the discussion—the body of the progress report. The audience for the discussion includes those readers who want a complete picture of the team's activities during the period covered; many readers, however, will not bother to read the discussion unless the summary highlights an unusual or unexpected development during the reporting period.

Of the several different methods of structuring the discussion section, perhaps the simplest is the past work/future work scheme. After describing the problem that motivated the project, the writer describes all of the work that has been completed in the present reporting period and then sketches in the work that remains to be done. The advantage of this scheme is that it is easy to follow and easy to write.

FIGURE 14-3

Discussion Section of a Progress Report

DISCUSSION

The Problem

Isolating and eliminating design and production errors has always been a top priority of our company. Recently, the sophistication of new computer systems we are producing has outpaced our testing capabilities. We have been asked to develop a testing device that can quickly and accurately debug our current and projected systems.

Currently, we have no technique for testing the individual logic modules of our new systems. Consequently, we cannot test the system until all of the modules are in place. When we do discover a problem, such as a timing error in one of the signals, we have to disassemble the system and analyze each of the modules through which the signal flows. Even after we have discovered a problem, we cannot know if that is the only problem until we reassemble the system and test it again. Although this testing method is effective--we are well within acceptable quality standards--it is very inefficient.

The solution to this problem is to develop a device for testing the logic modules individually before they are installed in the system. That is the overall objective of the project.

Another common structure for the discussion is based on the tasks involved in the project. If the project requires that the researchers work on several of these tasks simultaneously, this structure is particularly effective, for it enables the writer to describe, in order, what has been accomplished on each task. Often, the task-oriented structure incorporates the past work/future work structure:

III. Discussion

A. The Problem

B. Task I

Phase I of the project involved our determining the desired capabilities of this device.

Work Completed: Determining the Desired Characteristics of the Testing Device

The first characteristic of the testing device must be versatility. Over the next two years, we will be introducing three new systems. This rate of introduction is expected to continue at least through the 1980s. Although we cannot foresee the specific components of these new systems, of course, all are expected to incorporate the latest large-scale integration techniques, in conjunction with microprocessor control units. This basic structure will enable us to employ separate logic modules, each of which performs a specific function. The modules will be built on separate logic cards that can be tested and replaced easily. To achieve this versatility, the testing device should be able to "address" each module automatically and ask it to perform the desired task. Because the testing device will be asked to handle a large number of logic modules, it should be able to distinguish between the different modules in order to ask them to perform their appropriate functions.

The second characteristic of the testing device should be simplicity of operation: the ability to display for the operator

1. past work
2. future work

C. Task II
1. past work
2. future work

Figure 14–3 exemplifies the standard chronological progression—from the problem to past, present, and future work. Notice, however, that the writer uses a combination of generic and specific phrases in the headings.

FIGURE 14-3

(Continued)

the response from the module being tested. After the testing device transmits different command and data lines to the module, it should be able to receive a status or response word and communicate it to the operator.

Future Work

We are now at work on Phase II--designing the device to reflect these desired characteristics.

We are analyzing ways to enable the device to automatically "address" the various logic modules it will have to test. The most promising approach appears to be to equip each logic module with a uniform integrated circuit--such as the 45K58, a four-bit magnitude comparitor--that can be wired to produce a unique word that indicates the board address of that module.

The display capability appears to be a simpler problem. Once the module being tested has executed a command, it will generate a status word. The testing device will receive this status word by sending an enable signal to the status enable pin on the unit holding the module. Standard LCD indicators on the front panel of the testing device will display the status word to the operator.

CONCLUSION

A progress report is, by definition, a description of the present status of a project. The reader will receive at least one additional communication—the completion report—on the same subject. The conclusion of a progress report, therefore, is more transitional than final.

In the conclusion of a progress report, your task is to convey to the reader your evaluation of how the project is proceeding. In the broadest sense, you have one of two messages:

1. Things are going well.
2. Things are not going so well as anticipated.

Through a careful use of language, try to communicate your evaluation accurately.

If the news is good, convey your optimism, but avoid overstatement.

OVERSTATED

We are sure the device will do all that we ask of it, and more.

REALISTIC

We expect that the device will perform well and that, in addition, it might offer some unanticipated advantages.

Beware, too, of promising that the project will be completed early. Experienced writers know that such optimistic forecasts are rarely accurate, and of course it is always embarrassing to have to report a failure after you have promised success.

On the other hand, don't panic if the preliminary results are not so promising as had been anticipated, or if the project is behind schedule. Experienced readers are fully aware that the most sober and conservative proposal writers cannot anticipate all of the problems that can—and generally do—arise. As long as the original proposal contained no wildly inaccurate computations or failed to consider crucial factors, don't feel personally responsible. Just do your best to explain what happened and the current status of the work. If you suspect that the results will not match earlier predictions—or that the project will require more time, personnel, or equipment—say so, clearly. Don't give your reader an unduly optimistic picture, in the hope that you eventually will be able to work out the problems on your own or make up the lost time. If your news is not good, at least give your reader as much time as possible to deal with it effectively.

Figure 14–4 shows the conclusion for the computer-testing-device progress report.

Because the design of the testing device will affect the future design of the new computer systems, the writer has wisely decided to ask for technical assistance—from the Research and Development Department.

APPENDIXES In the appendixes to the report, include any supporting materials that you feel your reader might wish to consult: computations, printouts, schematics, diagrams, charts, tables, or a revised task

FIGURE 14–4

Conclusion of a Progress Report

CONCLUSION

Phase I of the project has been completed successfully and on schedule.

We hope to work out the basic schematic of the testing device within two weeks. The one aspect of Phase II that is giving us trouble is the question of versatility. Although we can equip our future systems with a uniform integrated circuit that can be wired to produce unique identifiers, this procedure will create future headaches for R&D. We have arranged to meet with R&D next week to discuss this problem.

schedule. Be sure to provide cross-references to these appendixes in the body of the report, so that the reader can consult them at the appropriate stage of the discussion.

SAMPLE PROGRESS REPORTS

Two sample progress reports are included here:

1. "Progress Report on Investigation of Electronic Security Systems for the EPA Region III Library" (Swift 1979)
2. "Progress Report: Removing Miscible Organic Pollutants from Water Effluents" (Charles 1983)

The first progress report was written as a follow-up on the proposal included in Chapter 13. (The completion report on this investigation is included in Chapter 15.) The author was a librarian for an Environmental Protection Agency library that was experiencing theft problems. Her project was to investigate the available security systems and, if appropriate, recommend one for her library.

The second progress report was written by a scientist employed by a chemical company. The subject of the report is a project to devise a method of removing miscible organic pollutants from water effluents.

Marginal notes have been added to both progress reports.

This purpose statement defines the purpose of the memo, identifies the period the progress report covers, and states the general subject of the investigation.

The first sentence of the summary suggests that the report will be structured according to the tasks.

Sentence two presents the conclusions of the first two parts of the investigation and sketches in how the conclusion of the third part will affect the overall recommendation. The last sentence tells the reader when the final report will be submitted.

The "Work Completed" section discusses each of the two parts in turn.

February 13, 1979

TO: John J. Sherman, Regional Administrator, U.S. EPA Libraries

FROM: Catherine M. Swift, Information Specialist, EPA Region III Library

SUBJECT: Progress report on Investigation of Electronic Security Systems for EPA Region III Library

PURPOSE: This memo reports the progress of the first half of my investigation of the eight currently available electronic security systems for possible use in the EPA Region III Library.

SUMMARY: So far, I have completed my analysis of the systems based on technical and management/maintenance criteria. My research suggests that the "Checkpoint Mark II," manufactured by Checkpoint Systems, Inc., fulfills both sets of criteria completely. None of the other systems satisfies all of our needs. If the Checkpoint system is cost-effective, it will clearly be the best system for our library; if it is not cost-effective, our choice will be more difficult. You will receive my report on February 20.

WORK COMPLETED: Here are the results of my analysis of

Throughout her discussion, the writer helps the reader understand her working method: to compare the available systems with the needs of her library, using the process of elimination.

the currently available electronic library security systems according to the technical and management/maintenance criteria.

Technical Criteria

Six of the eight systems are compact enough for our checkout area. All but one system--the Checkpoint Mark II--are magnetic-based; some magnetic-based systems have been subject to electronic interference. Currently, the medical community has not determined precisely the health risk posed by this type of radiation. For these reasons, the Checkpoint Mark II appears to satisfy best our technical criteria.

Management/Maintenance Criteria

Several of the systems were eliminated from consideration because they were too complicated to implement, given our personnel resources. Reliability studies rank three of the remaining systems as highly dependable; of these three, the "Checkpoint Mark II" was rated highest. In addition, the "Checkpoint Mark II" is adaptable to our various forms. For these reasons, the "Checkpoint Mark II"

The writer clearly describes the work she is now doing.

The conclusion states that the work is proceeding satisfactorily and suggests how the final phase of the investigation will determine the ultimate recommendation, which will be included in the completion report.

appears to satisfy best our management/maintenance criteria.

FUTURE WORK: I am now researching the financial criteria. An authoritative article by Bommer and Ford in College and Research Libraries provides a useful formula for calculating a cost-benefit ratio using book replacement costs, man-hour costs, and interlibrary loan costs.

CONCLUSION: The work is proceeding on schedule. If the cost-benefit analysis reveals that "Checkpoint Mark II" is within our financial resources, it will receive my clear recommendation. If we cannot afford it, we will have to look very carefully at the other seven systems, none of which will fulfill all of our criteria. The completed report, which you will receive on February 20, will contain my recommendation and the full documentation of my research.

The type of report.

The subject of the report.

PROGRESS REPORT:

REMOVING MISCIBLE ORGANIC
POLLUTANTS FROM WATER
EFFLUENTS

Submitted by: Martin Charles
Biologist Class I
Marshall Chemicals, Inc.

to: Dr. Helen Jenners
Chief, Biological Division
Marshall Chemicals, Inc.

May 11, 1983

The writer clearly defines the nature of this report and places this progress report within the sequence of reports.

The purpose of the study is defined.

The writer fills in the background of the problem.

Introduction

This progress report describes my findings after the first week of the investigation. The second progress report will follow in one week. The project is expected to be completed within two weeks, at which time a completion report will be submitted.

The purpose of this project is to devise a method for removing miscible organic pollutants from water effluents.

Discussion

The Problem

Currently, there are two methods of removing miscible organic pollutants from water effluents: column filtration and absorbent screens.

Column filtration, in which the effluent is pumped through an absorbent-packed column, is ineffective for rivers and streams, because of their great flowrates. The result is that a large percentage of the effluent remains unfiltered.

Absorbent screens, in which the effluent passes through an absorbent screen, are ineffective because the

The writer repeats the goal of the project.

This introductory paragraph forecasts the discussion that follows.

Notice how the writer uses chronology to emphasize the coherence of the discussion.

screens create resistance, causing the effluent to flow around, rather than through, them.

The goal of this project is to devise a more effective method than column filtration or absorbent screens for removing miscible organic pollutants from water effluents.

Work Completed

To date, we have devised the general principle for the new method--the use of floatable absorbent particles --and selected an absorbent.

The general principle behind the new method is to introduce floatable absorbent particles into the bottom of the stream or river. As the particles float to the surface, they come in contact with and absorb pollutants. Particle floatability is limited, to allow for resubmergence in turbulent water conditions. At a calm section of water downstream, the particles are surface-skimmed.

We then tested three absorbents--Absorbtex 115, Filtrasorb 553, and activated carbon--to determine cost-effectiveness. These data are listed below.

Absorbent	Cost/ Lb.	Absorption/ Lb. Absorbent	Lbs. Absorbed/ $1
1) Absorbtex 115	$9.50	20.9 lb. CCl_4/lb.	2.2 lb.
2) Filtrasorb 553	$7.65	18.6 lb. CCl_4/lb.	2.4 lb.
3) Activated Carbon	$2.89	10.9 lb. CCl_4/lb.	3.7 lb.

These data show that although activated carbon is the least-effective absorbent, it is--by far--the most cost-effective of the three.

Future Work

Here the writer explains how the nature of the problem affected his research plan.

In our preliminary tests, each of the three absorbents presented different technical problems. Because of the substantial cost advantage of activated carbon, we decided to address the technical problems associated with that absorbent before investigating the other two absorbents.

In our experiments with activated carbon, we discovered that it loses buoyancy after short periods of contact with water. The necessary flotation must be induced by attaching the carbon to a buoyant material. Currently we are testing two materials--polyethylene and paraffin --and one other alternative: affixing the carbon to the outside of glass spheres.

In this paragraph, as well as in the next, the writer uses chronology effectively.

The next phase of study will involve determining the optimum size for the carbon particles. The smaller the size, the greater the surface area, of course, per gram of carbon. On the other hand, smallness increases the risk of water saturation, which causes the particles to sink and makes recovery impossible.

Finally, a method of introducing the carbon into the stream or river must be devised. If coated carbon is the absorbent, weighted milk jugs with water-soluble caps will probably be effective.

Conclusion

The writer attempts to forecast the future of the project.

We foresee no special problems in completing the next phases of the project. The second progress report should answer the questions of coatings and particle size. We expect to conclude the project successfully on time.

WRITER'S CHECKLIST

Even though progress reports vary considerably in format and appearance, the basic strategy behind them remains the same. The purpose of this checklist is to help you make sure you have included the major elements of a progress report.

1. Does the introduction
   - a. Identify the document as a progress report? ____
   - b. Indicate the period the progress report covers? ____
   - c. Place the progress report within the sequence of any other progress reports? ____
   - d. State the objectives of the project? ____
   - e. Outline the major phases of the project? ____
2. Does the summary
   - a. Present the major accomplishments of the period covered by the report? ____
   - b. Present any necessary comments on the current work? ____
   - c. Direct the reader to crucial portions of the discussion section of the progress report? ____
3. Does the discussion
   - a. Describe the problem that motivated the project? ____
   - b. Describe all the work completed during the period covered by the report? ____
   - c. Describe any problems that arose, and how they were confronted? ____
   - d. Describe the work remaining to be done? ____
4. Does the conclusion
   - a. Accurately evaluate the progress on the project to date? ____
   - b. Forecast the problems and possibilities of the future work? ____
5. Do the appendixes include the supporting materials that substantiate the discussion? ____

EXERCISES

1. Write a progress report describing the work you are doing for the major assignment you proposed in Chapter 13.

2. The following progress report was written by an engineer working for the waste resources department of a medium-size city. The reader is the head of the writer's department. In an essay of 250–300 words, evaluate the report from the points of view of clarity, completeness, and writing style.

THE FUTURE OF MUNICIPAL

SLUDGE COMPOSTING

BY: Walter Prentice, P.E.

DEPARTMENT OF WASTE RESOURCES

CITY OF CORINTH

Introduction

Sludge composting is a 21-day process by which wastewater sludge is converted into organic fertilizer which is aesthetically acceptable, pathogen-free, and easy to handle. Composted sludge can be used to improve soil structure, increase the soil's water retention, and provide nutrients for plant growth.

Discussion

Sludge composting is essentially a two-step process:

1. Aerated-Pile Composting

   Dump trucks deliver the dewatered raw sludge to the compost site.

Approximately 10 tons of sludge is dumped on a 25 $yd.^3$ bed of bulking agent (usually woodchips). A front-end loader mixes the sludge into the bulking agent. The mixture is then placed on a compost pad and covered with a blanket of unscreened compost 1 ft. thick. This layer is applied to insulate the sludgebulking agent for ambient temperatures and for preventing the escape of odors from the pile. The air and odors are sucked out of the bulking agent base by an aeration system of pipes under the compost pad. After three weeks, the sludge in the aerated compost pile is essentially free of pathogens and stabilized.

2. Drying, Screening, and Curing the Composted Sludge
   After the aerated pile composting is completed, the pile is spread out and harrowed periodically until it is dry enough to screen. Screening is desirable, because it recovers 80% of the costly bulking agent for reuse with new sludge. The screened compost is stored for at least 30 days before being distributed for use. During the curing period, the compost continues to decompose, ensuring an odor- and pathogen-free product.

<u>Future Work</u>

The next step is to determine the cost of sludge composting. Sludge composting utilizing the aerated-compost-pile method is estimated to cost between $35 and $50 per dry ton ($35 for a 50-dry-ton per day operation, $50 for a 10-dry-ton per day operation). These estimates include all facilities, equipment, and labor necessary to compost at a site separate from the treatment plant. Not included are the costs of sludge dewatering, transportation to and from the site, and runoff treatment. These additional factors can raise the cost of composting to $160 per dry ton.

The breakdown of the capital costs for composting follows:

1. Site development--one acre of land is required for every three dry tons per day capacity. Half of the site should be surfaced. Asphalt paving costs about $60,000 per acre. In addition, the site requires electricity for the aeration blowers.
2. Equipment--front-end loader, trucks, tractors, screens, blowers, and pipes are required.

3. Labor--labor represents between a third and a half of the operating costs. Labor is estimated to cost $6 per hour, with five weeks of paid sick or vacation time.

However, a potential market exists for compost. Although the high levels of certain heavy metals in the compost restricts its use in some cases, compost can be used by businesses such as nurseries, golf courses, landscaping, and surface mining. Transportation costs can be high, however.

Conclusion

This information comes from published reports by federal agencies and journal articles. Since the government banned ocean sludge dumping in 1981, composting has become a viable method of waste treatment, and much has been written about it.

Before we can reach a final decision about whether composting would be economically justifiable for Corinth, we must add our numbers to the costs above. This process should take about two more weeks, provided that we can get all of the information.

3. The following progress report was written by a college student who worked part-time in a tavern/restaurant in a major city. In an essay of 250 to 300 words, evaluate the report from the points of view of clarity, completeness, and writing style.

HIRING ACCOUNTING HELP OR BUYING
AN ELECTRONIC REGISTER FOR THE CROW'S NEST:
A PROGRESS REPORT

Background

The arrival of the Saratoga for a 30-month renovation program has increased our business about 60%. As a result, the preparation of daily reports and inventory calculations has become a lengthy and cumbersome task. What used to take two hours a day now takes almost four. We have two alternatives: hire a part-time accountant (such as a local university student), or invest in an electronic cash register system. Following is a report on my first week's findings in the investigation of these two alternatives.

Work Completed

The going rate for an accounting student is about $4.50/hour.

At two hours per night, seven nights per week, our annual costs would be about $3,500, including the applicable taxes. If the student were to take over all of your bookkeeping tasks--about four hours per night--the cost would be about $7,000 annually.

Both local colleges have told me on the phone that we would have no trouble locating one or more students who would be interested in such work. Break-in time would probably be short; they could learn our system in a few hours.

The analysis of the electronic cash registers is more complicated. So far, I have figured out a way to compare the various systems and begun to gather my information. Five criteria are important for our situation:

1. overall quality
2. cost
3. adaptability to our needs
4. dealer servicing
5. availability of buy-back option

To determine which machines are the most reputable, several magazine articles were checked. Five brands--NCR, TEC, Federal, CASIO, and TOWA--were on everyone's list of best machines.

I am now in the process of visiting the four local dealers in business machines. I am asking each of them the same questions--about quality, cost, versatility, frequency-of-repair records, buy-back options, etc.

Work Remaining

Although I haven't completed my survey of the four local dealers, one thing seems certain: a machine will be cheaper than hiring a part-time accountant. The five machines range in price from $1,000 to $2,000. Yearly maintenance contracts are available for at least some of them. Also, buy-back options are available, so we won't be stuck with a machine that is too big or obsolete when the Saratoga repairs are complete.

I expect to have the final report ready by next Tuesday.

REFERENCE

Swift, C. M. 1979. Progress report on the investigation of electronic security systems for the EPA Region III Library. Unpublished.

CHAPTER FIFTEEN

COMPLETION REPORTS

A completion report is generally the culmination of a substantial research project. Two other reports often precede it. A proposal (see Chapter 13) argues that the writer or writers be allowed to begin and carry out a project. A progress report (see Chapter 14) describes the status of a project that is not yet completed; its purpose is to inform the sponsors of the project how the work is proceeding. The completion report, written when the work is finished, provides a permanent record of the entire project, including the circumstances that led to its beginning.

FUNCTIONS OF THE COMPLETION REPORT

A completion report has two basic functions. The first is immediate documentation. For the sponsors of the project, the report provides the necessary facts and figures, linked by narrative discussion, which enable them to understand how the project was carried out, what it found, and, most importantly, what those findings mean. All completion reports lead, at least, to a discussion of results. For example, a limited project might call for the writer to determine the operating characteristics of three competing models of a piece of lab equipment. The heart of that completion report will be the presentation of those results. Many completion reports call for the writer to go beyond the results and analyze the results and present conclusions. The writer of the report on the lab equipment might be authorized to inform the readers which of the three machines appears to be the most appropriate for his organization's needs. And finally, many completion reports go one step further and present recommendations: suggestions about how to proceed in light of the conclusions. The writer of the lab equipment report might have been asked to recommend which of the three machines—if any—should be purchased.

The second basic function of the completion report is to serve as a future reference. Three common situations send employees searching for old reports. The first such situation is a personnel change: a new employee is likely to consult filed reports to determine the kinds of projects the organization has completed recently. Reports are not only the best source of this information—they are often the only source, because the employees who participated in the project might have left the organization. Second, when the organization contemplates a major new project, it usually will want to determine how the new project would affect existing procedures or operations. The completion reports in the files will be the best source of this information, too. For example, if the owner of an office complex wants to computerize the temperature control of his buildings, he will bring in an expert to consult the reports on the electrical wiring and the heat and air-conditioning systems to determine whether computerization is technically feasible and economically justifiable. Third, and perhaps most important, if a problem develops after the project has been completed, employees will turn first to the project's completion report to try to figure out what went wrong. An analysis of a breakdown in a production line requires the technical description of the production line—the completion report that was written when the line was implemented. In these three situations, completion reports are valuable long after the projects they describe have been completed.

THE STRUCTURE OF THE COMPLETION REPORT

Like proposals and progress reports, completion reports must be self-sufficient: that is, they have to make sense without the authors there to explain them. The difficulty for you as a writer is that you can never be sure when your report will be read—or by whom. All you can be sure of is that some of your readers will be managers who are *not* technically competent in your field and who need only an overview of the project, and that others will be technical personnel who *are* competent in your field and who need detailed information. There will be yet other readers, of course—such as technical personnel in related fields—but in most cases the divergent needs of managers and of technical personnel are all you need to consider.

To accommodate these two basic types of readers, completion reports today generally contain an executive summary that precedes the body (the full discussion). These two elements overlap but remain independent; each has its own beginning, middle, and end. Most readers will be interested in one of the two, but probably not in both. As a formal report, the typical completion report will contain other standard elements:

title page
abstract
table of contents
list of illustrations
executive summary
glossary
list of symbols
body
appendix

This chapter will concentrate on the body of a completion report; the other elements common to most formal reports are discussed in Chapter 11.

THE BODY OF THE COMPLETION REPORT

The body of a typical completion report contains the following five elements:

1. introduction
2. discussion
3. results
4. conclusions
5. recommendations

FIGURE 15–1

Introduction to the Body of a Completion Report

The writer begins with the general background, then brings the readers up to date on the recent history of the process. The last sentence of this paragraph suggests the potential cost-effectiveness of employing the new technique.

The writer describes the problem: copper metalization works, but nobody knows how well it works.

INTRODUCTION

A large portion of the expense in the production of thick film micro-circuits is the cost of the metalization which is used to provide the conduction lines. In the past, thick film technology has relied on the use of expensive gold and platinum-gold metalizations. Recent advances in the field of micro-electronics have brought about the development of new low-cost copper metalizations. The copper metalizations can be obtained for about one-eighth the cost of gold and platinum-gold metalizations. Therefore, the use of copper metalizations could effectively reduce the cost of producing thick film micro-circuits.

A successful process has been developed for the screen printing and firing of the new copper metalizations; however, the quality of the finished circuits has not yet been determined. Gold circuits have long provided low resistance conduction paths and durability in the face of harsh environmental conditions. Therefore, in order for copper to replace gold and platinum-gold for thick film applications, several questions must be answered about the quality and reliability of circuits produced with this metalization.

THE INTRODUCTION

The first section of the detailed discussion is the introduction, which enables the readers to understand the technical discussion that follows. Usually, the introduction contains most or all of the following elements:

1. A statement of the problem that led to the project. What was not working, or not working well, in the organization? What improvements in the operation of the organization could be considered if more information were known?

2. A statement of the scope of the project. What aspects of the problem were included in the project and what aspects excluded? For example, a

The writer defines the scope of the project: the two questions posed here.

1. Will the resistance of the copper metalization be low enough to avoid large voltage drops in power applications?
2. Will the copper metalization be able to withstand severe environments such as salt spray, high temperature, and extreme moisture?

The writer defines the purpose of the project and the purpose of the report.

The purpose of this report is to document the results of an investigation aimed at answering in detail the two questions posed above. If these questions can be answered with positive results, then copper metalizations will be able to replace gold and platinum-gold for most thick film applications.

report on new microcomputers might be limited to those that cost less than $10,000, or those that have at least 64K of memory.

3. A statement of the purpose of the project. What exactly was the project intended to accomplish? What information was it intended to gather or create, or what action was it intended to facilitate?

Figure 15–1 shows an introduction to the body of a completion report. The subject of the report is an investigation to determine whether a new, inexpensive procedure can be used in the manufacture of thick film microcircuits, which are electronic components. The report, titled "Technical Evaluation of Copper Metalization in the Production of Thick Film Micro-circuits," was written by an electrical engineer working in the research and development division of a large electronics firm (Hall 1978). Marginal notes have been added.

THE DISCUSSION

In the discussion section of the report, you describe your methods—what you did. If you are reporting on a physical research project carried out in the lab or the field, the discussion section will closely resemble a traditional lab report. If several research methods were available to you, begin by describing why you chose the method(s) you did. The equipment and materials you used should be either listed before the description of the research or mentioned within the description itself. The preliminary listing is more common when some of the readers are going to duplicate the research. If they simply want to understand what you did, the listing is probably not necessary.

If you are reporting on an analysis of printed information, the nature of your subject will determine an appropriate structure. Begin with an explanation of the logic that guided you in carrying out the project. If, for example, you have been investigating the possibility of converting your manufacturing plant's energy source from oil to coal, the introduction will already have explained the problems involved in using oil as an energy source. In the discussion section, subheadings might be based on the following questions:

1. What are the advantages of using coal?
2. What are the disadvantages of using coal?
3. How much would it cost to convert from oil to coal?
4. How long would the conversion take?
5. To what extent would conversion interrupt our other operations?
6. What problems and benefits have similar plants experienced after conversion?

Another common type of report calls for an evaluation of alternative solutions to a problem. For instance, suppose that your organization is considering installing a word-processing system. In the introduction, you have already described the company's current system and its disadvantages. The discussion section might begin with an explanation of the logic behind your working methods. First, you analyzed your organization's needs, such as number of word-processing stations needed, the technical capabilities of each station, the cost, and the cost limitations. Then you gathered information on the available systems and compared each one to your needs. By a process of elimination, you narrowed your choice to three systems. After comparing the three in detail, you reached your conclusions and recommended one of the systems.

The structure of such a conclusion report might be based on the following questions:

1. What are the word-processing needs of our organization?
2. Within what cost limitations must we work?
3. What other factors should be considered in the evaluation?
4. What are the capabilities, costs, and other important characteristics of the available word-processing systems?
5. Of the available systems that meet our needs and fall within our budget, which is the best for us?
6. What would the system cost in the short run and in the long run? At what point would it pay for itself?
7. In what way would the system improve our operations?
8. Should we buy that system? Lease it? Do nothing?

The discussion section of the report on copper metalization begins with the excerpt shown in Figure 15–2.

Each of the three sets of experiments is discussed in turn. Figure 15–3 contains the first of the three—the line definition tests.

The writer uses the same structure—a general discussion followed by the test procedure—for each of the other two sets of tests.

THE RESULTS

The results are the data you observed, discovered, or created. You should present the results objectively, so that the readers can "experience" the methods just as you did. Save the interpretation of the results—the conclusion—for later. If you intermix results and conclusions, your readers might be unable to follow your reasoning process. Consequently, they will not be able to tell whether your conclusions are justified by the evidence—the results.

Just as the methods section answers the question, "What did you do?" the results section answers the question, "What did you see?"

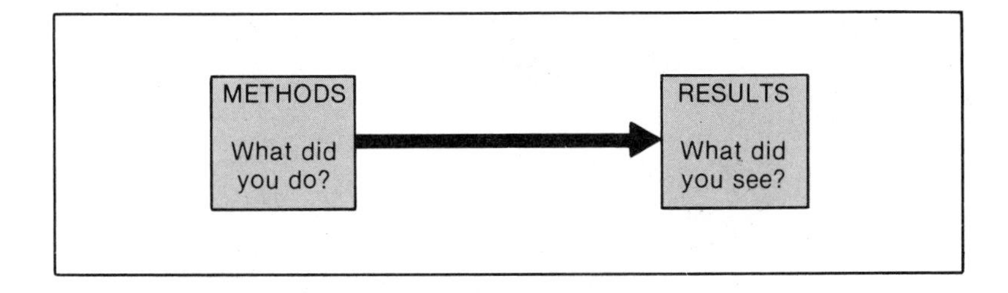

FIGURE 15-2

Excerpt from the Discussion Section in the Body of a Completion Report

The writer restates, in more detail, the general background of the project.

Here the writer expands the two properties—low resistance and durability—into three by dividing resistance into two properties—line definition and resistivity.

RESEARCH METHODS

In order for copper metalizations to replace gold and platinum-gold metalizations for thick film micro-circuit conduction lines, the copper must be able to meet low resistance requirements. Low resistance is most important in power bus and microwave applications where low loss conduction lines are critical for the maximum transfer of power. Also, these conduction lines must be able to maintain their low resistance if they are exposed to harsh environmental conditions.

Gold and platinum-gold metalizations have long been accepted for their low resistivity and durability. If copper metalizations are to replace gold and platinum-gold metalizations, the copper must exhibit the following properties:

1. High profile line definition
2. Low resistivity
3. Resistance to corrosion under harsh environmental conditions

The first two properties define the electrical resistance of the copper; the third property defines whether the copper metalizations will be durable enough to carry out their intended uses.

The writer introduces his description of the test methods.

To investigate whether copper has these properties, a series of tests was performed comparing copper to gold and to platinum-gold metalizations.

FIGURE 15-3

Excerpt from the Discussion Section in the Body of a Completion Report

The writer begins by explaining the importance of line definition. Then he describes the factors that affect line definition.

Here the writer describes the test procedures he used.

LINE DEFINITION

What Determines Line Definition?

Line definition is very important because even if the metalization possesses a low resistivity, the resistance of the line could still be high if the line has a low profile. So it is desirable to print lines that are as thick as possible. Line thickness is controlled by printing techniques, such as emulsion thickness, squeegee pressure, and breakaway. Even with these printing techniques, the line thickness still varies widely with the metalization used and is quite uncontrollable for some metalizations.

Line Definition Analysis

To evaluate and compare the line definitions of copper, gold, and platinum-gold metalization, the following procedure was used:

1. A test circuit was printed for each metalization.
   a. The pattern for the circuit (as seen in Appendix I, page 13) contains one 50 mil wide line 29.05 inches long.
   b. The printing techniques used were those specified by Military Standard 817.

c. All circuits were fixed at 850°C. The gold and platinum-gold circuits were fired in an air furnace.

2. Each circuit was evaluated by a Surfanalyzer Model #4280A. The Surfanalyzer contains a diamond stylus which travels across each conduction line. The stylus sends an electric signal to a graphing unit, which makes a cross sectional profile graph of the conduction line. From this graph, the shape and dimensions of lines as small as 0.0001 inches thick can be obtained.

FIGURE 15–4

Excerpt from the Results Section in the Body of a Completion Report

The writer presents his principal results and refers his readers to the full data.

The writer restates the relationship between line definition and resistance.

Results of the Line Definition Analysis

The results of the Surfanalyzer graphs showed that the copper and the platinum-gold printed the thickest, both at 0.78 mils thick, with the gold metalization printing the lowest profile at 0.48 mils thick. A table of results is shown in Appendix II, page 14.

During repeated measurements, the copper showed a constant thickness and a fairly flat cross section, while the gold and platinum-gold showed varying thicknesses and a cross section that was concave in the middle. This tended to suggest that the surface forces of the copper were holding up, while the surface forces of the gold and platinum-gold were breaking down during the firing cycle.

Therefore, since the copper printed with the thickest conduction lines, all that was now needed was proof of its low resistivity in order to assure low resistance conduction lines.

The nature of the project will tell you how to structure the results. If the project requires that you perform several unrelated tasks simultaneously, the results will probably be easiest to understand if they are presented together, following the discussion of the methods. If, however, the results of one of the tasks determine how or whether to carry out the next task, the results will probably be clearer if they are presented separately, at the end of the discussion of each task.

The results section of the copper metalization report includes the results of the line-definition analysis, as shown in Figure 15–4.

THE CONCLUSIONS The conclusions are the implications—the "meaning" of the results. Drawing valid conclusions from results requires great care. Suppose, for example, that you work for a company that manufactures and sells clock radios. Your records tell you that in 1983, 2.3% of the clock radios your company produced were returned as defective. An analysis of company records over the previous five years yields these results:

YEAR	% RETURNED AS DEFECTIVE
1982	1.3
1981	1.6
1980	1.2
1979	1.4
1978	1.3

One obvious conclusion can be drawn: a 2.3% defective rate is a lot higher than the rate for any of the last five years. And that conclusion is certainly a cause for concern.

But do those results indicate that your company's clock radios are less well made than they used to be? Perhaps—but in order to reach a reasonable conclusion from these results, two other factors must be considered. First, you must account for consumer behavior trends. Perhaps consumers were more sensitive to quality in 1983 than they had been in previous years. A general increase in awareness—or a widely reported news item about clock radios—might account in part or in whole for the increase in consumer complaints. (Presumably, other manufacturers of similar products have experienced similar patterns of returns if general consumer trends are at work.) A second factor to examine is your company's policy on defective clock radios. If a new, broader policy was instituted in 1983, the increase in number of returns might imply nothing about the quality of the product. In fact, the clock radios sold in 1983 might even be better than the older models. In other words, beware of drawing hasty conclusions from your results. Examine all the relevant information.

Just as the results section answers the question, "What did you see?" the conclusions section answers the question, "What does it mean?"

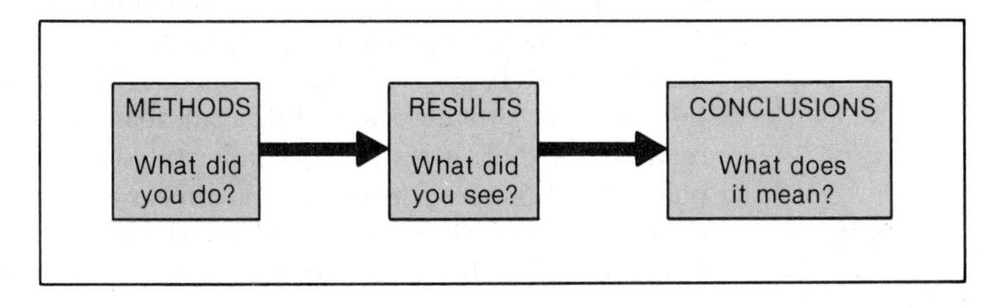

The conclusion to the line definition test in the copper metalization report is the following simple sentence:

```
The copper metalization showed excellent line-definition proper-
ties.
```

THE RECOMMENDATIONS

Recommendations are statements of action. Just as the conclusions section answers the question, "What does it mean?" the recommendations section answers the question, "What should we do now?"

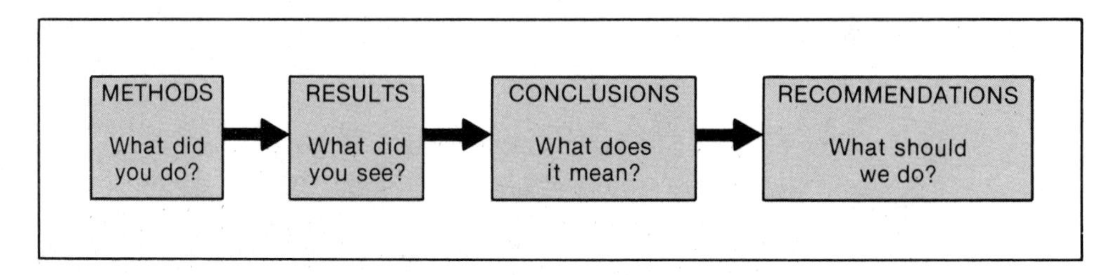

The recommendations section is always placed at the end of the discussion; because of its importance, however, the recommendations section is often summarized—or inserted verbatim—after the executive summary.

If the conclusion of the report leads to more than one recommendation, use a numbered list. If the report leads to only one recommendation, traditional paragraphs are appropriate.

Of more importance than the form of the recommendations section are its content and tone. Remember that when you tell your readers what you think they ought to do next, you must be clear, complete, and polite. If the project you are describing has been unsuccessful, don't simply recommend that your readers "try some other alternatives." Be specific: What other alternatives do you recommend, and why?

And keep in mind that when you recommend a new course of action, you might well be rejecting a previous course of action. Thus you run the risk of offending whoever formulated the earlier course. Do not write that your new direction will "correct the mistakes" that have been made recently. Instead, write that your new direction "offers great promise for success." If the previous direction was not proving successful, your readers will probably already know that. A restrained, understated tone is not only more polite but also more persuasive: you appear to be interested only in the good of your company, not in personal rivalries.

Figure 15–5 shows the recommendations section of the copper

FIGURE 15–5

The Recommendations Section in the Body of a Completion Report

The writer introduces his recommendations.

Here the writer fills in the background for what he hopes will be the next research project.

The writer explains the promise that copper metalization holds.

RECOMMENDATION

It is the recommendation of this report that a new series of evaluations be performed to determine the feasibility of using conformal coatings to protect the copper metalization for environmental use.

Several conformal coatings exist on today's market. The most accepted one is Parylene, a vaporized wax which is condensed on the circuit and forms an air-tight coating. Two problems arise with the use of Parylene. One problem is that the whole circuit gets coated, including terminals where exterior electrical connections must be made. This results in extra man-hours required to scrape off unwanted Parylene in order to make electrical contacts. The other problem is that the vaporized Parylene can work its way into the crystal lattice and raise the resistivity of the metalization.

A new coating on the market is Sylgard 180, a Dow Corning product. The Sylgard is in liquid form when applied and, therefore, will not diffuse into the crystal lattice and raise the resistivity. It is a plastic and is easily scraped off unwanted areas where electrical contact must be made.

With further experimentation, the possibility of using a conformal coating as a protective device against corrosion can

metalization report. The results of the tests showed that copper failed the environmental-durability test.

Notice that this writer states his recommendations directly and objectively. The fact that the conclusions derived from the project were essentially negative—that copper metalization would not be an effective, low-cost substitute at this time—does not reflect negatively on the writer. Don't feel that your supervisors expect "good news" in every completion report. Most projects yield recommendations like the ones in Figure 15–5 ("more work needs to be done") or wholly negative recommendations ("let's abandon the idea"). If the problem you investigated were easy to solve, it probably would have been solved long ago. Give the best advice you can in the recommendations, even if the advice is to do nothing.

FIGURE 15–5
(Continued)

become a reality. Once this copper coating system is successfully developed, copper will become the principal metalization used for all applications. With the combination of copper's properties of low cost, low resistivity, and high profile line definition, and the durability provided by a conformal coating, copper could be far superior to any other known metalization now in use.

SAMPLE COMPLETION REPORT

The following completion report was written by the librarian at the EPA library that was experiencing theft problems (Swift 1979). Her project was to investigate the available security systems and, if appropriate, recommend one for her library. The proposal and the progress report that preceded this completion report are included in Chapters 13 and 14, respectively. Marginal comments have been added.

ENVIRONMENTAL PROTECTION AGENCY
REGION III LIBRARY

March 5, 1979

Mr. John J. Sherman
Regional Administrator
U.S. EPA Libraries
1960 Walnut Street
Philadelphia, PA 19101

Dear Mr. Sherman:

Purpose of the report. General conclusion.

Presented in this report is my investigation of, and recommendation for, the purchase of an electronic security system for the EPA Region III Library. The installation of such a system would greatly reduce the incidence of theft in the library.

Methods. Results.

For this investigation, I established quality criteria for the selection of a system. The "Checkpoint Mark II" security system, manufactured by Checkpoint Systems, Inc., ranked consistently higher in all areas than any of the other systems.

Conclusion.

The Checkpoint system will provide safe, efficient security for the EPA Region III Library. At a cost of about $4,400, the system will provide savings of about $3,000-$4,000 annually.

Polite conclusion to the letter.

If you have any questions concerning this report, please feel free to contact me. I shall be happy to discuss them with you, at your convenience.

Sincerely,

Catherine M. Swift

Catherine M. Swift

Information Specialist
EPA Region III Library

Purpose and subject of the report.

RECOMMENDATION OF AN ELECTRONIC SECURITY SYSTEM
FOR THE
EPA REGION III LIBRARY

Name and title of writer and reader.

Submitted to: Mr. John J. Sherman
Regional Administrator
U.S. EPA Libraries

Submitted by: Catherine M. Swift
Information Specialist
EPA Region III Library

Date of submission.

March 5, 1979

ABSTRACT

Title of the report.

"Recommendation of an Electronic Security System for the EPA Region III Library," Catherine M. Swift, Information Specialist, U.S. EPA Region III Library

Name and position of the writer.

Background of the problem.

During the last two years, the EPA Region III Library, a classified "Special Library," has experienced a dramatic increase in the incidence of library thefts. Currently, no formal security system is employed by the EPA Library, and recent studies by leading library journals have indicated substantial savings at libraries which have installed an electronic security system. There are eight systems available, and, of these, the "Checkpoint Mark II" system, manufactured by Checkpoint Systems, Inc., is the most adaptable to the needs of the EPA Library. This system operates on radio frequencies and electronically detects all "checked" materials which are passed through a designated detection area without proper authorization. The implementation of the "Checkpoint Mark II" system would greatly reduce the incidence of library theft and produce substantial savings of approximately $3,000-$4,000 annually for the EPA Library and possibly for other Special Libraries experiencing similar security problems.

Conclusion.

Principle of operation.

Recommendation.

i

Prefatory pages are numbered with lowercase Roman numerals, centered horizontally.

The executive summary.

In the body of this report, the writer will combine her results with the discussion of each of the three sets of criteria by which she judges the antitheft systems.

CONTENTS

I. SUMMARY

Problem.

The EPA Region III Library in 1978 suffered theft losses of $2,000 in the legal collection and $2,300 in the general collection. With an annual acquisitions budget of only $40,000, the library cannot afford such losses. The 1978 theft losses represent an increase of more than 100% over the 1977 figures.

Methods and conclusion.

Currently, there are eight electronically operated library security systems on the market. They average an 80-90% reduction in resource losses (2:226). In accordance with my "Proposal to Investigate Electronic Security Systems for EPA Region III Library," dated February 7, 1979, I have investigated the eight systems, using three criteria: technical, management/maintenance, and financial. Based on my analysis, the "Checkpoint Mark II" system is the best for our needs.

Costs and benefits of the recommended system.

It can be adapted to our small checkout area, is immune to electronic interference, and poses no safety risk to library patrons or personnel. It is relatively simple to use, and has a good reliability and user satisfaction record. At $4,400, it is from $1,800 to $4,150 less expensive than the other systems. A cost-benefit analysis shows that almost three dollars of benefits are gained for every dollar expended on the system over a ten-year period and that it will pay for itself in less than a year. The "Checkpoint Mark II" system is the best solution to the theft problem at the EPA Region III Library, and perhaps at other EPA libraries.

II. CHOOSING AN ELECTRONIC SECURITY SYSTEM FOR THE EPA REGION III LIBRARY

A. Introduction

Problem.

Purpose of report.

Purpose of investigation.

Methods, scope.

Last year the EPA Region III Library suffered $4,300 in theft losses. This report describes the project intended to determine whether any of the currently available library anti-theft systems would effectively and efficiently combat this theft problem. To carry out this analysis, I judged the available anti-theft systems according to technical, management/maintenance, and financial criteria.

B. Technical Criteria

Introduction to discussion of technical criteria.

In the analysis of the prospective security systems, three aspects of their technical characteristics were emphasized: (1) personal safety standards, (2) spatial limitations, (3) electronic device interference. These three criteria placed many limitations on the available systems; however, this form of selection process eliminated systems that potentially would not comply with the EPA Library's service standards.

1. Personal Safety Standards

Discussion of first aspect of technical criteria.

The most important aspect of the technical criteria used in the evaluation of the system was personal safety standards. Within the last two years, theft detection systems have increasingly come under federal regulation, primarily from the U.S. Food and Drug Administration. At a public hearing sponsored by the FDA on June 25, 1975, the hazards of theft detection systems to cardiac pacemakers were discussed. The hearing concluded with the request "that the manufacturers of cardiac pacemakers and antitheft devices attempt to work out a program for determining whether or not antitheft devices can have adverse effects on cardiac pacemakers" (3:581).

During investigation of the available systems adaptable to the EPA Library, it was discovered that the "Knogo Mark II" system could "possibly" interfere with the operation of cardiac pacemakers and

Elimination of potentially unsafe system.

hearing aids. This system was immediately eliminated from consideration because it was thought a possible danger to the personal safety of the library users and staff. (See Appendix A, page 10.)

There also has been debate recently on health hazards associated with the electromagnetic radiation utilized by most of the theft prevention systems. (The FDA does not address the issue in its electromagnetic radiation guidelines.) The "Checkpoint Mark II" is the only system which isn't operated on a magnetic basis; it therefore eliminates potentially hazardous electromagnetic radiation exposure among library patrons and personnel.

Safety of "Checkpoint" system.

The "Checkpoint Mark II" system operates on radio frequency transmissions in coordination with "checked" materials. In this system, all library resources are "checked" with a small printed circuit which is affixed to the protected materials. When "checked" material is carried through the designated detection area ("Checkpoint"), the sensing screens pick up the small circuit frequency. One of the sensing screens is a low-power radio frequency transmitter; the other sensing screen is a radio receiver. The receiver detects all "checked" materials which have not been "de-sensitized" and activates an audio and visual signal which then locks the exit turnstile in place, prohibiting unauthorized removal of the library material. This system poses no threat to the personal safety of library users.

Principle of operation of "Checkpoint" system.

2. Spatial Limitations

Introduction to discussion of second technical criterion.

The area designated as the future charging area in the EPA Library measures approximately 6 ft. long and 10 ft. wide. This space is very small in comparison with charging areas in larger libraries, such as public or university libraries, where standard checkout areas measure approximately 12 ft. long and 15 ft. wide. Because of space limitations, many of the larger, bulkier systems, such as the "Gaylord/Magnavox" and "Knogo Mark II," were eliminated from consider-

Elimination of systems that won't fit.

Physical dimensions of the "Checkpoint" system.

ation. The "Checkpoint Mark II" detection unit, which is small and suitable for this area, consists of two parallel screens between which patrons exit from the library. The near screen (receiver) is floor-mounted against the discharging desk; the remote screen (transmitter) stands across the exit aisle from the receiver. The overall dimensions of each screen are 66 inches high, 14 1/2 inches wide, and 5 1/2 inches deep. The suggested aisle width for the "Checkpoint" is 36 inches (3:602).

The "Checkpoint Mark II" system is the most compact of the available systems, in addition to being physically attractive; the sensing screens will blend with the library's decor.

3. Electronic Device Interference

Discussion of third technical criterion.

The Region III Library exit is located approximately 5 ft. from an elevator. Unfortunately, some of the magnetic-based systems have reported incidences of false alarm triggering due to nearby electrical disturbances, escalators, elevators, vacuum cleaners, electrical storms, etc. The "Checkpoint Mark II," the only non-magnetic system currently available, operates on radio frequency transmissions that are not affected by the operation of any electrical systems. Therefore, in the case of the EPA Library, the implementation of the "Checkpoint" system would reduce the potential for false alarms.

Interference potential of the "Checkpoint" system.

C. Management/Maintenance Criteria

Introduction to management/maintenance criteria.

Because of limited manpower in the EPA Library, management/maintenance criteria were established to reflect the special needs of this library. Unlike a large university or public library where various cataloging and maintenance functions are distributed to distinct departments, the EPA Library Staff equally participate in the operation of all library functions. Although many of the available library security theft-prevention systems satisfied this category's re-

Justification for these criteria.

"Checkpoint" system's satisfaction of these criteria.

quirements, the "Checkpoint Mark II" system consistently ranked higher in reliability and maintenance standards.

1. Implementation Requirements

Elimination of systems that are difficult to implement.

The eight available systems were studied for implementation requirements. Because of its small staff, the EPA Library cannot afford more than 20 man-hours/week implementing a Library Security System. This stipulation limited the number of systems adaptable to the EPA Library. For example, the "Tattle-Tape" and "Spartan" systems require a minimum of eight steps for the manual insertion of the magnetized sensing strips. In contrast, the "Checkpoint" system requires the use of an automatic labeling device which inserts labels neatly in less than three steps. In addition, Baker & Taylor, a security system processing company, can insert the sensitive labeling in most of our books, eliminating much of the need for manual implementation by our staff. We, however, shall process journals and other forms.

"Checkpoint" system's implementation requirements.

2. Maintenance and Repair

Introduction to a discussion of maintenance and repair.

Limited staffing in the EPA Library also undercuts the effectiveness of any system which would require continual maintenance and repair. Data provided in the November 1976 issue of Library Technology Reports provided valuable information regarding users' experiences in terms of system reliability. Most of the users involved in this study had had at least one year of experience with their chosen system and answered various questions dealing with the system's dependability and quality.

Elimination of an unsatisfactory system.

"Downtimes" (amount of time in which the system was inoperable) were reported by the users. "Bookmark" appeared to be the most ineffective. Libraries using the "Bookmark" system had repeatedly experienced breakdowns resulting from a variety of mechanical and operational errors. All of the remaining systems were satisfactorily

"Checkpoint" system's maintenance and repair record.

effective, with the "Checkpoint," "Tattle-Tape," and "Spartan" systems having a higher degree of dependability.

The "Checkpoint Mark II" users reported a minimum amount of downtime, and most of this time was spent making slight adjustments. All of the users relied upon Checkpoint Systems, Inc., for maintenance repairs and reported fast and efficient service for their systems. "Tattle-Tape" and "Spartan" systems, manufactured by 3M Corporation, also received favorable service and dependability reports from their users.

3. Adaptability to Different Forms

Discussion of adaptability.

In addition to high maintenance and repair standards, the EPA Library also demanded a system which would be adaptable to the varied forms in the collection. The selected theft-prevention system was required to be equally effective in the protection of books, periodicals, microfilm, microfiche, cassettes, and phonodiscs. The "Checkpoint Mark II" system was easily adaptable to all of these forms because of the simplicity of the radio frequency labeling technique.

"Checkpoint" system's adaptability.

D. <u>Financial Criteria</u>

Summary of previous discussions.

According to my analysis of the technical and management/maintenance criteria, the "Checkpoint Mark II" system is clearly the best solution for the theft problem at the EPA Region III Library. However, no system would be recommended unless it is cost effective. A cost/benefit analysis, using Bommer and Ford's model, was performed on the "Checkpoint Mark II" system.

Introduction to financial criteria.

1. Bommer and Ford's Cost-Benefit Analysis Model

Michael Bommer and Bernard Ford have developed a cost-benefit analysis model in studying the problem of lost and stolen items from the University of Pennsylvania Van Pelt Library (1:270).

Detailed discussion of cost/benefit model.

Bommer and Ford's model defines the total costs as the installation cost of the system and the yearly operating costs. Benefits are defined as the savings in materials that are not lost or stolen and in reduced expenses in tracking down or borrowing replacement materials. In addition, Bommer and Ford calculate the subjective benefits, both to the library and the user, of having the material accessible (e.g., users' confidence, reduced time delays, and increased potential for scholarship). The cost-benefit model can be adjusted to any "planning horizon": that is, to any length of time that the system would be in use. Also, the model can easily determine the length of time it takes for the system to pay for itself.

2. Bommer and Ford's Model Applied

Method of applying the cost/benefit model.

To apply Bommer and Ford's Model, I gathered data from our 1978 Resource Inventory, 1978 Acquisitions Listing, and the 1978 interlibrary loan forms.

The EPA Library staff estimated that approximately 300 documents were missing. Of these, 100 documents were eliminated for reacquisition because they had become outdated or were seldom requested by users.

Two hundred of the missing documents, however, did warrant reacquisition. Of these 200, the majority consisted of legal material which had "disappeared" from the library's shelves soon after its purchase. Since the legal collection would be grossly inadequate without the missing material, a high priority was placed on reordering these items.

The 1978 Resource Inventory revealed that approximately $2,000 worth of material was missing from the legal collection and approximately $2,300 of material was missing from the general collection. The actual "out of pocket" costs to the library, in terms of replacement, would be $4,300.

The writer does not address the question of the costs involved in inserting the "sensitive pieces" in the library's collection.

Another "out-of-pocket" cost is the man-hours used in the tracking down and reacquiring missing library materials. The minimum wage of the Regional Library Staff members is $4.50/hour. If a half-hour is allocated for each missing document, and the total number of documents to be reordered is 200, the potential savings in this area amounts to $450.00.

In addition, Interlibrary Loan costs accrue for the ordering and photocopying of materials which are requested from other libraries because they are missing or have been stolen from the EPA Library collection. According to 1978 estimates, approximately $150 was expended for this reason.

Appendix B, page 11, shows a cost-benefit analysis performed on the "Checkpoint Mark II" system for the EPA Region III Library over a projected ten-year period. According to this analysis, $2.71 worth of benefits is gained for every dollar spent on the system. The system would pay for itself in less than a year.

Result of the calculations.

E. Survey of User Satisfaction

Discussion of the user-satisfaction survey.

The user survey whose results were published in the 1976 Library Technology Reports included one question of particular importance: Would you purchase another system from the same manufacturer? The author of the questionnaire offers several precautions in interpreting the answers to this question (3:576). First, the questionnaire was sent to current users of the system, and therefore does not include users who have discontinued using it. Second, the names of the users were supplied by the manufacturers, who might not have provided complete lists. But despite these precautions, the survey shows (see Appendix C, page 12) that the "Checkpoint Mark II" received, by far, the highest proportion of affirmative answers.

Results of the survey.

F. Conclusion

Basic conclusion.

Library theft is a problem which is not peculiar to the EPA Region III Library (2:224). Many other special libraries have experienced a sharp increase in theft within the last ten years. The advent of electronic security systems has alleviated many of the problems in the large public libraries, and recently the attention of security specialists has turned to small special libraries.

The EPA Library has a small yet valuable environmental collection --a collection that warrants the protection and security provided by a theft prevention system. The "Checkpoint Mark II" system can easily adapt to the needs of the EPA Library and safely provide the security so desperately needed.

G. Recommendation

Recommendation.

I recommend that we ask a Checkpoint representative to visit our library so that we can discuss the details of purchase and installation.

REFERENCES

1. Bommer, M., and B. Ford. 1974. A cost-benefit analysis for determining the value of an electronic security system. College and Research Libraries 35(July):270-79.
2. Griffith, J. W. 1978. Library theft: A problem that won't go away. American Libraries 9(April):224-27.
3. Knight, N. H. 1976. Theft-detection systems--a survey. Library Technology Reports 12(November):575-690.

APPENDIX A FORM PROTECTION, POTENTIAL ELECTRONIC INTERFERENCE, AND COST DATA FOR EACH SYSTEM

Because so little information is contained in this table, the writer might have expressed the pertinent facts in the body of the report.

Part 1 Survey of Form Protection and Items Potentially Affected by Electronic Interference

System	Forms NOT Protected	Items Potentially Affected by Electronic Interference
Book Mark	None	None
Checkpoint Mark II	None	None
Gaylord/Magnavox	Phonodiscs	None
Knogo Mark II	None	Cardiac pacemakers, hearing aids
Sentronic S-76	None	None
S-64	None	None
Tattle Tape	None	None
Spartan	None	None

Part 2 Cost Data for Each System (3:579-80)

System	Purchase Price	Lease Price (annual)	Service Contract (annual)[a]	Installation	Sensitive Pieces[b]
Book Mark	$6,550	NA	$750	Varies	10¢ to 14¢
Checkpoint Mark II	4,400	$1,100	308	$75-150[c]	9½¢ to 10½¢
Gaylord/ Magnavox	6,850	2,397.50	240	750 per unit[d]	9¢ to 11¢
Knogo Mark II	7,600[e]	3,000[f]	220	740	9¢ to 14¢
Sentronic S-76	7,950	2,400	360	500	8¢ to 25¢
S-64	6,750	2,100	360	500	8¢ to 25¢
Tattle-Tape	8,550	3,120	220	750	10¢ to 14¢
Spartan	6,200	2,400	220	375	10¢ to 14¢

Notice that those lettered notes that refer to all of the systems are placed in the appropriate column headings.

[a]None for first year.
[b]Depending on quantity.
[c]Depending on complexity of installation; no charge if library maintenance people perform the work under Checkpoint supervision.
[d]Additional units installed at same time in same location, $325 each.
[e]Includes charge/discharge unit with verifier; $6,900 is cost of system without a verifier on the charge/discharge unit.
[f]Includes charge/discharge unit with verifier; $2,700 is annual lease without a verifier on the charge/discharge unit.

APPENDIX B COST-BENEFIT ANALYSIS OF "CHECKPOINT MARK II" (BASED ON 1:277-78)

Cost-benefit analysis over a ten-year period

1. $$\frac{\text{Benefit}}{\text{Cost}} = \frac{n(M + L + U)}{I + n(A)}$$

Where: n = number of years the system is to be in place

M = monetary benefits accrued

L = subjective benefits to the library

U = subjective benefits to the users

I = installation cost

A = annual operating cost (maintenance plus cost of "sensitizing" the library materials)

2. $$\frac{\text{Benefit}}{\text{Cost}} = \frac{10(5{,}500 + 2{,}000 + 2{,}000)}{5{,}000 + 10(3{,}000)} = \frac{95{,}000}{35{,}000} = 2.71$$

For every dollar expended in the system, $2.71 in benefits would be gained.

Payback period

The payback period is the time at which the costs equal the benefits. Therefore, "n" becomes the variable as the numerator of the original equation is equated with the denominator.

1. $n(M + L + U) = I + n(A)$
2. $n(9{,}500) = 5{,}000 + n(3{,}000)$
3. $n(6{,}500) = 5{,}000$
4. $n = 0.769$

The system would pay for itself in 0.769 years.

11

APPENDIX C SURVEY OF USER SATISFACTION (3:575-690)

System	Number of Users Responding	Number/Percentage of Users Who Would Purchase Another System by the Same Manufacturer
Book Mark	6	3/50%
Checkpoint Mark II	14	12/85.7%
Gaylord/Magnavox	3	1/33.3%
Knogo Mark II	6	4/66.6%
Sentronic (S-76 and S-64)	12	3/25%
Tattle Tape/Spartan	20	11/55%

12

WRITER'S CHECKLIST

The following checklist applies only to the body of the completion report. The checklist pertaining to the other report elements is included in Chapter 11.

1. Does the introduction
   a. Identify the problem that led to the project? _____
   b. Identify the scope of the project? _____
   c. Identify the purpose of the project? _____

2. Does the discussion
   a. Provide a complete description of your methods? _____
   b. List or mention the equipment or materials? _____

3. Are the results presented clearly and objectively, and without interpretation? _____

4. Are the conclusions
   a. Presented clearly? _____
   b. Drawn logically from the results? _____

5. Are the recommendations stated directly, diplomatically, and objectively? _____

EXERCISES

1. Write the completion report for the major assignment you proposed in Chapter 13.

2. Following is a report written by a chemist working at a brewery (DeMedio 1980). In a 300-word essay, evaluate the effectiveness of the report.

American Brewing Company

May 28, 1980

Mr. William Miller
Brewmaster
American Brewing Company
Center and Elm Streets
Philadelphia, PA 19123

Dear Mr. Miller:

This report covers the observations and results of the near beer yeast propagation procedure, originally proposed April 19, 1980. The procedure was found to be highly effective in preventing yeast infection.

The procedure uses the logarithmic growth pattern of the yeast, as well as certain sterilization procedures fitted to the brewery, to grow yeast aseptically from test tube slants to the eight barrel pitching amount.

The propagation is very inexpensive: $840.00 for labor and equipment. It is also a rapid procedure and can be started in a day's notice. The procedure should save the brewery any troubles associated with infected near beer brews.

If you have any questions concerning this report, please feel free to contact me. I shall be happy to discuss them with you, at your convenience.

Sincerely,

William DeMedio

William DeMedio
Chemist
Brewing Laboratory

A PROCEDURE FOR THE ASEPTIC
PROPAGATION OF NEAR BEER YEAST

Submitted to: Mr. William Miller
Brewmaster
American Brewing Company

Submitted by: William J. DeMedio
Chemist
Brewing Laboratory
American Brewing Company

May 28, 1980

ABSTRACT

The brewing laboratory was given the task of developing a method for the aseptic propagation of near beer yeast from test tube slants. Vague details of the propagation were sent from Switzerland, but it was the task of the brewing laboratory to develop a method that would fit the needs of the brewery. A highly successful and inexpensive method was developed, which involved specialized sterilization procedures and the utilization of the logarithmic growth pattern of the yeast. The total cost of the method (material and labor) was \$840.00. If this is compared to the cost of an infected brew, which would be a direct consequence of an improper yeast propagation, one sees a considerable savings. In my opinion, the method of yeast propagation developed in the brewing laboratory is the safest and least expensive way to grow yeast.

William DeMedio
Chemist
Brewing Laboratory

CONTENTS

I. EXECUTIVE SUMMARY

If a near beer yeast culture becomes infected, serious problems can result. First, an entire new brew would be ruined, resulting in a loss of the range $700-$1,000 for materials alone. Second, the company would have to go through government clearance procedures in order to dispose of the brew. Third, draining the beer would increase the sewerage costs somewhat. Finally, if infected near beer were canned and sold, the company would be liable to severe fines.

An infected brew can be prevented by pitching it with infection-free yeast.

The brewing laboratory was given the task of developing a procedure for the aseptic propagation of near-beer yeast from test tube slants to an amount feasible for pitching a commercial brew. A successful and inexpensive method was developed, utilizing the growth patterns of yeast. The procedure was shown to be extremely effective in preventing costly infection.

As well as being safe, the method was also very inexpensive. Total labor costs amounted to $640.00, while equipment costs were $200.00. (Please refer to Appendix 1, on page 12.) Also, the equipment that was purchased is a permanent investment, since it can be used in repeated propagations.

The procedure takes eight days to produce an amount of yeast feasible for pitching a 160-barrel brew. This is enough beer to can 5,000 cases. The cost of propagation is a small fraction (6%) of the total cost of production of near beer.

The propagation is based on the vague details sent from Switzerland. (Please see Appendix 2, on page 13.) It was the task of the brewing laboratory to develop a method that both meets the requirements and fits the needs of the brewery.

II. THE PROPAGATION PROCEDURE

1. Background

The method of propagation contained four major steps. The first step was the transfer of a loopful of each of the two yeast strains into individual 100 ml. bottles of wort, each containing 5% sucrose (saccharose). After two days of fermentation time, each 100 ml. sample of yeast culture was transferred into individual 1,000 ml. wort samples, containing 5% sucrose. Again, these were allowed to ferment for two days. In the third step of the propagation, each 1,000 ml. sample of yeast culture was transferred into individual 10,000 ml. wort samples, containing 5% sucrose. Finally, in the fourth step of the propagation, the two strains of yeast were mixed in eight barrels of plain wort. This was allowed to ferment for two days to obtain the objective: eight barrels of aseptic pitching yeast. The procedure obviously involved taking advantage of the logarithmic growth pattern of yeast.

In order to prevent infection, standard aseptic technique was used in all yeast transfers. (Please see H. J. Benson, Microbiological Applications, W. C. Brown and Co., 1979, p. 78, for an example of standard aseptic technique.)

Three germicides were used in order to sterilize utensils: heat, iodophor, and anhydrous isopropyl alcohol. First, each utensil was rinsed in a 50 PPM iodophor solution. Then, each was rinsed clean with anhydrous isopropyl alcohol in sparing amounts. Finally, the utensils were placed in an oven at 100°C to maintain sterility. Presterilized absorbent cotton was used to plug all flasks in the propagation.

In order to maintain the pH at the bacteria-inhibiting 4.0-4.5 level, water corrective was employed (85% phosphoric acid) in the final steps of the propagation. A negligible amount was used.

The wort used was ordinary lager kettle wort. In order to differentiate between strains, one was called strain "A" and the other was called strain "B."

2. Step One

   a) Preparation: First, we sterilized three 250 ml. centrifuge bottles. Then, we cleared sufficient bench space and lit two bunsen burners and a propane torch for a hood effect. We then got all materials ready: two 7 g. packets of sucrose, 2 yeast slants in a test tube stand, inoculating loops, towels. Towels were used to handle all hot material.

   b) Wort Obtaining: Two technicians were needed for this procedure. First, one technician took a 2½ gallon bucket, with a 12-foot rope attached, to the kettle. With another technician assisting, he then lowered the bucket into the hot (214°F) wort. The bucket was allowed to fill with wort, and then the technicians carefully raised it out of the kettle. This was the wort to be used as a medium for growing the yeast. The technicians then took this hot wort up to the laboratory, where they placed it on the stand. Since the wort was boiling, it was already sterile.

   c) Preparation of 100 ml. Wort Media: Two technicians were needed for this procedure. One technician carefully placed 7 g. of sucrose into each of two centrifuge bottles, marked "A" and "B." A third bottle, marked "C," was used as a control. After placing the sucrose into the sterile bottles, the technician placed a sterile ½" X 5" stainless steel funnel into the mouths of bottles "A" and "B." Another technician dipped hot wort fresh from the kettle and filled first the "A" bottle, and then the "B" bottle up to the 100 ml. mark. The "B" bottle funnel was then placed in the mouth of the "C" bottle and this was filled to the 100 ml. mark. A sterile thermometer was placed in the "C" bottle, and all three bottles were placed in the refrigerator. When the "C" bottle registered 70°F, all three were taken from the refrigerator.

d) Inoculation: One slant was marked "A" and the other "B" to differentiate them. A loopful of the yeast from the "A" slant was transferred into the bottle marked "A." The same was done with the slant marked "B." The bench area had been cleaned in advance with a 50 PPM iodophor solution. After inoculation, all of the bottles were placed in a dark place at 70°F. The uninoculated "C" bottle was used to determine whether any infection could have invaded the "A" or "B" bottles, with the idea that if the "C" bottle remained uninfected, so would the "A" and "B" bottles.

e) Fermentation: The bottles were allowed to ferment for two days. Bubbles appeared in the "A" bottle after 24 hours and in the "B" bottle after 28 hours. The bottles were periodically swirled to disperse the yeast.

Each bottle had a very clean smell. After two days, no growth was seen in the "C" bottle, nearly confirming the asepticness of bottles "A" and "B." A simple taste test showed no infection present in the "C" bottle.

3. Step Two
   a) Preparation: We cleared and prepared a bench area as in "Step One--Preparation," on page 3, except for obtaining the slants, test tube stand, and inoculating loops. Also, we used three 2,000 ml. Erlenmeyer flasks and two 1" X 7" stainless steel funnels in this step.
   b) Wort Obtaining: The wort was obtained in exactly the same way as in "Step One--Wort Obtaining," on page 3.
   c) Preparation of 1,000 ml. Wort Media: The method of preparation was the same as in "Step One--Preparation of 100 ml. Wort Media," on page 3, except that 2,000 ml. Erlenmeyer flasks, 1,000 ml. wort samples, 70 g. packets of sucrose, and 1" X 7" funnels were used.
   d) Inoculation: Each 100 ml. yeast culture was transferred into its respective flask, marked "A" and "B." This was done aseptically by simply pouring the contents of each bottle into the other flask.
   e) Microbial Yeast Examination: A wet mount slide was prepared from the residue in each bottle marked "A" and "B." The slide was observed under high power. A good number (¾) of the yeast cells showed signs of budding. There was no sign of any infection microbes.
   f) Fermentation: The observations on maintenance of the flasks were exactly the same as in "Step One--Fermentation," on page 4. The only difference was size.

4. Step Three

   a) Preparation: The preparation for Step Three was the same as that mentioned in "Step Two--Preparation," on page 5, except that 12,000 ml. Florence flasks, 10,000 ml. wort samples, and 700 g. packets of sucrose were obtained. Also, 1½" X 12" stainless steel funnels were used.

   b) Wort Obtaining: The wort was obtained in the same way as mentioned in "Step One--Wort Obtaining," on page 5. Several trips to the kettle were necessary, in order to get enough. It was found that one 2½ gallon bucketful of wort is approximately 10,000 ml. of wort.

   c) Preparation of 10,000 ml. Wort Media: The preparation of media for step three is the same as in "Step One--Preparation of 100 ml. Wort Media," on page 5, except 12,000 ml. Florence flasks, 700 g. sucrose packets, and 1½" X 12" funnels were used.

   d) Inoculation: We transferred the 1,000 ml. yeast cultures into their respective flasks as in the method described in "Step Two--Inoculation," on page 5.

   e) Microbial Yeast Examination: The same wet mount procedure was performed here as in "Step Two--Microbial Yeast Examination," on page 5. Again, we found much yeast activity and little infection.

   f) Fermentation: The same procedure and observations were made here as in "Step Two--Fermentation," on page 5.

5. Step Four

   a) Preparation: An eight barrel yeast tank was filled with wort by pumping it directly from the kettle. The lines, pump, and yeast tank were sterilized with 50 PPM iodophor in advance. The wort was adjusted to pH 4.3 by adding 4 ounces of 85% H_3PO_4 in a slurry. (Please refer to Appendix 3, on page 9, to see the experiment used to determine this figure.)

b) Inoculation: The two 10,000 ml. yeast cultures were taken to the yeast tank and poured directly into it together. The tank was then covered. No sucrose was added to the wort.

c) Microbial Yeast Examination: We periodically observed wet mounts of the wort-yeast mixture in the tank. No signs of infection were observed, and most yeast cells showed activity.

d) Fermentation: The inoculated wort in the yeast tank was allowed to ferment for two days at 70°F. At the end of two days, we had reached our objective: eight barrels of aseptic near beer yeast. This was then given to the fermenting room men, who pitched a 160-barrel brew with it.

III. CONCLUSION

According to our results, we can conclude that the procedure we have developed is very effective in preventing the infection of near beer yeast. This procedure is also very inexpensive, involving little cost in labor and materials. The propagation takes only eight days to perform and can be started with a day's notice. Overall, we conclude that the procedure is safe, effective, inexpensive, and fits the needs of the brewery well.

APPENDIX 1. Costs of Equipment

ITEM	PRICE
Case of four 12,000 ml. Florence Flasks	$ 50.00
6 Stainless Steel Funnels	20.00
1 Carton Wax Pencils	3.00
Case of four 2,000 ml. Erlenmeyer Flasks	25.00
Platinum Inoculating Loops	102.00
TOTAL	$200.00

APPENDIX 2. Original Propagation Instructions

BIRELL--Yeast Propagation

- The two yeast strains A and B must be propagated in separate flasks up to the 10 liter stage under strict sterile conditions.

- The following procedure is recommended:

- Inoculate from the agar slant tube into 100 ml. of hopped wort, containing 5% sucrose.

- After two days at 20°C, pour the 100 ml. culture into flasks containing 1 liter of wort with 5% sucrose.

- Pour the 1 liter culture, after two days at 20°C, into flasks containing 10 liters of wort with 5% sucrose.

- Pitch 5 ml of wort (as sterile as possible) with the 10 liter cultures of strains A and B.

Dr. K. Hammer

APPENDIX 3. Data from Experiment to Change the pH of Lager Wort from 5.10 to 4.30

Water corrective (85% H_3PO_4 (aq)) was added in varying amounts to one gallon of lager wort, and the pH was measured. The purpose was to find out how much corrective is required to lower the pH of lager wort from 5.10 to 4.3-4.5 range.

Trial	Amt. 85% H_3PO_4 Added	pH of Wort	Comments
1	0.00	5.10	Start
2	0.50	4.47	Considerable change
3	1.0	4.30	OK

CONCLUSION

We must add 1.00 ml. corrective to one gallon lager wort to change the pH from 5.10 to 4.3-4.5, or we must add four ounces to eight barrels of lager wort.

10

REFERENCES

DeMedio, W. 1980. A procedure for the aseptic propagation of near beer yeast. Unpublished.

Hall, W. 1978. Technical evaluation of copper metalization in the production of thick film micro-circuits. Unpublished.

Swift, C. M. 1979. Recommendation of an electronic security system for the EPA Region III Library. Unpublished.

PART FOUR

OTHER TECHNICAL WRITING APPLICATIONS

CHAPTER SIXTEEN

TECHNICAL ARTICLES

Many popular magazines, from *Reader's Digest* to *Popular Mechanics* and *Personal Computing*, print journalistic articles about technical subjects. This chapter discusses technical articles, which in contrast might be defined as essays that are written by specialists—for other specialists—and that appear in professional journals. Technical articles are written principally to inform their readers, not to entertain them, and most of the readers of these articles read them for the information they convey.

TYPES OF TECHNICAL ARTICLES

Technical articles fall into three categories: (1) research articles, (2) review articles, and (3) conference papers. A research article reports on a research project. In most cases, the research is carried out by the writer in the laboratory or the field. A typical research article might describe the effect of a new chemical on a strain of bacterium, or the rate of contamination of a freshwater aquifer. A review article, on the other hand, is an analysis of the published research on a particular topic. It might be, for example, a study of the advances in chemotherapy research over the last two years.

The purpose of a review article is not merely to provide a bibliography of the important research, but also to classify and evaluate the work and perhaps to suggest fruitful directions for future research. The third kind of technical article—the conference paper—is the text of an oral presentation that the author gave at a professional conference. Because conferences provide a useful forum for communicating tentative or partially completed research findings, a paper printed in the "proceedings"—the published collection of the papers given at the meeting—is generally less authoritative and prestigious than an article published in a scholarly journal. However, proceedings do give the reader a good idea of the research currently being done in the field.

WHO WRITES TECHNICAL ARTICLES, AND WHY?

Technical articles are written by the people who actually perform the research or, in the case of review articles, by specialists who can evaluate the research of others. Often, the only characteristic these authors share is that they themselves generated or gathered the information they communicate: a physicist at a research organization writes up her experiment in laser technology, just as a hospital administrator writes up his research on the impact of health maintenance organizations on the private hospital industry.

Technical articles are the basic means by which professionals communicate with each other. Most professional organizations hold national conferences (and some hold regional and international ones, too) at which members present papers and discuss their collective interests. But meetings occur infrequently, and not all interested parties can attend; therefore, conferences are not the principal means by which professionals stay current with the day-to-day changes in their fields. Only technical journals can fulfill that function, for they provide a convenient and inexpensive way to transmit recent, authoritative information. A technical article can be in a reader's hand less than two months after the writer sent it off to the journal; most books, which deal with broader subjects and contain more information than articles can, require at least two years and therefore are not useful for communicating new findings. And because it has been accepted for publication by a group of referees—an editorial board of specialists who evaluate its quality—a technical article is likely to be authoritative.

Because technical articles are a vital communications tool, their authors are highly valued and frequently rewarded by employers and professional organizations. In some professions—notably university teaching and research—regular and substantial publication is virtually a job requirement. In many other professions, occasional articles enhance the reputations and hasten the promotion and advancement of their authors.

TECHNICAL ARTICLES AND THE STUDENT

As an undergraduate you are unlikely to write a technical article that would make an important contribution to the scholarship in your field, although you might well write a journalistic article for a local publication about some subject you're studying in an advanced course or on a project one of your professors is working on. However, within a few years—probably far fewer than you think—you may find yourself applying for a job that requires publication in a professional journal. Or you may realize while working on a project that you have perfected a new technique for carrying out a particular task and that your technique warrants publication.

In the meantime most of your contact with technical articles will be as a reader. Nonetheless, when you read technical articles, don't limit your attention to the information that they convey. Notice the conventions and strategies that they employ, keeping in mind that at some point your résumé might benefit from an article or two.

When the occasion for writing a technical article arises, you will want to know how to approach your task. At such a point, you will probably not have a great deal of time to investigate your options, and for you to understand from the start the fundamentals of publication is important. Even for a reader of technical articles, some familiarity with publication procedures can be helpful: recognizing the editorial positions of the journals in your field and understanding how an author's article has come to be published may shed some light on apparent mysteries; moreover, knowing the standard structure of technical articles will help you skim possible sources for research.

CHOOSING A JOURNAL

Having determined the topic of a possible article and perhaps having written a rough outline, a writer must think of a journal to which to submit the article. You might ask, "Why should anyone worry about placing an article that hasn't been written yet—

wouldn't it be more logical to write it first and then find an appropriate journal?" The answer would be "yes" if all journals operated in the same way. But they don't.

Most professional journals are run by volunteers. Although two or three persons at a journal might be salaried employees, the bulk of the work usually is done by eminent researchers and scholars whose principal goal is to strengthen the work in their fields. Consequently, these editors and their assistants are extremely serious about their work; they draw up and publish careful and comprehensive statements of purpose and editorial policies. They see themselves as addressing a particular audience, in a particular way, for a particular reason.

Editors expect a prospective author to take the journals they publish seriously and to follow their editorial policies. An article that does not follow stated editorial policies suggests that its author is careless, indifferent, or a bit smug. If, for instance, the journal uses one style of documentation and an article conforms to a different one, the author should not be surprised if the article is rejected without comment or explanation.

Sheer numbers contribute to this situation. Publication is so highly valued these days that few editors of reputable journals ever have a shortage of good articles from which to choose. The acceptance rate might be as low as 3 or 4%; rarely is it above 20%. Understandably, editors are unwilling to take the time or use their tiny staffs to make an article conform to their preferred style when they already have a dozen equally good articles that don't need stylistic revisions.

For these reasons, then, an author greatly enhances his or her chances of success by choosing a journal *before* starting to write. Only by doing so can an author tailor an article to its first audience—the editorial board of a particular journal.

WHAT TO LOOK FOR

If you plan to write a professional article, start in the library. You already know the leading journals in your field. Get a few recent issues of each journal and find a quiet place to study them.

First, check the masthead of each journal. The masthead, which is generally located on one of the first few pages, is the statement of ownership and editorial policy. The editorial board is often listed on the masthead as well. The editor is likely to have a column—also near the front of the issue—that sometimes discusses the kinds of articles the journal is (or isn't) looking for. Study the editor's comments and the journal's editorial policy carefully; these sources are likely to provide crucial information that will save you considerable time, trouble, and expense. For instance, every once in a while a journal stops accepting submissions for a specified period (sometimes up to a year or two), because of a backlog of good articles. Of course, you would not write your article for that journal. Or you might learn that a particular journal does not accept unsolicited manuscripts: that is, that it commissions all of its articles from well-known authorities. Again, this is not the journal for you. Or maybe a journal is planning a special issue devoted entirely to one subject. Some journals print *only* special issues, and they announce the subjects in advance. If you choose one of these journals, make sure your idea fits the subject before you write the article.

Not all the information in the editorial policy or the editor's column is negative, of course. Most journals *do* accept unsolicited articles and are happy to explain their requirements: the explanation saves everyone a lot of frustration. Often, they will provide guidelines for sending in manuscripts (how to mail them, how many copies to send, etc.). Perhaps the most pertinent piece of information journals convey is the name of their preferred style sheet: the *Council of Biology Editors Style Manual*, *The Chicago Manual of Style*, the U.S. Government style sheet, or any other style guide. If the journal specifies a style sheet, find it and follow it. Editors notice these things immediately. Figure 16–1 provides an example of an editorial policy statement.

After you have studied a journal's editorial policies, turn to what it actually publishes: the articles themselves. Read a number of articles and try to determine as much as you can about the following matters:

1. *Article length.* Sometimes, journals only print articles that fall within a certain range, such as 4,000 to 6,000 words. Even if a range is not stated explicity in the journal, the editors might have one in mind as they read your article. If your article is going to be either much shorter or longer than the average length of the articles the journal publishes, you might not want to write it for that journal.

2. *Level of technicality.* Are the articles moderately technical or extremely technical? Do they include formulas, equations, figures?

3. *Prose style.* How long are the paragraphs? Do the authors use the passive voice ("The mixture was added . . .") or active voice ("I added the mixture . . .")? Is the writing formal or somewhat informal?

4. *Formal requirements.* Are the titles purely informative or are they "catchy"? Are subtitles included? Abstracts? Biographical sketches? Are the articles written with American or British spelling and punctuation?

THE QUERY LETTER

Once you are fairly certain of the journal you want to shape your article for, it's a good idea to find out what the editor thinks. Some

MANUSCRIPT PREPARATION

Send all manuscripts to:
Jerold F. Lucey, MD
Editor
Pediatrics Editorial Office
Mary Fletcher Hospital
Colchester Avenue
Burlington, VT 05401

In view of the Copyright Revision Act of 1976, effective January 1, 1978, transmittal letters to the editor should contain the following language: "In consideration of the American Acaαemy of Pediatrics taking action in reviewing and editing my submission entitled _______, also known as _______, the author(s) undersigned hereby transfers, assigns, or otherwise conveys all copyright ownership to the AAP in the event that such work is published by the AAP." We regret that transmittal letters not containing the foregoing language signed by all authors of the submission will delay review of the manuscripts.

Manuscripts should be prepared in the manner described in *Manual for Authors & Editors* © 1981 by the American Medical Association. See also "Uniform Requirements for Manuscripts Submitted to Biomedical Journals." A current issue of PEDIATRICS should be consulted for general style.

Three complete copies of the manuscript including tables and illustrations must be supplied. All material should be typed on white bond paper, 21.6 × 27.9 cm (8½ × 11 in). Use double spacing throughout, including title page, abstract, text, acknowledgments, references, tables, and legends for illustrations.

The author's style will be respected; however, writing should conform to acceptable English usage and syntax, and American Medical Association style preferences will be observed. Titles should be concise and clear, subtitles avoided. Terminology should follow *Standard Nomenclature of Diseases and Operations*. Give authors' full names and professional degrees, principal author's address, and name of institution(s) where work was done; omit departmental appointments unless necessary for special reasons. Slang, medical jargon, obscure abbreviations, and abbreviated phrasing should be avoided. Mathematical terms, formulas, abbreviations, and units of measurement must conform to usage in PEDIATRICS, based on standards in *Science* 120:1078, 1954. The metric system will be used; equivalent measurement in the English system may be included in parentheses. Name of chemical compounds—not formulas—should be given. Proprietary names, if unavoidable, will be indicated by capitalization of the first letter. Conversions to accepted standards and terms should be made before the manuscript is submitted.

Authors are requested to furnish (in addition to the full title) a condensed title for the cover, not exceeding 60 spaces, and a running foot of not more than 35 spaces. Original articles should be accompanied by an abstract of 200 words or less, as well as up to five key words under which the paper should be indexed. Authors should also supply an alphabetical list of any unusual abbreviations used and their definitions.

Manuscripts should include a clear introductory statement of purpose; a historical review when desirable; a description of the technique and the scope of the experiments or observations (previously published procedures require only references to the original); a full presentation of the *Results* obtained; a brief *Comment or Discussion* on the significance of the findings and any correlation with those of other workers; a paragraph headed *Speculation and Relevance*, or *Implications*; and a *Summary*, in brief, logical résumé which may include conclusions.

References must be numbered consecutively according to their citation in the text. Abbreviations for journals should be those listed in *Index Medicus*. The following reference style (a modified form of that shown in "Uniform Requirements for Manuscripts Submitted to Biomedical Journals") will appear in the journal effective with volume 71 (January 1983 issue):

Journal (list first three authors then et al):
1. Starzl TE, Klintmalm GBG, Porter KA, et al: Liver transplantation with use of cyclosporin A and prednisone. *N Engl J Med* 1981;305: 266–269

Book
1. Kavet J: Trends in the utilization of influenza vaccine: An examination of the implementation of public policy in the United States, in Selby P (ed): *Influenza: Virus, Vaccines, and Strategy*. New York, Academic Press Inc, 1976, pp 297–308

Tables must be comprehensible to the reader without reference to the text and typed (double-spaced) rather than photographed. Each table should be typed on a separate sheet, be numbered consecutively, and have a brief title. Care should be taken to make tables as concise and brief as possible.

Illustrations—Photographs of line drawings and any other figures that are not composed simply of letters, numerals, and routine symbols must be furnished. Do not send original artwork or printed forms. A reasonable number of black-and-white illustrations will be printed from black-and-white glossies or film without charge.

Each illustration should be identified on its back, indicating the number, author's name, and "top." They should be keyed in the text. If unessential, their omission may be requested. The prints should not be stapled, clipped together, mounted, or trimmed. Details to be emphasized or crop marks should be indicated on a tissue overlay, not on the illustration itself. Illustrations of poor quality may be returned for improvement. Photographs of patients should be submitted *only* when written parental permission has been obtained. It is the responsibility of the authors to obtain this permission and to keep it in their files. If a figure has been published, acknowledge the original source and obtain written permission for its use from the coyright holder. Use cardboard inserts to protect illustrations in the mail. Legends for figures are to be on a separate sheet.

Color illustrations and other special processing involve extra costs that are usually borne by the author. Manuscripts containing such materials will not be processed until arrangements for payment, on the basis of estimated prices, are made. Color work requires one month longer for production.

Revised, March 1982

PEDIATRICS (ISSN 0031 4005) is owned and controlled by the American Academy of Pediatrics. It is published monthly by the American Academy of Pediatrics, Pediatrics, P.O. Box 1034, Evanston, IL 60204.

Subscription price per year: U.S., Mexico, Canada, Central and South America, $37.00; other countries, $47.00. Special rates for medical students, hospital residents and fellows in full time training in U.S., Mexico, Canada, Central and South America, $25.00 per year. Renewal at special rate beyond two years will require a letter from an appropriate authority stating the individual's eligibility. Air mail delivery available outside U.S. and Canada for an additional $55 per year. Please allow 6–8 weeks for delivery of first issue. Single issues $5.00. Payment must accompany order.

Second-class postage paid at EVANSTON, ILLINOIS 60204 and at additional mailing offices.

ABC PENDING

FIGURE 16–1

A Typical Editorial Policy Statement

FIGURE 16–2

Query Letter for an Article

Century Plastics

ALBION, MISSISSIPPI 34213

March 21, 19--

Dr. William Framingham
Editor
Plastics Technology
1913 Vine Street
Fowler, MS 34013

Dear Dr. Framingham:

I would like to know if you would be interested in considering an article for Plastics Technology.

My work at Century Plastics over the last two years has concentrated on laminate foamboard premixes. As you know, one of the major problems with these premixes is the loss of the blowing agent. In my research, I devised a new compound that effectively retards these losses.

We are currently in the process of patenting the compound, and I would now like to share my findings. I think this compound will reduce fabrication costs by 15%, as well as improve the quality of the premix.

The article, which will be a standard report of a laboratory procedure, will run about 2,500 words. Currently I am a Senior Chemist at Century Plastics and the author of fourteen articles in the leading plastics journals (including Plastics Technology, Winter 1972, Fall 1976, and Summer 1979).

If you think this article would interest your readers, please contact me at the above address or at (316) 437-1902.

Sincerely yours,

Anson Hawkins

Anson Hawkins, Ph.D.
Senior Chemist

editors will talk to you on the phone, but most prefer that you write a query letter.

A query letter is brief—usually less than a page. Your purpose is to answer briefly the following questions:

1. What is the subject of the article?
2. Why is the subject important?
3. What is your approach to the subject (that is, will the article be the report of a laboratory procedure, a rebuttal of another article, etc.)?
4. How long will the article be?
5. What are your credentials?
6. What is your phone number?

Figure 16–2 shows an example of a query letter.

THE STRUCTURE OF THE ARTICLE

Most of this text has been devoted to an explanation of multiple-audience documents—reports addressed both to people who know your subject and want to read all of the details and to people who want to know only the basics. Although some readers of a technical article will skip one or several of the sections, for the most part you will be writing to fellow specialists who will read the whole article carefully. Because technical articles are intended for a national or international audience, a uniform structure has been developed to assist both writers and readers.

This structure is essentially the same as that of the *body* of the average report, except that an abstract generally precedes the article itself, with no other material intervening. The components of this structure are as follows:

1. title of article and name of author
2. abstract
3. introduction
4. materials and methods
5. results
6. discussion
7. references

The following discussion describes each of these components and includes representative sections of a brief technical article from the journal *Stain Technology* as illustration.

TITLE AND AUTHOR

Technical articles tend to have long (and clumsy) titles, because of the need for specificity. Abstracting and indexing journals classify articles according to the key words in the title; therefore, you should be sure to include in your title those terms that will enable the readers to find your article.

Following the title comes your name (and the name of any co-authors). In addition, you should include your institutional affiliation, as shown in Figure 16–3.

Notice that every word in this title includes important information. The first three words—"A Simplified Method"—not only convey the *type* of article (a description of a laboratory method) but also hint at the reason the research was undertaken (to simplify an existing method). Notice also how the rest of the title conveys specific information: "Mast cells," not just "cells"; and an identification of the type of stain. A simpler title—"A Method for Staining Cells," for example—would be much less informative.

A SIMPLIFIED METHOD FOR STAINING MAST CELLS WITH ASTRA BLUE

DAVID M. BLAIES AND JEFFREY F. WILLIAMS, *Department of Microbiology and Public Health, College of Veterinary Medicine, Michigan State University, East Lansing, Michigan 48824*

FIGURE 16–3

Title and Author Listing of a Technical Article

SOURCE David M. Blaies and Jeffrey F. Williams, "A Simplified Method for Staining Mast Cells with Astra Blue," *Stain Technology*, 56. Copyright 1981, the Williams & Wilkens Co. Reproduced by permission.

ABSTRACT. **The copper phthalocyanin dye astra blue has been used to stain differentially mast cells of the intestine; however, the procedure has not been used widely because of the difficulty in preparing and using the dye solution. Described here is a simple, reliable, and consistent method for selectively staining mast cells using a dye solution that may be prepared in any laboratory without the aid of sophisticated pH metering equipment. Astra blue is mixed with an alcoholic solution containing $MgCl_2 \cdot 6H_2O$ and the pH indicator pararosaniline hydrochloride. Concentrated hydrochloric acid is added dropwise, changing the dye mixture from purple to violet and then to blue. In this low range the weakly ionizing ethanol provides a more stable hydrogen ion concentration than the corresponding aqueous solutions used previously. Alcoholic acid fuchsin is a convenient counterstain, and this simple procedure then provides good contrast between the blue staining mast cell granules and the red tissue background.**

FIGURE 16–4

Abstract of a Technical Article

ABSTRACT

Technical articles contain abstracts for the same reasons that reports do: to enable readers to decide whether to read the whole discussion. Most journals today prefer informative abstracts to descriptive ones (see Chapter 11 for a discussion of these two basic types). Figure 16–4 shows the abstract for the article on mast cells.

This is an excellent abstract. The first sentence effectively defines the problem: although astra blue has been used to stain mast cells, the procedure has been difficult. The second sentence summarizes the article, suggesting the major advantages of the new method. The remaining sentences list the principal steps in the procedure and the important application of the method.

INTRODUCTION

The purpose of an introduction to a technical article is to define the problem that led to the investigation. To do so, the introduction generally must (1) fill in the background and (2) isolate the particular deficiencies that currently exist. Often, the introduc-

Use of the copper phthalocyanin dye astra blue for the demonstration of mast cells in tissue sections was first described by Bloom and Kelly (1960). Since that time the stain has been used by many investigators to characterize the distinct mast cell population in the rat which develops at mucosal surfaces, especially in response to intestinal infections with parasitic helminths (Enerback 1966, Miller and Walshaw 1972, Murray *et al.* 1968). A modified procedure has found application in the detection of basophil granules in human peripheral blood and bone marrow smears (Inagaki 1969). Under appropriate conditions the use of astra blue results in highly selective staining of sulfated mucopolysaccharides in mucosal mast cell granules, provided these have been retained through proper fixation. The corresponding absence of background color offers considerable advantage over the less selective Bismarck brown (Gottlieb *et al.* 1961) or thiazine metachromatic dyes such as azure A and toluidine blue (Enerback 1966).

A major disadvantage of the above procedure, which impairs markedly the reproducibility of the results, is the critical dependence of selective staining on a pH environment in the range 0.2–0.3. Solutions with this specific degree of acidity are difficult to produce with satisfactory consistency, and pH at this low level cannot be monitored accurately with most commonly used laboratory metering equipment. By contrast, the method which we describe obviates the need for sophisticated pH detection devices. The success of this technique depends on the use of alcoholic solutions of astra blue and the pH dependent color change of pararosaniline hydrochloride in acid alcohol in the presence of magnesium. It incorporates the principle of critical electrolyte concentration in mucopolysaccharide staining (Scott and Dorling 1965).

FIGURE 16–5

Introduction to a Technical Article

tion requires a brief review of the literature: that is, an annotated summary of the major research, either on the background of the problem or on the problem itself. This literature review provides a context for the discussion that is to follow. Perhaps more importantly, the literature review gives credence to the article by demonstrating that the author is thoroughly familiar with the pertinent research.

Figure 16–5 shows the introduction to the article on mast cells.

This introduction begins with a brief literature review that describes the history of the use of astra blue as a dye for mast cells. The first three sentences of the opening paragraph identify the major applications of the dye. The remaining two sentences describe why astra blue was an advance over the dyes used previously. Paragraph one, then, answers the reader's basic questions about the use of astra blue as a dye for mast cells: (1) What is the history of the procedure? and (2) Why was it developed?

The second paragraph of the introduction explains why the authors decided to try to refine the procedure: the solution is difficult to formulate and monitor. Notice, in the middle of the paragraph, how effectively the authors introduce the principle behind their new method: "By contrast, the method which we describe. . . ."

MATERIALS AND METHODS

"Materials and methods" is a standard phrase used to describe the actual procedure the researcher followed. Basically, this section of the article has the same structure as a common recipe: these are the things you will need, and this is what you will do. Generally, the materials and methods section is clearly divided into its two halves, and the steps of the procedure are numbered and expressed in the imperative mood (for example, "Mix the . . .").

Figure 16–6 shows the materials and methods section of the article.

This materials and methods section is divided into "Reagents" and "Staining Procedure." The reagents section describes how to prepare the "ingredients" for the staining procedure itself. Notice that this section carefully tells the readers where to obtain the exact substances to use if they want to reproduce the procedure themselves. It also gives advice at several points about how to correct common mistakes the reader might make in doing the preparation.

RESULTS

The results of the procedure are what happened. Keep in mind that the word *results* refers only to the observable or measurable effects of the methods; it does not attempt to explain or interpret those effects.

DISCUSSION

The discussion section answers two questions: (1) Why did the results happen? and (2) What are the implications of the results? The two halves of the typical discussion section of an article are thus similar to the conclusion and recommendation sections of a report.

Figure 16–7 provides an example of a combined results and discussion section. Such a combination is common in brief technical articles.

The first paragraph of Figure 16–7 describes the results of the procedure: "Mast cell granules stain intensely blue. . . ." The next paragraph is the first half of the discussion—the "why" of the procedure. In this case, the researchers explain why their procedure does *not* cause the undesirable effect that characterizes the old procedure. The final two paragraphs suggest the usefulness of the new technique: it ". . . facilitates detection and counting of mast cells . . ." and is ". . . very attractive for routine use in histology laboratories."

STAIN TECHNOLOGY

MATERIALS AND METHODS

Reagents

1. Dissolve 2 g $MgCl_2 \cdot 6H_2O$ and 0.1 g pararosaniline hydrochloride C.I. 42500 (Sigma Chemical Co., St. Louis, MO) in 80 ml 95% ethanol in an Erlenmeyer flask.
2. Dissolve 0.5 g astra blue FM (Chroma 1B163, Chroma-Gesellschaft, Schmid Co., W. Germany. U.S. distributor Roboz Surgical Instr. Co., 1000 Connecticut Avenue, Washington, D.C.) in 10 ml glass distilled water.
3. Slowly add the astra blue solution to the alcoholic mixture while stirring constantly.
4. Add 12 N HCl dropwise, dispensing from a burette, until the color of the solution changes from purple to violet, then to royal blue. The amount needed is approximately 9 ml; however, adding several drops more will not affect the stain adversely because in the weaker ionizing ethanol environment low hydrogen ion concentrations are less liable to fluctuate widely than in wholly aqueous solutions.
5. After allowing the solution to settle for 1 hour, filter through Whatman #114v paper. This stain may be stored at 22 C for months, but it should always be refiltered before use. If the color shifts toward violet on storage a few drops of 12 N HCl will restore the royal blue tint.
6. Prepare 5% (v/v) stock HCl in 70% ethanol for the rinsing solution. The final acid concentration should be 0.6–0.7 M.

If acid fuchsin C.I. 42685 (Matheson, Coleman and Bell, Norwood, OH) is to be used as a counterstain, prepare a stock by adding 0.25 g to 100 ml of 1% (v/v) HCl in 95% ethanol. Further dilute this solution 1:100 in 1% acid ethanol to obtain a working stain containing 0.0025% (w/v) acid fuchsin.

Staining Procedure

1. Bring paraffin tissue sections to 95% ethanol.
2. Stain in astra blue solution for 30 to 60 minutes.
3. Rinse in the 5% acid ethanol solution until all the excess stain is removed.
4. Counterstain in acid fuchsin solution for 5 minutes. The solution is dilute enough so that the exact color intensity may be varied by lengthening or reducing the time.
5. Wash in 95% ethanol.
6. Dehydrate, clear, and mount.

FIGURE 16-6

Materials and Methods Section of a Technical Article

SIMPLIFIED ASTRA BLUE PROCEDURE

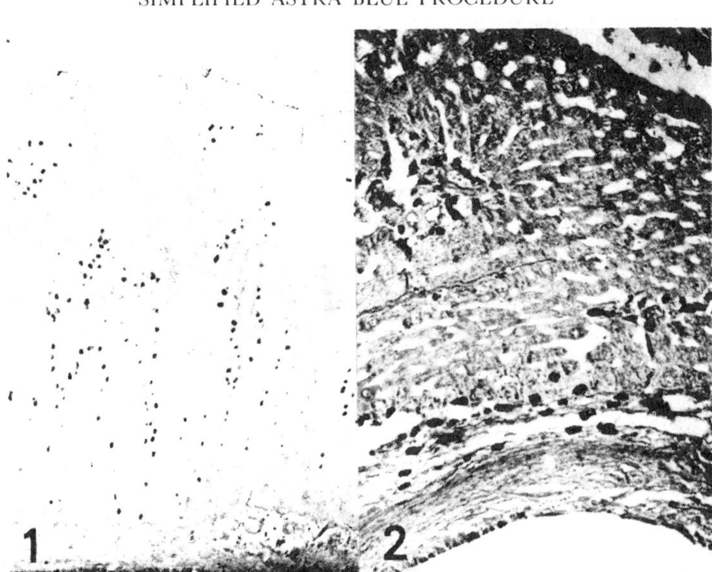

Fig. 1. (*left*) Duodenum from rat infected with *Taenia taeniaeformis*. Tissue was fixed in Carnoy's fluid and stained with astra blue in the presence of sufficient Mg^{++} to prevent mucus staining. BG-38 filter. × 55.

Fig. 2. (*right*) Liver from rat infected with *Taenia taeniaeformis* showing mast cells at periphery of host capsule. Formalin fixed, astra blue. × 100.

An additional advantage of the procedure is that it eliminates loss of tissue sections from slides, which frequently occurs with traditional astra blue methods. In these procedures, the low pH of the stain leads to hydrolysis of protein adhesives; when the aqueous solvent is replaced by alcohol for dehydration, surface tension changes often cause sections to float off. In our method the dye is used in an alcoholic environment and there is no abrupt solvent change to produce this effect.

The contrast between the deeply stained granules and the clear or lightly counterstained background facilitates detection and counting of mast cells, especially in circumstances where pathological increases are encountered. Mast cells in the intestinal lamina propria of both rats and humans stain well when the tissues are fixed in either Carnoy's fluid or lead subacetate. Rat liver mast cells are clearly identifiable, even when formalin is used as fixative (Fig. 2).

The simplicity and reproducibility of the procedure combined with its compatibility with a variety of tissue fixatives make it very attractive for routine use in histology laboratories.

FIGURE 16–7

Results and Discussion Section of a Technical Article

REFERENCES There are different styles for listing references, but each method has the same objective: to enable readers to locate the cited source, should they wish to read it. Journals either specify that writers use a particular style sheet (which in turn defines a style for references) or provide their own guidelines in the discussion of editorial policies near the front of the issue. Following is the editorial statement on references from *Stain Technology*, the journal that printed the mast cells article.

> *References:* Alphabetical listings (not numbered) at the end of the paper; text citations given by dates. Complete names, titles, and first and last pages are required; for example: Lyons, W. R. and Johnson, Ruth E. 1953. Embedding stained mammary glands in plastic. Stain Technol. *28:* 201–204. References to books should include date, edition, publisher and publisher's city.

Figure 16–8 shows the references section of the article.

This format highlights the names of the authors and the dates of publication of their works. Notice that journal titles are abbreviated. The italicized number that follows each title is the volume number; after the colon are the page numbers of the article. References to books are not common in technical articles, because most research is printed in journals.

REFERENCES

Bloom, G. and Kelly, J. W. 1960. The copper phthalocyanin dye "Astrablau" and its staining properties, especially the staining of mast cells. Histochemie *2:* 48–57.

Enerback, L. 1966. Mast cells in rat gastrointestinal mucosa. 2. Dye-binding and metachromatic properties. Acta Pathol. Microbiol. Scand. *66:* 303–312.

Gottlieb, L. S., Robertson, R. and Zamcheck, N. 1961. A method for staining mouse mastocytoma. Amer. J. Clin. Pathol. *36:* 336–338.

Inagaki, S. 1969. Cytochemical reactions of basophil leukocytes. VIII. Astra blue staining. Acta Haematol. Jap. *32:* 148–153.

Miller, H. R. P. and Walshaw, R. 1972. Immune reactions in mucous membranes. IV. Histochemistry of intestinal mast cells during helminth expulsion in the rat. Amer. J. Pathol. *69:* 195–208.

Murray, M., Miller, H. R. P. and Jarrett, W. F. H. 1968. The globule leukocyte and its derivation from the subepithelial mast cell. Lab. Invest. *19:* 222–234.

Scott, J. E. and Dorling, J. 1965. Differential staining of acid glycosaminoglycans (mucopolysaccharides) by Alcian blue in salt solutions. Histochemie *5:* 221–233.

FIGURE 16–8

References Section of a Technical Article

The article on dyeing mast cells includes another section that has not been mentioned yet:

ACKNOWLEDGMENT

This work was supported by NIH grant AI-10842. This is journal article No. 9499 from the Michigan Agricultural Experiment Station.

An acknowledgment section, which generally precedes the references, gives you the opportunity to express your appreciation to persons or institutions that have assisted you in your research.

The sample article that appears in this chapter demonstrates the usefulness of the basic article structure. You might not have understood the details of the discussion, but the logical flow of the argument should be clear: the existing method of using astra blue dye to stain mast cells works, but it is difficult to carry out, so here are the directions for our new technique, which will give good results without any of the problems associated with the old technique. This basic structure makes it easier for the writer to develop the article—and easier for the reader to understand it.

PREPARING AND MAILING THE MANUSCRIPT

Preparing and mailing the manuscript in a professional manner will not guarantee that an article is accepted for publication, of course. However, an unprofessional approach will almost certainly ensure that an article is *not* accepted.

TYPING THE ARTICLE

The article must be typed. Whether or not *you* type it, keep in mind that editors are like any other readers: they don't like sloppy manuscripts. And they particularly don't like articles that violate the preferred stylistic guidelines. One reason is that it costs the journal money to revise a manuscript to make it conform stylistically. Perhaps a more important reason is that it is rude.

If the journal has stated its preferences, follow them to the letter. Most journals have clear guidelines about margins, pagination, spacing, number of copies, and so on. Look, for example, at the following discussion from *Stain Technology*:

Preparation of Manuscript: Use letter-size paper, not larger than 8½ × 11 inches. Type *all* material in double-spaced lines (including footnotes, references, legends, *etc.*) on only one side of the paper. Allow a margin of at least 1 inch at the top, bottom, and *both sides* of the page. On the first 4 pages, place and mark items as follows: page 1, short title (running head), the name and address of the person to whom proof is to be sent, and key words to assist in indexing; page 2, full title only; page 3, names and addresses, including institutional affiliations, of author or authors as they are to appear in the journal; and page 4, abstract only. Use page 5 and succeeding pages exclusively for the textual material of the article, without footnotes or other material which is to be set in small type. The latter material consists of: references, footnotes, tables, and legends. Combine these on pages separate from the text proper as may be convenient; *e.g.*, a separate page for each legend or each footnote is not required. Legends should not be attached to photomicrographs or other illustrative material. Charts should have a typed legend, not one that is an integral part of the chart. Usually, a page for each table is advantageous for both author and typesetter. Submit manuscripts in duplicate.

By conforming with these specifications, authors can greatly facilitate both editing and typesetting. Manuscripts which do not conform may be either returned to the authors for retyping, or cut and rearranged when the deviations are minor.

The second paragraph is a polite way of saying that the journal will correct minor stylistic problems, but the writer will have to retype the whole article if the deviations are great. What this statement does not say—but what any editor will admit—is that any manuscript that blatantly violates the journal's specifications will probably be rejected outright.

If the journal has provided no typing guidelines, double-space everything and type on only one side of 8½- by 11-inch nonerasable bond paper. Use a good typewriter (preferably electric) and a fresh ribbon. Send a photocopy with the original, and *make sure you keep a photocopy.* The percentage of manuscripts lost in the mail or in the journal's office is probably quite small, but when it happens to people who didn't keep a copy, they start to ask fundamental questions about life.

WRITING A COVER LETTER

An article should not arrive alone in an envelope. Enclose a cover letter. This letter should be brief, for your purpose is not to "sell" the article: any boasting would probably work against you. The abstract will tell your readers everything they want to know.

If the editor has responded to a query letter, you should work this fact into the first paragraph. If you didn't inquire first, simply state that you are enclosing an article that you would like the editor to consider. In the next paragraph, you might wish to define briefly the subject or approach of your article and give its title if you haven't done so already. Then conclude the letter politely. Figure 16–9 shows a sample cover letter.

FIGURE 16-9
Cover Letter for a Technical Article

Century Plastics
ALBION, MISSISSIPPI 34213

April 3, 19--

Dr. William Framingham
Editor
Plastics Technology
1913 Vine Street
Fowler, MS 34013

Dear Dr. Framingham:

Thank you for replying promptly to my inquiry about the article on laminate foamboard premixes. As you requested, I am submitting the article for your consideration.

The article provides a step-by-step procedure for using Century Plastics' new compound in reducing blowing agent losses.

I look forward to learning your decision.

Sincerely yours,

Anson Hawkins

Anson Hawkins, Ph.D.
Senior Chemist
(316) 437-1902

MAILING THE ARTICLE

Follow the journal's mailing guidelines—about the number of copies to include, for example. Some journals request that you enclose a self-addressed, stamped envelope (SASE); others ask for loose postage. You might wish to state, in your cover letter, that you are enclosing an SASE or postage.

To keep the article from getting mangled, it's a good idea to enclose it between pieces of cardboard. Be especially careful if you are enclosing photographs or other artwork. If you are sending a total of about 100 pages or more, use a padded envelope (a "Jiffy Bag"), which can be obtained from the post office.

Some journals state that they acknowledge received manuscripts. If the journal you are writing to doesn't, enclose a self-addressed postcard and ask the editor to mail it when he or she receives your article.

RECEIVING A RESPONSE FROM THE JOURNAL

When your article arrives at a journal, it is read by the editor. If he or she believes that the subject matter is within the scope of the journal and that the article appears authoritative and is reasonably consistent with the specified stylistic guidelines, it will be sent out to two or three reviewers who are specialists in your subject.

WHAT THE REVIEWERS LOOK FOR

There is little mystery about what reviewers look for when they read an article. They want to determine, basically, if the subject of the article is worthy of publication in the journal, if the article is technically sound, and if it is reasonably well written. These three questions might be expanded as follows.

First, does the article fall within the scope of the journal? If it doesn't, the article will be rejected, regardless of its quality. If the article does fall within the journal's scope, the reviewers will want to see a convincing case that the subject is important enough to warrant publication.

Second, is the article sound? Is the problem defined clearly, and is the literature review comprehensive? Would the readers be able to reproduce your work based on the methods and materials section? Are the results clearly expressed, and do they flow logically from the methods? Is the discussion clear and coherent? Have you drawn the proper conclusions from the results, and have you clearly suggested the implications of your research?

Third, is the article well written? Does it conform to the journal's stylistic requirements? Is it free of grammar, punctuation, and spelling errors?

THE EDITOR'S RESPONSE

While the reviewers are studying your article, you wait—anywhere from three weeks to six months or a year. When you receive an acknowledgment from the editor, he or she will usually tell you how long the review process should take. If you haven't received an acknowledgment within three or four weeks, call or write the editor, asking if your manuscript arrived.

Finally, you will receive a response from the editor. Most likely, the response will take the form of an acceptance, a request to revise the article and resubmit it, or a rejection.

1. *Acceptance.* Very few articles—usually fewer than 5% of those submitted—are accepted outright. If you are one of the select few, congratulations. The acceptance letter might tell you the issue the article will appear in. Some journals edit the author's manuscripts; you might be asked to review the edited manuscript and to proofread the galleys—the first stage of typeset proof, which is later corrected and cut into pages. Other journals don't edit the manuscripts or ask the authors to proofread.

2. *Request to revise the article.* The majority of articles that eventually get published were revised according to the reviewers' recommendations. Sometimes reviewers call for major revisions: a total rethinking of the article. Sometimes the repairs are minor. Some journals call this response a provisional acceptance: they will accept your article if you make the specified revisions. Other journals do not commit themselves, but merely invite you to send it in again. If the reviewers' comments seem reasonable to you, revise the article and send it in. If they don't, you might want to consider trying another journal. Keep in mind, however, that another journal might require other revisions.

3. *Rejection.* The majority of articles submitted to most journals are simply rejected. A rejection means that the journal will not print it and does not want to see another version of it. All rejections hurt. The worst are the curt letters that offer no explanation: "We do not feel your article is appropriate for our journal." These rejections are frustrating, because you wonder if anyone read the article that you put so much time and effort into. Some journals offer explanations, either by including the reviewers' comments or by excerpting them. If the criticism is valid, you don't feel too bad. If it isn't, you feel miserable.

Most experienced writers would advise that you develop a thick skin—but not too thick a skin. Many important articles are rejected by several journals before they are accepted. On the other hand, you should try to learn from the rejections. If your article has been rejected, study the reviewers' comments (if you received any) and try to react objectively. Consider it free advice, not a personal insult. If you haven't received any comments, take the article apart piece by piece, and see if you can discover any flaws. If you do, correct them and submit the article to another journal.

The first article is the hardest to write, and the one most likely to be rejected. After you publish two or three, you will develop a system that works well for you and learn to accept rejection as an inevitable part of publication. Knowing that most articles never get printed, you can take understandable pride in your successes.

WRITER'S CHECKLIST

1. Does the article fall within the scope of the journal? ____

2. Is the subject of the article sufficiently important to justify publication in the journal? ____

3. Does the abstract effectively summarize the problem, methods, results, and conclusions? ____

4. Is the article technically sound?
   a. Is the problem clearly defined? ____
   b. Is the literature review comprehensive? ____
   c. Is the methods and materials section complete and logically arranged? Would it be sufficient to enable researchers to reproduce the technique? ____
   d. Are the results clear? Do they seem to follow logically from the methods? ____
   e. Is the discussion clear and coherent? Has the writer drawn the proper conclusions from the results and clearly suggested the implications of the research? ____

5. Is the article reasonably well written?
   a. Does it conform (at least basically) to the journal's stylistic requirements? ____
   b. Is the writing clear and unambiguous? Is it free of basic errors of grammar, punctuation, and spelling? In short, can it be understood easily? ____

EXERCISES

1. Choose a major journal in your field and write a 500-word essay analyzing the degree to which its contents correspond to its stated editorial policy. First, using the editorial policy statement, identify what the editor says the journal publishes. Consider such questions as range of subjects covered, type of article, writing style (paragraph and sentence

length, technicality, etc.), and format (type of abstract, documentation style, etc.). Then, survey two or three articles in the issue. How closely do the articles reflect the editorial policy statement?

2. Write a 300-word essay evaluating an article in your field.

REFERENCE

Blaies, D. M., and J. F. Williams. 1981. A simplified method for staining mast cells with astra blue. *Stain Technology* 56, no. 2: 91–94.

CHAPTER SEVENTEEN

ORAL PRESENTATIONS

TYPES OF ORAL PRESENTATIONS

Although you will occasionally be called on for impromptu oral reports on some aspect of your work, most often you will have the opportunity to prepare what you have to say. This chapter covers two basic kinds of oral presentations that are prepared in advance: the extemporaneous presentation and the presentation read verbatim from manuscript. In an extemporaneous presentation, you might refer to notes or an outline, but you actually make up the sentences as you go along. Regardless of how much you have planned and rehearsed the presentation, you create it as you speak. An oral presentation read from manuscript is a talk that is completely written in advance; you simply read your text.

This chapter will not discuss the memorized oral presentation. Memorized presentations are not appropriate for most technical subjects, because of the difficulty involved in trying to remember technical data. In addition, few people other than trained actors are able to memorize oral presentations of more than a few minutes in length.

The extemporaneous presentation is the preferred type for all but the most formal occasions. At its best, it combines the virtues of clarity and spontaneity. If you have planned and rehearsed your presentation sufficiently, the information will be accurate, complete, logically arranged, and easy to follow. And if you can think well on your feet without grasping for words, the presentation will have a naturalness that will help your audience concentrate on what you are saying, just as if you were speaking to each person individually.

The presentation read from manuscript sacrifices naturalness for increased clarity. Most people sound stilted when they read from a text. But, obviously, having a complete text ensures that you will say precisely what you intend to say.

THE ROLE OF ORAL PRESENTATIONS

In certain respects, oral presentations are inefficient. For the speaker, preparing and rehearsing the presentation generally take more time than writing a report would. For the audience, physical conditions during the presentation—such as noises, poor lighting or acoustics, or an uncomfortable temperature—can interfere with the reception of information.

Yet oral presentations have one big advantage over written presentations: they permit a dialogue between the speaker and the audience. The listeners can offer alternative explanations and viewpoints, or simply ask questions that help the speaker clarify the information.

Oral presentations are therefore an increasingly popular means of technical communication. You can expect to give oral presentations to three different types of listeners: clients and customers, fellow employees in your organization, and fellow professionals in your field.

In representing your product or service to clients and customers, you will find oral presentations to be a valuable sales technique. Whether you are trying to interest them in a silicon chip or a bulldozer, you will have to make a presentation of the features of your product or service and the advantages it has over the competition's. Then, after the sale is made, you will likely give oral presentations to the client's employees, providing detailed operating instructions and outlining maintenance procedures. In both cases, extemporaneous presentations would probably be the more effective.

This same kind of informational oral presentation is necessary within an organization, too. If you are the resident expert on a certain piece of equipment or a procedure, you will be expected to instruct your fellow workers, both technical and nontechnical. After you return from an important conference or work on an out-of-town project, your supervisors will, whether or not you submit a written trip report, want an opportunity to ask questions: you must anticipate these questions and prepare answers. Another occasion for an oral presentation to your fellow workers comes up after you submit an informal proposal. If you have an idea for improving operations at your organization, you probably will first write a brief memo outlining your idea to your supervisor. If he or she agrees with you that the idea is promising, you might then be asked to present it orally to a small group of managers. This face-to-face meeting will give them an opportunity to question you in detail. In an hour or less, they can determine whether it is prudent to devote resources to studying the idea.

Oral presentations to fellow professionals generally are given at technical conferences and are often read from manuscript. You might speak on a research project with which you have been personally involved. Your subject might be a team project carried out at your organization. Or you may be invited to speak to professionals in other fields. If you are an economist, for example, you might be invited to speak to an association of realtors on the subject of interest rates.

The situations that demand oral presentations, therefore, are numerous, and they increase as you assume greater responsibility within an organization. You might not have had much experience in public speaking, and perhaps your few attempts have been difficult. But oral presentations on technical subjects are essentially the spoken form of technical writing. Just as there are a few writers who can produce effective reports without outlines or rough drafts, you might know a natural speaker who can talk to groups "off the cuff" effortlessly. For most persons, however, an oral presentation requires deliberate and careful preparation.

PREPARING TO GIVE AN ORAL PRESENTATION

Preparing to give an oral presentation requires four steps:

1. assessing the speaking situation
2. preparing an outline
3. preparing the graphic aids
4. rehearsing the presentation

ASSESSING THE SPEAKING SITUATION

The first step in preparing an oral presentation is to assess the speaking situation: audience and purpose are as important to oral presentations as they are to written reports.

Analyzing the audience requires careful thought. Who are the people who make up the audience? How much do they know about your subject? You must answer these questions in order to determine the level of technical vocabulary and concepts appropriate to the audience. Speaking over an audience's level of expertise is ineffective; oversimplifying and thereby appearing condescending might be just as bad. If the audience is relatively homogeneous—for example, a group of landscape architects or physical therapists—it is not difficult to choose the correct level of technicality. If, however, the group contains individuals of widely differing backgrounds, you should try in your presentation to accommodate everyone's needs by providing brief parenthetical definitions of those technical words and concepts that some of your listeners are not likely to know. Ask yourself the same kinds of questions you would ask about a group of readers: Why is the audience there? What do they want to accomplish as a result of having heard your presentation? Are they likely to be hostile, enthusiastic, or neutral toward your topic? A presentation on the virtues of free trade, for instance, will be received one way if the audience is a group of conservative economists and another way if it is a group of American steelworkers.

Equally important, analyze your purpose in making the presentation. Are you attempting to inform your audience, or both to inform as well as persuade them? If you are explaining how windmills can be used to generate power, you have one type of presentation. If you are explaining how *your* windmills are an economical means of generating power, you have another type of presentation. A successful oral presentation requires that you begin with a clear understanding of your purpose.

Finally, determine how the audience and purpose affect the strategy—the content and the form—of your presentation. You might have to emphasize some aspects of your subject or eliminate some altogether. You might have to arrange the elements of the presentation in a particular order to accommodate the audience's needs.

As you are planning, don't forget the time limitations established for your speech. At most professional conferences, the organizers clearly state a maximum time, such as 20 or 30 minutes, for each speaker. If you know that the question-and-answer period is part of your allotted time, plan accordingly. Even at an informal presentation, you probably will have to work within an unstated time limitation—a limit that you must figure out from your understanding of the speaking situation. Speaking too long is ineffective. Claiming more than your share of an audience's time is rude and egotistical, and eventually the audience will simply stop paying attention.

PREPARING AN OUTLINE

After assessing the audience, purpose, and strategy of the presentation, you should prepare an outline, regardless of whether you plan to bring it with you and consult it as you speak. In preparing the outline, keep in mind the difference between an oral presentation and a written one. An oral presentation is, in general, simpler than a written version of the same material, for the listeners cannot stop and reread a paragraph they do not understand. Statistics and equations, for example, generally should be kept to a minimum.

The oral presentation must be structured logically. The organizational patterns presented in Chapter 4 are as useful for structuring an oral presentation as they are for structuring a written document.

The introduction to the oral presentation is very important, for your opening sentences must gain and keep the audience's attention. An effective introduction might define the problem that led to the project, offer an interesting fact that the audience is unlikely to know, or present a brief quotation from an authoritative figure in the field or a famous person not generally associated with the field (for example, Abraham Lincoln on the value of American coal mines). All of these techniques should lead into a clear statement of the scope and purpose of the presentation. If none of these techniques offers an appropriate introduction, a speaker can begin directly by defining the scope and purpose of the presentation. A forecast of the major points is useful for long or complicated presentations. Don't be fancy. Use the words *scope* and *purpose*. And don't try to enliven the presentation by adding a joke. Humor is usually inappropriate in technical presentations.

The conclusion, too, is crucial in an oral presentation, for it summarizes the major points of the talk and clarifies their relationships with one another. Without a summary, many otherwise effective oral presentations would sound like a jumble of unrelated facts and theories.

With these points in mind, write an outline just as you would for a written report. Your own command of the facts will determine the degree of specificity necessary in the outline. Many people prefer a sentence outline (see Chapter 4) because of its specificity; others feel more comfortable with a topic outline. Figure 17–1

FIGURE 17–1

Sentence/topic Outline for an Oral Presentation

OUTLINE

DESCRIBING, TO A GROUP OF CIVIL ENGINEERS, A NEW METHOD OF TREATMENT AND DISPOSAL OF INDUSTRIAL WASTE.

I. Introduction
   A. The recent Resource Conservation Recovery Act places stringent restrictions on plant engineers.
   B. With neutralization, precipitation, and filtration no longer available, plant engineers will have to turn to more sophisticated treatment and disposal techniques.

II. The Principle Behind the New Techniques
   A. Waste has to be converted into a cementitious load-supporting material with a low permeability coefficient.
   B. Conversion Dynamics, Inc., has devised a new technique to accomplish this.
   C. The technique is to combine pozzolan stabilization technology with the traditional treatment and disposal techniques.

III. The Applications of the New Technique
   A. For new low-volume generators, there are two options.
      1. Discussion of the San Diego plant.
      2. Discussion of the Boston plant.
   B. For existing low-volume generators, Conversion Dynamics offers a range of portable disposal facilities.
      1. Discussion of Montreal plant.
      2. Discussion of Albany plant.
   C. For new high-volume generators, Conversion Dynamics designs, constructs, and operates complete waste-disposal management facilities.
      1. The Chicago plant now processes up to 1.5 million tons per year.
      2. The Atlanta plant now processes up to 1.75 million tons per year.

shows a combined sentence/topic outline, and Figure 17–2 shows a topic outline. The speaker is a specialist in waste-treatment facilities. The audience is a group of civil engineers who are interested in gaining a general understanding of new developments in industrial waste disposal. The speaker's purpose is to provide this information and to suggest that his company is a leader in the field.

Notice the differences in both content and form between these two versions of the same outline. Whereas some speakers prefer to have full sentences before them, others find that full sentences interfere with the spontaneity of the presentation. (For presenta-

```
        D.  For existing high-volume generators, Conversion Dy-
            namics offers add-on facilities.
            1.  The Roanoke plant already complies with the new
                RCRA requirements.
            2.  The Houston plant will be in compliance within
                six months.

    IV. Conclusion
        A.  The Resource Conservation Recovery Act will necessi-
            tate substantial capital expenditures over the next
            decade.
```

tions read from manuscript, of course, the outline is only a preliminary stage in the process of writing the text.)

PREPARING THE GRAPHIC AIDS

Graphic aids fulfill the same purpose in an oral presentation that they do in a written one: they clarify or highlight important ideas or facts. Statistical data, in particular, lend themselves to graphical presentation, as do representations of equipment or processes. The same guidelines that apply to graphic aids in text apply here: they should be clear and self-explanatory. The audience should know immediately what each graphic aid is showing. In addition,

FIGURE 17–2

Topic Outline for an Oral Presentation

OUTLINE

DESCRIBING, TO A GROUP OF CIVIL ENGINEERS,
A NEW METHOD OF TREATMENT AND DISPOSAL OF INDUSTRIAL WASTE.

Introduction:
- Implications of the RCRA

Principle Behind New Technique
- reduce permeability of waste
- pozzolan stabilization technology and traditional techniques

Applications of New Technique
- for new low-volume generators (San Diego, Boston)
- for existing low-volume generators (Montreal, Albany)
- for new high-volume generators (Chicago, Atlanta)
- for existing high-volume generators (Roanoke, Houston)

Conclusion

the material conveyed should be simple. In making up a graphic aid for an oral presentation, be careful not to overload it with more information than the audience can absorb. Each graphic aid should illustrate a single idea. Remember that your listeners have not seen the graphic aid before, and that they, unlike readers, do not have the opportunity to linger over it.

In choosing a medium for the graphic aid, consider the room in which you will give the presentation. The people seated in the last row and near the sides of the room must be able to see each graphic aid clearly and easily (a flip chart, for instance, would be ineffective in an auditorium). Following is a listing of the basic media for graphic aids, with the major features cited.

1. *Slide projector:* projects previously prepared slides onto a screen.

ADVANTAGES:

Very professional appearance.
Versatile—can handle photographs or artwork, color or black-and-white.
With a second projector, the pause between slides can be eliminated.
During the presentation, the speaker can easily advance and reverse the slides.

DISADVANTAGES:

Slides are expensive.
Room has to be kept relatively dark during the slide presentation.

2. *Overhead projector:* projects transparencies onto a screen.

ADVANTAGES:

Transparencies are inexpensive and easy to draw.
Speakers can draw transparencies "live."
Overlays can be created by placing one transparency over another.
Lights can remain on during the presentation.

DISADVANTAGES:

Not as professional-looking as slides.
Each transparency must be loaded separately by hand.

3. *Opaque projector:* projects a piece of paper onto a screen.

ADVANTAGES:

Can project single sheets or pages in a bound volume.
Requires no expense or advance preparation.

DISADVANTAGES:

Room has to be kept dark during the presentation.
Cannot magnify sufficiently for a large auditorium.
Each page must be loaded separately by hand.

4. *Poster:* a graphic drawn on oak tag or other paper product.

ADVANTAGES:

Inexpensive.
Requires no equipment.
Posters can be drawn or modified "live."

DISADVANTAGES:

Ineffective in large rooms.

5. *Flip chart:* a series of posters, bound together at the top like a looseleaf binder; generally placed on an easel.

ADVANTAGES:

Relatively inexpensive.
Requires no equipment.
Speaker can easily flip back or forward.
Posters can be drawn or modified "live."

DISADVANTAGES:

Ineffective in large rooms.

6. *Felt board:* a hard, flat surface covered with felt, onto which paper can be attached using doubled-over adhesive tape.

ADVANTAGES:

Relatively inexpensive.
Particularly effective if speaker wishes to rearrange the items on the board during the presentation.
Versatile—can handle paper, photographs, cutouts.

DISADVANTAGES:

Informal appearance.

7. *Blackboard:*

ADVANTAGES:

Almost universally available.
Speaker has complete control—can add, delete, or modify the graphic easily.

DISADVANTAGES:

Complicated or extensive graphics are difficult to create.
Ineffective in large rooms.
Very informal appearance.

A word of advice: before you design and create any graphic aids, make sure the room in which you will be giving the presentation has the equipment you need. Don't walk into the room carrying a stack of transparencies only to learn that there is no overhead projector. Even if you have arranged beforehand to have the necessary equipment delivered, check to make sure it is there; or, if possible, bring it with you.

REHEARSING THE PRESENTATION

Even the most gifted speakers have to rehearse their presentations. It is a good idea to set aside enough time to be able to rehearse your speech several times. For the first rehearsal of an extemporaneous presentation, simply sit at your desk with your outline before you. Don't worry about posture or voice projection. Just try to compose your presentation out loud. Many times speakers cut corners when they construct their outlines. In this first rehearsal, your goal is to see if the speech makes sense—if you can explain all of the points you have listed and can forge effective transitions from point to point. If you have any trouble, stop and try to figure out the problem. If you need more information, get it. If you need a better transition, create one. You might well find that you have to revise your outline. This is very common and no cause for alarm. Pick up where you left off and continue through the presentation, stopping again where necessary to revise the outline. When you have finished your first rehearsal, put the outline away and do something else.

Come back again when you are rested. Try the presentation once more. This time, it should flow more easily. Make any necessary revisions in the outline. Once you have complete control over the structure and flow, check to see if you are within the time limits for the presentation.

After a satisfactory rehearsal, try the presentation again, under more realistic circumstances. If it is possible to rehearse in front of people, take advantage of the opportunity. The listeners will be able to offer constructive advice about parts of the presentation they didn't understand or about your speaking style. If people aren't available, use a tape recorder and then evaluate your own delivery. If you can visit the site of the presentation to get the feel of the room and rehearse there, you will find giving the actual speech a little bit easier.

Once you have rehearsed your presentation a few times and are satisfied with it, stop. Don't attempt to memorize it—if you do, you will surely panic the first time you forget the next phrase. During the presentation, you must be thinking of your subject, not about the words you used during the rehearsals.

Rehearse a written-out presentation in front of people, too, if possible, or use a tape recorder. After a few rehearsals, you might want to signal the pauses—the points at which you will take a breath. For this purpose, speakers often use a slash (/).

GIVING THE ORAL PRESENTATION

Most professional actors freely admit to being nervous before a performance, and so it is no wonder that occasional speakers are nervous. You might well fear that you will forget everything you have to say or that nobody will be able to hear you. These fears are universal. But signs of nervousness are much less apparent to the audience than to the speaker. And it is true that after a few minutes most speakers are able to relax and concentrate effectively on the subject.

All of this sage advice, however, is unlikely to make you feel much better as you wait to give your presentation. When the moment comes, don't jump up to the lectern and start speaking quickly. Walk up slowly and arrange your text, outline, or note cards before you. If water is available, take a sip. Look out at the audience for a few seconds before you begin. It is polite to begin formal presentations with "Good morning" (or "Good afternoon," "Good evening") and to include a reference to the officers and dignitaries present. If your name has not been mentioned by the introductory speaker, identify yourself. In less formal contexts, just begin your presentation.

Your goal, of course, is to have the audience listen to you and have confidence in what you are saying. Try to project the same image that you would in a job interview: restrained self-confidence. Show your listeners that you are interested in your topic and that you know what you are talking about. In giving the presentation, this sense of control is conveyed chiefly through your voice and your body.

YOUR VOICE

Inexperienced speakers encounter problems with five aspects of vocalizing: volume, speed, pitch, clarity of pronunciation, and verbal fillers.

1. VOLUME The acoustics of rooms vary greatly, and so it is impossible to be sure how well your voice will carry in a room until you have heard someone speaking there. In some well-constructed auditoriums, speakers can use a conversational tone. Other rooms require a real effort at voice projection. An annoying echo effect plagues some rooms. These special circumstances aside, it can be said in general that more people speak too softly than too loudly. After your first few sentences, ask if the people in the back of the room can hear you. When people speak into microphones, they tend to bend down toward the microphone and end up speaking too loudly. Glance at your audience to see if you are having volume problems.

2. SPEED Nervousness makes people speak more quickly. Even if you think you're speaking at the right rate, you might be going a little too fast for some of your audience. Remember, you know where you're going—you know the material well. Your listeners, however, are trying to understand new information. For particularly difficult points, slow down your delivery, for emphasis. After finishing a major point, try to pause before beginning the next point.

3. PITCH In an effort to achieve voice control, many speakers end up flattening their pitch. The resulting monotone is boring—and, for some listeners, actually distracting. Try to let the natural rhythm of your speech come through. Let the pitch of your voice go up or down as it would in a normal conversation. In fact, experienced speakers often exaggerate pitch variations.

4. CLARITY OF PRONUNCIATION The nervousness that goes along with an oral presentation tends to accentuate sloppy pronunciation. If you want to say the word "environment," don't say "envirament." Say "nuclear," not "nucular." Don't drop final g's. Say "contaminating," not "contaminatin'." A related pronunciation problem concerns technical words and phrases—especially the important ones. When a speaker uses a phrase over and over, it tends to get clipped and becomes difficult to understand. Unless you articulate carefully, "Scanlon Plan" will end up as "Scanluhplah," or "total dissolved solids" will be heard as "toe-dizahved sahlds."

5. VERBAL FILLERS Avoid such meaningless fillers as "you know," "okay," "right," "uh," and "um." These phrases do not disguise the fact that the speaker isn't saying anything; they call attention to it. It is better to have a thoughtful pause than a nervous and annoying verbal tic.

YOUR BODY

The audience will be listening to what you say, but they will also be looking at you. Effective speakers know how to take advantage of this fact by using their bodies to help the listeners follow the presentation.

Perhaps most important in this regard are your eyes. People will be looking at you as you speak, and it is only polite to look back at them. This is called "eye contact." For small groups, look at each listener in turn; for larger groups, be sure to look at each segment of the audience during your speech. Do not stare at your notes or at the floor or the ceiling or out the window. Even if you are reading your presentation, you should be well enough rehearsed to allow frequent eye contact. Eye contact is not only polite; it also gives you a valuable indication of how the audience is receiving your presentation. You will be able to tell, for instance, if the listeners in the back are having trouble hearing you.

Your arms and hands also are important. Watch practiced speakers to see how they use their arms and hands to signal pauses and to emphasize important points. Gestures are particularly useful when you are referring to graphic aids. Use your arm to direct the audience's attention to an area of the graphic.

Watch out for mannerisms—those physical gestures that serve no useful purpose. Don't play with your jewelry or the coins in your pocket. Don't tug at your beard or fix your hair. These nervous gestures can quickly distract an audience from what you are saying. Like verbal mannerisms, physical mannerisms are often unconscious. Constructive criticism from friends can help you to pinpoint physical mannerisms.

AFTER THE PRESENTATION

On all but the most formal occasions, an oral presentation is followed by a question-and-answer period. In fielding questions, first make sure that everyone in the audience has heard the question. If there is no moderator to do this job, you should ask if everyone has heard the question. If they haven't, repeat or paraphrase it yourself. Sometimes this can be done as an introduction to your response: "Your question about the efficiency of these three techniques . . ."

If you hear the question but don't understand it, ask for a clarification. After responding, ask if you have understood the question correctly.

If you understand the question but don't know the answer, tell the truth. Only novices believe that they ought to know all the answers. If you have some ideas about how to find out the answer—by checking a certain reference text, for example—share them. If the question is obviously important to the person who asked it, you might offer to meet with him or her after the question-and-answer period to discuss ways to give a more complete response, such as through the mail.

If you are unfortunate enough to have a belligerent member of the audience who is not content with your response and insists on restating his or her original point, a useful technique is to offer politely to discuss the matter further after the session. This will prevent the person from boring or annoying the rest of the audience.

If it is appropriate to stay after the session to talk individually with members of the audience, offer to do so. Don't forget to thank them for their courtesy in listening to you.

SPEAKER'S CHECKLIST

This checklist covers the steps involved in preparing to give an oral presentation.

1. Have you assessed the speaking situation—the audience and purpose of the presentation? _____

2. Have you determined the content of your presentation? _____

3. Have you shaped the content into a form appropriate to your audience and purpose? _____

4. Have you prepared an outline? _____

5. Have you prepared graphic aids that are

   a. Clear and easy to understand? _____

   b. Easy to see? _____

6. Have you made sure that the presentation room will have the necessary equipment for the graphic aids? _____

7. Have your rehearsed the presentation so that it flows smoothly? _____

8. Have you checked that the presentation will be the right length? _____

EXERCISES

1. Prepare a five-minute oral version of your proposal for a final report topic.

2. After the report is written, prepare a five-minute oral presentation based on your findings and conclusions.

3. Write a memo to your instructor in which you analyze a recent oral presentation of a guest speaker at your college or a politician on television.

4. Prepare a five-minute extemporaneous presentation on a term or concept in your major field of study. Address the presentation to an audience that is unfamiliar with that term or concept.

5. Prepare a five-minute extemporaneous presentation explaining how to use a piece of equipment or how to carry out a procedure in your major field of study. Address the presentation to an audience that is unfamiliar with that piece of equipment or procedure.

CHAPTER EIGHTEEN

CORRESPONDENCE

The letter is the basic means of communication between two organizations; millions of letters are written each working day. Although the use of the telephone is constantly increasing, letters remain the basic communication link, because they provide documentary records. Often, phone conversations and transactions are immediately written up in the form of a letter, to become a part of the records of both organizations. Therefore, even as a new employee, you can expect to write letters regularly. And as you progress to positions of greater responsibility, you will write even more letters, for you will be representing your organization more often.

To a greater extent than any other kind of technical writing, letters represent the dual nature of the working world. On the one hand, a letter must be every bit as accurate and responsible as a legal contract. On the other hand, a letter is an individual communication between two people. Your reader will form an opinion of your organization based on the impression that you as a writer make. Regarding your reader as an individual person—while at the same time representing your organization effectively—is the challenge of writing good business letters.

THE "YOU ATTITUDE"

Like any other type of technical writing, the business letter should be clear, concise, accessible, correct, and accurate. The information it contains must be conveyed in a logical order. It should not contain small talk: the first paragraph should get directly to the point without wasting words. And to enable the reader to locate the information in the letter quickly and easily, the writer must make careful use of topic sentences at the start of the paragraphs. Often, writers use headings and indentation to make information more accessible, just as they would in a report. In fact, some writers use the term *letter report* to describe a technical letter of more than, say, two or three pages. In substance, it is a report; in form, it is a letter, containing all of the letter's traditional elements.

Moreover, because it is a personal communication from one person to another, a letter must also convey a courteous, positive tone. The key to this tone is sometimes called the "you attitude." This term means looking at the situation from your reader's point of view and adjusting the content, structure, and tone to meet that person's needs. The "you attitude" is largely common sense. If, for example, you are writing to a supplier who has failed to deliver some merchandise to you on the agreed-upon date, the "you attitude" dictates that you not discuss problems you are having with other suppliers—those problems don't concern your reader. Rather, you should concentrate on explaining clearly and politely to your reader that he or she has violated your agreement and that not having the merchandise is costing you money. Then you should propose ways to expedite the shipment of the merchandise.

Looking at things from the other person's point of view would be simple if both parties always saw things the same way. They don't, of course. Sometimes the context of the letter is a dispute. Despite the specifics of the situation, good letter writers always maintain a polite tone. Civilized behavior is good business.

Following are examples of thoughtless sentences, each followed by an improved version that exhibits the "you attitude."

EGOTISTICAL

Only our award-winning research and development department could have devised this revolutionary new sump pump.

BETTER

Our new sump pump features significant innovations.

BLUNT

You wrote to the wrong department. We don't handle complaints.

BETTER

Your letter has been forwarded to the Customer Service Division.

ACCUSING

You must have dropped the engine. The housing is badly cracked.

BETTER

The badly cracked housing suggests that your engine must have fallen onto a hard surface from some height.

SARCASTIC

You'll need two months to deliver these parts? Who do you think you are, the Post Office?

BETTER

A two-month delay for the delivery of the parts is unacceptable.

BELLIGERENT

I'm sure you have a boss, and I doubt if he'd like to hear about how you've mishandled our account.

BETTER

I would prefer to settle our account with you rather than having to bring it to your supervisor's attention.

CONDESCENDING

Haven't you ever dealt with a major corporation before? A 60-day payment period happens to be standard.

BETTER

We had assumed that you honored the standard 60-day payment period.

OVERSTATED

Your air-filter bags are awful. They're all torn. We want our money back.

BETTER

Nineteen of the 100 air-filter bags we purchased are torn. We would therefore like you to refund the purchase price of the 19 bags: $190.00

After you have drafted a letter, look back through it. Put yourself in your reader's place. How would you react if you received it? A calm, respectful, polite tone always makes the best impression and therefore increases your chances of achieving your goal.

THE ELEMENTS OF THE LETTER

Almost every business letter has a heading, inside address, salutation, body, complimentary close, signature, and reference initials. In addition, some letters contain one or more of the following notations: attention, subject, enclosure, and copy. These elements are discussed in the following paragraphs in the order in which they would ordinarily appear in a letter. Six common types of letters will be discussed in detail later in this chapter.

HEADING The typical organization has its own stationery, with its name, address, and phone number—the letterhead—printed on the top. The letterhead and the date on which the letter will be sent (typed two lines below the letterhead) make up the heading. When typing on blank paper, use your address (without your name) as the letterhead. Note that only the first page of any letter is typed on letterhead stationery. Type the second and all subsequent pages on blank paper, with the name of the recipient, the page number, and the date in the upper left-hand corner. For example:

```
Mr. John Cummings
Page 2
July 3, 19--
```

Do not number the first page of any letter.

INSIDE ADDRESS The inside address is your reader's name, position, organization, and business address. If your reader has a courtesy title, such as *Dr.*, *Professor*, or—for public officials—*Honorable*, use it. If not, use *Mr.* or *Ms.* (unless you know the reader prefers *Mrs.* or *Miss*). If your reader's position can be fit conveniently on the same line as his or her name, add it after a comma; otherwise, place it on the line below. Spell the name of the organization the way it does: for example, International Business Machines calls itself IBM. Include the complete mailing address: the street, city, state (using the Post Office abbreviations listed in Appendix C), and zip code.

ATTENTION LINE Sometimes you will be unable or unwilling to address the letter to a particular person. If you don't know (and cannot easily find out) the person's first name or don't know the person's name at all, use the attention line:

```
Attention: Technical Director
```

The attention line is also useful if you want to make sure that the organization you are writing to responds to your letter even if the person you would ordinarily write to is unavailable. In this case, the first line of the inside address contains the name of the organization or of one of its divisions:

```
Operations Department
Haverford Electronics
117 County Line Road
Haverford, MA 01765

Attention: Charles Fulbright, Director
```

SUBJECT LINE The subject line contains either a project number (for example, "Subject: Project 31402") or a brief phrase defining the subject of the letter (for example, "Subject: Price Quotation for the R13 Submersible Pump").

```
Operations Department
Haverford Electronics
117 County Line Road
Haverford, MA 01765

Attention: Charles Fulbright, Director

Subject: Purchase Order #41763
```

SALUTATION Assuming there is no attention line or subject line, the salutation is placed two lines below the inside address. The traditional salutation is *Dear* followed by the reader's courtesy title and last name. Use a colon after the name, not a comma. If you are fairly well acquainted with your reader, use *Dear* followed by the first name. When you do not know the name of the person to whom you are writing, use a general salutation:

```
Dear Technical Director:

Dear Sir or Madam:
```

When you are addressing the letter to a group of people, use one of the following salutations:

```
Ladies and Gentlemen:

Gentlemen: (if all the readers are male)

Ladies: (if all the readers are female)
```

Or you can tailor the salutation to your readers:

```
Dear Members of the Restoration Committee:

Dear Members of Theta Chi Fraternity:
```

This same strategy is useful for sales letters without individual inside addresses:

```
Dear Homeowner:

Dear Customer:
```

BODY The body is the substance of the letter. Although in some cases it might be only a few words long, generally it will be three or more paragraphs in length. The first paragraph introduces the subject of the letter, the second elaborates the message, and the third concludes it.

COMPLIMENTARY CLOSE After the body of the letter, include one of the traditional closing expressions: *Sincerely, Sincerely yours, Yours sincerely, Yours very truly, Very truly yours.* Notice that only the first word in the complimentary close is capitalized, and that all such expressions are followed by a comma. Today, all of the phrases have lost whatever particular meanings and connotations they once possessed. They can be used interchangeably.

SIGNATURE Type your full name on the fourth line below the complimentary close. Your signature, in ink, is written above the typewritten name. Some organizations prefer that you add, beneath your typed name, your position. For example:

```
Very truly yours,

Chester Hall

Chester Hall
Personnel Manager
```

REFERENCE LINE If someone else types your letters, the reference line identifies, usually by initials, both you and the typist. It appears a few spaces below the signature line, along the left margin. Generally, the writer's initials—which always come first—are capitalized, and the typist's initials are lowercase. For example, if Marjorie Connor wrote a letter that Alice Wagner typed, the standard reference notation would be MC/aw.

ENCLOSURE LINE If the envelope contains any documents other than the letter itself, identify the number of enclosures:

FOR ONE ENCLOSURE

```
Enclosure
```

OR

```
Enclosure (1)
```

FOR MORE THAN ONE ENCLOSURE

```
Enclosures (2)
Enclosures (3)
```

In determining the number of enclosures, count only the separate items, not the number of pages. A three-page memo and a ten-page report constitute only two enclosures. Some writers like to identify the enclosures by name:

```
Enclosure: 1984 Placement Bulletin

Enclosures (2): "This Year at Ammex"

                1983 Annual Report
```

COPY LINE If you want the reader of your letter to know that other people are receiving a copy of it, use the symbol *cc* (for "carbon copy"—even though photocopies have replaced carbon copies in many organizations) followed by the names of the other recipients (listed either alphabetically or according to organizational rank). If you do not want your reader to know about other copies, type *bcc* ("blind carbon copy") on the copies only—not on the original.

THE FORMAT OF THE LETTER

There are three popular formats used for business letters: modified block, modified block with paragraph indentations, and full block. Figures 18–1, 18–2, and 18–3 show letters written in these three formats. Read the letters carefully; they provide details on spacing, margins, and so forth.

TYPES OF LETTERS

No two communication situations in the technical world are identical. There are literally dozens of kinds of letters meant for specific occasions. This chapter could not possibly cover all of the types of letters that are sent routinely. It focuses, instead, on the six basic types of letters written most frequently in the technical

FIGURE 18–1

Modified Block Format

CENTERVILLE LIONS

CENTERVILLE, PA 15316

August 13, 19--

Mr. Albert Lusha
Director of Operations
Holiday Bookings, Inc.
119 Market Street
Orleans, MA 02653

Subject: China Tour #314

Dear Mr. Lusha:

This format is called the modified block form. All elements except the date, the complimentary close, and the signature begin at the left margin.

All letters should have margins of at least an inch on both sides and the bottom, and one line of space between the letterhead and the date. Letters are single-spaced, with double spacing between all elements except where shown on this letter.

The date should not extend into the right margin. Many writers begin the date at the horizontal center of the page. The complimentary close generally is aligned with the date.

Sincerely yours,

William Murphy

William Murphy
Centerville Lions

WM/mk

Enclosure

cc: Alice Nostrand
Fred Shipple

FIGURE 18–2

Modified Block Format, with Paragraph Indentations

CENTERVILLE LIONS

CENTERVILLE, PA 15316

August 13, 19--

Holiday Bookings, Inc.
119 Market Street
Orleans, MA 02653

Attention: Mr. Alan Hawkins

Subject: China Tour #314

Ladies and Gentlemen:

The name of this format is self-explanatory: the only modifications are the paragraph indentations. Each paragraph is indented five spaces.

Notice in this letter the use of the attention line, with a line of space above and below it, and the salutation. The letter is addressed to the company; therefore, the salutation is addressed to the group of people who work there.

Sincerely yours,

William Murphy

William Murphy
Centerville Lions

WM/mk

Enclosure

cc: Alice Nostrand
Fred Shipple

FIGURE 18-3
Full Block Format

CENTERVILLE LIONS
CENTERVILLE, PA 15316

August 13, 19--

Mr. Albert Lusha
Director of Operations
Holiday Bookings, Inc.
119 Market Street
Orleans, MA 02653

Dr. Mr. Lusha:

This is the full block format. All elements begin at the left margin. Obviously, it is the easiest format for the typist. However, it is not as popular as the two modified block formats, because it can look asymmetrical, especially if the body of the letter is brief.

Sincerely yours,

William Murphy

William Murphy
Centerville Lions

WM/mk

Enclosure

cc: Alice Nostrand
Fred Shipple

world: order, inquiry, response to inquiry, sales, claim, and adjustment. The transmittal letter is discussed in Chapter 11. The job-application letter is discussed in Chapter 19. For a more detailed discussion of business letters, consult one of the several full-length books on the subject.

THE ORDER LETTER

Perhaps the most basic form of business correspondence is the order letter, written to a manufacturer, wholesaler, or retailer. When writing an order letter, be sure to include all of the infor-

FIGURE 18–4
Order Letter

WAGNER AIRCRAFT
116 North Miller Road
Akron, OH 44313

September 4, 19--

Franklin Aerospace Parts
623 Manufacturer's Blvd.
Bethpage, NY 11741

Attention: Mr. Frank DeFazio

Gentlemen:

Would you please send us the following parts by parcel post. All page numbers refer to your 19-- catalog.

Quantity	Model No.	Catalog Page	Description	Price
2	36113-NP	42	Seal fins	$ 34.95
1	03112-Bx	12	Turbine bearing support	19.75
5	90135-QN	102	Turbine disc	47.50
1	63152-Bx	75	Turbine bearing housing	16.15
			Total Price:	$118.35

Yours very truly,

Christopher O'Hanlon

Christopher O'Hanlon
Purchasing Agent

mation your reader will need to identify the merchandise. Include the quantity, model number, dimensions, capacity, material, price, and any other pertinent details. Also specify the terms of payment (if other than payment in full upon receipt of the merchandise) and method of delivery. A typical order letter is shown in Figure 18–4. (Notice that the writer of this order letter uses an informal table to describe the parts he wishes to purchase.)

Many organizations have preprinted forms, called purchase orders, for ordering products or services. A purchase order calls for the same information that appears in an order letter.

THE INQUIRY LETTER

Your purpose in writing an inquiry letter is to obtain information from your reader. The difficulty of writing an inquiry letter is determined by whether the reader is expecting your letter.

If the reader is expecting the letter, your task is easy. For example, if a company that makes institutional furniture has advertised that it will send its 48-page, full-color brochure to prospective clients, you need write merely a one-sentence letter: "Would you please send me the brochure advertised in *Higher Education Today*, May 13, 19--?" The manufacturer knows why you're writing and, naturally, wants to receive letters such as yours, so no explanation is necessary. Similarly, a technical question, or set of questions, about any product or service for sale can be asked quickly. For instance, an inquiry letter might begin, "We are considering purchasing your new X-15 self-correcting typewriters for an office staff of 35 and would like some further information. Would you please answer the following questions?" The detail about the size of the potential order is not necessary, but it does make the point that the inquiry is serious and that the sale could be substantial. An inquiry letter of this kind will get a prompt and gracious reply.

If your reader is not expecting your letter, your task is more difficult. At times, you will want information from someone who will not directly benefit from supplying it to you. You must ask your reader simply to do you a favor. This situation requires careful, persuasive writing, for you must make your reader *want* to respond despite the absence of the profit motive.

In the first paragraph of this kind of letter, state why you decided to write to *this* person or organization, rather than any other organization that could supply the same or similar information. Subtle flattery is useful at this point—for example, "I was hoping that as the leader in solid-state electronics, you might be able to furnish some information about. . . ." Then explain why you want the information. Obviously, a company will not furnish information to a competitor. You have to show that your interests are not commercial—for instance, "I will be using this information in a senior project in electrical engineering at Illinois State University. I am trying to devise a. . . ." If you need the information by a certain date, this might be a good place to mention it: "The project is to be completed by April 15, 19--."

The bulk of the inquiry letter should be composed of a numbered list of the specific questions you want answered. Readers are understandably annoyed by thoughtless requests to send "everything you have" on a topic. They much prefer a set of carefully thought-out technical questions that show that the writer has already done substantial research. "Is your Model 311 compatible with Norwood's Model B12?" is obviously much easier to respond to than "Would you please tell me about your Model 311?" If your questions can be answered in a small space, it is a good idea to leave room for your reader's reply after each question or in the margin.

Because you are asking someone to do something for you, it is natural to offer to do something in return. In many cases, all you can offer are the results of your research. If possible, explicitly state that you would be happy to send your reader a copy of the report you are working on. And finally, express your appreciation. Don't say, "Thank you for sending me this information." Such a statement is presumptuous, because it assumes that the reader is both willing and able to meet your request. A statement such as the following is more effective: "I would greatly appreciate any help you could give me in answering these questions." Finally, if the answers will be brief, enclose a stamped self-addressed envelope for your reader's reply.

You should of course write a brief thank-you note to someone who has taken the time and trouble to respond to your inquiry letter.

Figure 18–5 provides an example of a letter of inquiry.

THE RESPONSE TO AN INQUIRY

If you ever receive an inquiry letter, keep the following suggestions in mind. If you can provide the information the writer asks for, do so graciously. If the questions were numbered, number your responses correspondingly. If you cannot answer the questions, either because you don't know the answers or cannot divulge them because the information is proprietary, explain this to the writer and offer to try to be of assistance with other requests. Figure 18–6 shows a response to the inquiry letter in Figure 18–5.

THE SALES LETTER

A large, sophisticated sales campaign costs millions of dollars—for marketing surveys and consulting fees, printing, postage, and promotions. This kind of campaign is beyond the scope of this book. However, it is not unusual for an employee to draft a sales letter for a product or service.

The "you attitude" is crucial in sales letters. Your readers are not interested in why you want to sell your product or service. They want to know why they should buy it. You are asking your readers to spend valuable time studying the letter. Provide clear, specific information to help them understand what you are selling and how it will help them. Be upbeat and positive in tone, but never forget that your readers are looking for facts.

A sales letter generally has a four-part structure. First, you

FIGURE 18-5
Inquiry Letter

14 Hawthorne Ave.
Belleview, TX 75234

November 2, 19--

Dr. Andrew Shakir
Director of Technical Services
Orion Corporation
721 West Douglas Avenue
Maryville, TN 31409

Dear Dr. Shakir:

I am writing to you because of Orion's reputation as a leader in the manufacture of adjustable X-ray tables. I am a graduate student in biomedical engineering at the University of Texas working on an analysis of diagnostic equipment. Would you be able to answer a few questions about your Microspot 311?

1. Can the Microspot 311 be used with lead oxide cassettes, or does it accept only lead-free cassettes?

2. Are standard generators compatible with the Microspot 311?

3. What would you say is the greatest advantage, for the operator, in using the Microspot 311? For the patient?

My project is due January 15. I would greatly appreciate your assistance in answering these questions. Of course, I would be happy to send you a copy of my report when it is completed.

Yours very truly,

Albert K. Stern

must gain the reader's attention. Unless the opening sentence seems either interesting or important, the reader will put the letter aside. To attract the reader, use interesting facts, quotations, or questions. In particular, try to identify a problem that will be of interest to your reader. A few examples of effective openings follow.

How much have construction costs risen since your plant was built? Do you know how much it would cost to rebuild at today's prices?

The Datafix copier is better than the Xerox--and it costs less, too. We'll repeat: it's better and it costs less!

FIGURE 18-6

Response to an Inquiry

ORION

721 WEST DOUGLAS AVE.
MARYVILLE TN 31409

(615) 619-8132
TECHNICAL SERVICES

November 7, 19--

Mr. Albert K. Stern
14 Hawthorne Ave.
Belleview, TX 75234

Dear Mr. Stern:

I would be pleased to answer your questions about the Microspot 311. We think it is the best unit of its type on the market today.

1. The 311 can handle lead oxide or lead-free cassettes.

2. At the moment, the 311 is fully compatible only with our Duramatic generator. However, special wiring kits are available to make the 311 compatible with our earlier generator models--the Olympus and the Saturn. We are currently working on other wiring kits.

3. For the operator, the 311 increases the effectiveness of the radiological procedure while at the same time cutting down the amount of film used. For the patient, it cuts down the number of repeat exposures and therefore reduces the total dose.

I am enclosing our brochure on the Microspot 311. If you would like copies, please let me know. I would be happy to receive a copy of your analysis when it is complete. Good luck!

Sincerely yours,

Andrew Shakir, M.D.
Director of Technical Services

AS/le

Enclosure
cc - Robert Anderson, Executive Vice-President

If you're like most training directors, we bet you've seen your share of empty promises. We've heard all the stories, too. And that's why we think you'll be interested in what Fortune said about us last month.

Second, describe the product or service you are trying to sell. What does it do? How does it work? What problems does it solve?

The Datafix copier features automatic loading, so your people don't waste time watching the copies come out. Datafix copies from a two-sided original--automatically! And Datafix can turn out 30 copies a minute--which is 25% faster than our fastest competitor. . . .

Third, convince your reader that your claims are accurate. Refer to users' experience, testimonials, or evaluations performed by reputable experts or testing laboratories.

In a recent evaluation conducted by Office Management Today, more than 85% of our customers said they would buy another Datafix. The next best competitor? 71%. And Datafix earned a "Highly Reliable" rating, the highest recommendation in the reliability category. All in all, Datafix scored higher than any other copier in the desk-top class. . . .

Finally, tell your reader how to find out more about your product or service. If possible, provide a postcard that the reader can use to request more information or arrange for a visit from one of your sales representatives. Make it easy to proceed to the next step in the sales process. Figure 18–7 provides an example of a sales letter.

THE CLAIM LETTER

A claim letter is a polite and reasonable complaint. If as a private individual or a representative of an organization you purchase a defective or falsely advertised product or receive inadequate service, your first recourse is a claim letter.

The purpose of the claim letter is to convince your reader that you are a fair and honest customer who is justifiably dissatisfied. If you can do this, your chances of receiving an equitable settlement are good. Most organizations today are very accommodating toward reasonable claims, because they realize that unhappy customers are bad business. In addition, claim letters provide a valuable index of the strong and weak points of an organization's product or service.

The claim letter has a four-part structure:

1. A specific identification of the product or service. List the model numbers, serial numbers, sizes, and any other pertinent data.
2. An explanation of the problem. State explicitly the symptoms. What function does not work? What exactly is wrong with the service?
3. A proposed adjustment. Define what you want the reader to do: for example, refund the purchase price, replace or repair the item, improve the service.
4. A courteous conclusion. Say that you trust that the reader, in the interest of fairness, will abide by your proposed adjustment.

Tone, the "you attitude," is just as important as content in a claim letter. You must project a calm and rational tone. A com-

FIGURE 18-7
Sales Letter

DAVIS TREE CARE
1300 Lancaster Avenue
Berwin, PA 19092

May 13, 19--

Dear Homeowner:

Do you know how much your trees are worth? That's right--your trees. As a recent purchaser of a home, you know how much of an investment your house is. Your property is a big part of your total investment.

Most people don't know that even the heartiest trees need periodic care. Like shrubs, trees should be fertilized and pruned. And they should be protected against the many kinds of diseases and pests that are common in this area.

At Davis Tree Care, we have the skills and experience to keep your trees healthy and beautiful. Our diagnostic staff is made up of graduates of major agricultural and forestry universities, and all of our crews attend special workshops to keep current with the latest information in tree maintenance. Add this to our proven record of 43 years of continuous service in the Berwyn area, and you have a company you can trust.

May we stop by to give you an analysis of your trees--absolutely without cost or obligation? A few minutes with one of our diagnosticians could prove to be one of the wisest moves you've ever made. Just give us a call at 865-9187 and we'll be happy to arrange an appointment at your convenience.

Sincerely yours,

Daniel Davis III
President

plaint such as "I'm sick and tired of being ripped off by companies like yours" leaves your reader no way of knowing whether your claim is accurate. If, however, you write, "I am very disappointed in the performance of my new Eversharp Electric Razor," you sound like a responsible adult. There is no reason to show anger in a claim letter, even if the other party has made an unsatisfactory response to an earlier claim letter of yours. Calmly explain what you plan to do, and why. Your reader then will be much more likely to try to see the situation from your perspective. Figure 18–8 provides an example of a claim letter.

FIGURE 18-8
Claim Letter

ROBBINS CONSTRUCTION, INC.
255 Robbins Place Centerville, MO 65101 (417) 934-1850

August 19, 19--

Mr. David Larsen
Larsen Supply Company
311 Elmerine Avenue
Anderson, MO 63501

Dear Mr. Larsen:

As steady customers of yours for over 15 years, we came to you first when we needed a quiet pile driver for a job near a residential area. On your recommendation, we bought your Vista 500 Quiet Driver, at $14,900. We have since found, much to our embarrassment, that it is not substantially quieter than a regular pile driver.

We received the contract to do the bridge repair here in Centerville after promising to keep the noise to under 90 db during the day. The Vista 500 (see enclosed copy of bill of sale for particulars) is rated at 85db, maximum. We began our work and, although one of our workers said the driver didn't seem sufficiently quiet to him, assured the people living near the job site that we were well within the agreed sound limit. One of them, an acoustical engineer, marched out the next day and demonstrated that we were putting out 104 db. Obviously, something is wrong with the pile driver.

I think you will agree that we have a problem. We were able to secure other equipment, at considerable inconvenience, to finish the job on schedule. When I telephoned your company that humiliating day, however, a Mr. Meredith informed me that I should have done an acoustical reading on the driver before I accepted delivery.

THE ADJUSTMENT LETTER

In an adjustment letter, you respond to a customer's claim letter. The purpose of the adjustment letter is to tell the customer how you plan to handle the situation. Whether you are granting the customer everything that the claim letter proposed, part of it, or none of it, your purpose remains the same: to show the customer that your organization is fair and reasonable, and that you value his or her business.

If you are able to grant the customer's request, the letter will be simple to write. Express your regret about the situation, state the adjustment you are going to make, and end on an upbeat note

Mr. David Larsen
Page 2
August 19, 19--

I would like you to send out a technician--as soon as possible--either to repair the driver so that it performs according to specifications or to take it back for a full refund.

Yours truly,

Jack Robbins, President

JR/lr

Enclosure

that will encourage the customer to continue doing business with you.

If you cannot grant the customer's request, you must try to salvage as much goodwill as you can from the situation. Obviously, your reader is not going to be happy. If your letter is carefully written, however, he or she might at least believe that you have acted reasonably. In denying a request, you are attempting to explain your side of the matter and thus to educate your reader about how the problem occurred and how to prevent it in the future.

This more difficult kind of adjustment letter generally has a four-part structure:

1. An attempt to meet the customer on some neutral ground. Often, an expression of regret—never an apology!—is used. Never admit that the customer is right in this kind of claim letter. If you write, "We are sorry that the engine you purchased from us is defective," the customer would have a good case against you if the dispute ended up in court. Sometimes, the writer will even thank the customer for bringing the matter to the attention of the company.
2. An explanation of why your company is not at fault. Most often, you explain to the customer the steps that led to the failure of the product or service. Do not say, "You caused this." Instead, use the less blunt passive voice: "The air pressure apparently was not monitored. . . ."
3. A clear statement that your company, for the above-mentioned reasons, is denying the request. This statement must come later in the letter. If you begin with it, most readers will not finish the letter, and therefore you will not be able to achieve your twin goals of education and goodwill.
4. An attempt to create goodwill. Often, an organization will offer a special discount if the customer buys another, similar product. A company's profit margin on any one item is almost always large enough that the company can offer the customer very attractive savings as an inducement to continue doing business with it.

Figures 18–9 and 18–10 show examples of "good news" and "bad news" adjustment letters. The first letter is a reply to the claim letter shown in Figure 18–8.

WRITER'S CHECKLIST

The following checklist covers the basic letter format and the six types of letters discussed in the chapter.

LETTER FORMAT

1. Is the first page of the letter typed on letterhead stationery? _____
2. Is the date included? _____
3. Is the inside address complete and correct? Is the appropriate courtesy title used? _____
4. If appropriate, is an attention line included? _____
5. If appropriate, is a subject line included? _____
6. Is the salutation appropriate? _____

FIGURE 18-9
"Good News" Adjustment Letter

Larsen Supply Company
311 Elmerine Avenue
Anderson, MO 63501

August 21, 19--

Mr. Jack Robbins, President
Robbins Construction, Inc.
255 Robbins Place
Centerville, MO 65101

Dear Mr. Robbins:

I was very unhappy to read your letter of August 19 telling me about the failure of the Vista 500. I regretted most the treatment you received from one of my employees when you called us.

Harry Rivers, our best technician, has already been in touch with you to arrange a convenient time to come out to Centerville to talk with you about the driver. We will of course repair it, re-place it, or refund the price. Just let us know your wish.

I realize that I cannot undo the damage that was done on the day that a piece of our equipment failed. To make up for some of the extra trouble and expense you incurred, let me offer you a 10% discount on your next purchase or service order with us, up to $1,000 total discount.

You have indeed been a good customer for many years, and I would hate to have this unfortunate incident spoil that relationship. Won't you give us another chance? Just bring in this letter when you visit us next, and we'll give you that 10% discount.

Sincerely,

Dave Larsen

Dave Larsen, President

7. Is the complimentary close typed with only the first word capitalized? Is the complimentary close followed by a comma? _____

8. Is the signature legible and the writer's name typed beneath the signature? _____

9. If appropriate, are the reference initials included? _____

10. If appropriate, is an enclosure line included? _____

11. If appropriate, is a copy line included? _____

12. Is the letter typed in one of the standard formats? _____

FIGURE 18-10
"Bad News" Adjustment Letter

PETERSON SNOWMOBILES
600 Midway Drive
Sawyer, NE 69101

February 3, 19--

Mr. Dale Devlin
1903 Highland Avenue
Glenn Mills, NE 69032

Dear Mr. Devlin:

Thank you for writing to us about the problem with your Eskimo Snowmobile.

Our Service Department found water in the fuel line of your snowmobile. Apparently some of the gasoline was bad. While we guarantee our snowmobile for a period of one year against defects in workmanship and materials, we cannot assume responsibility for problems caused by bad gasoline. Therefore, we are unable to grant your request to repair the snowmobile free of charge.

Fortunately, no serious harm was done to the snowmobile. We would be happy to flush the fuel line at cost, $30. Your Eskimo would then give you many years of trouble-free service.

If you will authorize us to do this work, we will have your snowmobile back to you within four working days. Just fill out the enclosed card and drop it in the mail.

Yours very truly,

Bart Peterson

Bart Peterson
Peterson Snowmobiles

TYPES OF LETTERS

1. Does the order letter
   a. Include the necessary identifying information, such as quantities and model numbers? ______
   b. Specify, if appropriate, the terms of payment? ______
   c. Specify the method of delivery? ______
2. Does the inquiry letter
   a. Explain why you chose the reader to receive the inquiry? ______
   b. Explain why you are requesting the information and to what use you will put it? ______
   c. Specify by what date you need the information? ______

d. List the questions clearly and, if appropriate, provide room for the reader's response? _____

e. Offer, if appropriate, the product of your research? _____

3. Does the response to an inquiry letter answer the reader's questions or explain why they cannot be answered? _____

4. Does the sales letter

a. Gain the reader's attention? _____

b. Describe the product or service? _____

c. Convince the reader that the claims are accurate? _____

d. Encourage the reader to find out more about the product or service? _____

5. Does the claim letter

a. Identify specifically the unsatisfactory product or service? _____

b. Explain the problem(s) clearly? _____

c. Propose an adjustment? _____

d. Conclude courteously? _____

6. Does the "good-news" adjustment letter

a. Express your regret? _____

b. Explain the adjustment you will make? _____

c. Conclude on a positive note? _____

7. Does the "bad-news" adjustment letter

a. Meet the reader on neutral ground, expressing regret but not apologizing? _____

b. Explain why the company is not at fault? _____

c. Clearly deny the reader's request? _____

d. Attempt to create goodwill? _____

EXERCISES

1. Write an order letter to John Saville, general manager of White's Electrical Supply House (13 Avondale Circle, Los Angeles, CA 90014). These are the items you want: one SB11 40-ampere battery backup kit, at $73.50; twelve SW402 red wire kits, at $2.50 each; ten SW400 white wire kits, at $2.00 each; and one SB201 mounting hardware kit, at $7.85. Invent any reasonable details about methods of payment and delivery.

2. Secure the graduate catalog of a university offering a graduate program in your field. Write an inquiry letter to the appropriate representative, asking at least three questions about the program the university offers.

3. You are the marketing director of the company that publishes this book. Draft a sales letter that might be sent to teachers of the course you are presently taking.

4. You are the marketing director of the company that makes your bicycle (calculator, stereo set, running shoes, etc.). Write a sales letter that might be sent to retailers to encourage them to sell the product.

5. You are the recruiting officer for your college. Write a letter that might be sent to juniors in local high schools to encourage them to apply to the college when they are seniors.

6. You purchased four "D"-size batteries for your cassette player, and they didn't work. The package they came in says that the manufacturer will refund the purchase price if you return the defective items. Inventing any reasonable details, write a claim letter asking for not only the purchase price, but other related expenses.

7. A thermos you just purchased for $8.95 has a serious leak. The grape drink you put in it ruined a $15.00 white tablecloth. Inventing any reasonable details, write a claim letter to the manufacturer of the thermos.

8. The gasoline you purchased from New Jersey Petroleum contained water and particulate matter; after using it, you had to have your automobile tank flushed at a cost of $50. You have a letter signed by the mechanic explaining what he found in the bottom of your tank. As a credit-card customer of New Jersey Petroleum, write a claim letter. Invent any reasonable details.

9. As the recipient of one of the claim letters described in Exercises 6 through 8, write an adjustment granting the customer's request.

10. You are the manager of a private swimming club. A member has written saying that she lost a contact lens (value $55) in your pool. She wants you to pay for a replacement. The contract that all members sign explicitly states that the management is not responsible for loss of personal possessions. Write an adjustment letter denying the request. Invent any reasonable details.

11. As manager of a stereo equipment retail store, you guarantee that you will not be undersold. If a customer who buys something from you can prove within one month that another retailer sells the same equipment for less money, you will refund the difference in price. A customer has written to you, enclosing an ad from another store showing that it is selling the equipment he purchased for $26.50 less than he paid at your store. The advertised price at the other store was a one-week sale that began five weeks after the date of his purchase from you. He wants his $26.50 back. Inventing any reasonable details, write an adjustment letter denying his request. You are willing, however, to offer him a blank cassette tape worth $4.95 for his equipment if he would like to come pick it up.

12. The following letters could be improved in both tone and substance. Revise them to increase their effectiveness, adding any reasonable details.

a.

Modern Laboratories, Inc.
DEAUVILLE, IN 43504

July 2, 19--

Adams Supply Company
778 North Henson Street
Caspar, IN 43601

Gentlemen:

Would you please send us the following items:

> one dozen petri dishes
> one gross pyrex test tubes
> three bunsen burners

Please bill us.

Sincerely,

Corey Dural

CD/kw

---

b.

1967 Sunset Avenue
Rochester, N. Y. 06803

November 13, 19--

Admissions Department
University of Pennsylvania Law School
Philadelphia, Pa. 19106

Gentlemen:

I am a senior considering going to law school. Would you please answer the following questions about your law school:

1. How well do your graduates do?
2. Is the LSAT required?
3. Do you have any electives, or are all the courses required?
4. Are computer skills required for admission?

A swift reply would be appreciated. Thank you.

Sincerely yours,

Eileen Forster

Eileen Forster

c.

14 Wilson Avenue
Wilton, ME 04949

November 13, 19--

Union Pacific Railroad
100 Columbia Street
Seattle, WA 98104

Attention: The President

Dear Sir:

I'm a college student doing a report on the future of mass transportation in the United States. Would you please tell me how many passenger miles your railroad traveled last year, the number of passengers, whether this was up or down from the year before, and the socioeconomic level of your riders?

My report is due in a week and a half. If you sent your answer in the next two days or so I'd be able to include your information in my report. Thank you in advance.

Very truly yours,

Jon Radley

Jonathan Radley

d.

Cutlass Vacuums, Inc.
Ridge Pike / Speonk, OR 97203

January 13,19--

Dear Service Station Owner:

I don't have to tell you that an indispensable tool for your garage is a good industrial-strength vacuum cleaner. It not only makes your garage better looking, but it makes it safer, too.

We at Cutlass are proud to introduce our brand-new Husky 450, which replaces our model 350. The 450 is bigger, so it has a greater suction power. It has a bigger receptacle, too, so you don't have to empty it as often.

I truly believe that our Husky 450 is the shop vacuum you've been looking for. If you would like further information about it, don't miss our ad in leading car magazines.

Yours truly,

Bob Wheeler

Bob Wheeler, President
Cutlass Vacuums, Inc.

e.

GUARDSMAN PROTECTIVE EQUIPMENT, INC.

3751 PORTER STREET
NEWARK, DE 19304

April 11, 19--

Dear Smith Family:

A rose is a rose is a rose, the poet said. But not all home protection alarms are alike. In a time when burglaries are skyrocketing, can you afford the second-best alarm system?

Your home is your most valuable possession. It is worth far more than your car. And if you haven't checked your house insurance policy recently, you'll probably be shocked to see how inadequate your coverage really is.

The best kind of insurance you can buy is the Watchdog Alarm System. What makes the Watchdog unique is that it can detect intruders before they enter your home and scare them away. Scaring them away while they're still outside is certainly better than scaring them once they're inside, where your loved ones are.

At less than two hundred dollars, you can purchase real peace of mind. Isn't your family's safety worth that much?

If you answered yes to that question, just mail in the enclosed postage-paid card, and we'll send you a 12-page, fact-filled brochure that tells you why the Watchdog is the best on the market.

Very truly yours,

Jerry Wexler

Jerry Wexler, President

f.

19 Lowry's Lane
Morgan, TN 30610

April 13, 19--

Sea-Tasty Tuna
Route 113
Lynchburg, TN 30563

Gentlemen:

I've been buying your tuna fish for years, and up to now it's been OK.

But this morning I opened a can to make myself a sandwich. What do you think was staring me in the face? A fly. That's right, a housefly. That's him you see taped to the bottom of this letter.

What are you going to do about this?

Yours very truly,

Seth Reeves

Seth Reeves

g.

Handyman Hardware, Inc.
Millersville, AL 61304

December 4, 19--

Hefty Industries, Inc.
19 Central Avenue
Dover, TX 76104

Gentlemen:

I have a problem I'd like to discuss with you. I've been carrying your line of hand tools for many years.

Your 9" pipe wrench has always been a big seller. But there seems to be something wrong with its design. I have had three complaints in the last few months about the handle snapping off when pressure is exerted on it. In two cases, the user cut his hand, one seriously enough to require nineteen stitches.

Frankly, I'm hesitant to sell any more of the 9" pipe wrenches, but I still have over two dozen in inventory.

Have you had any other complaints about this product?

Sincerely yours,

Peter Arlen, Manager
Handyman Hardware

PA/sc

h.

Sea-Tasty Tuna
Route 113
Lynchburg, TN 30563

April 21, 19--

Mr. Seth Reeves
19 Lowry's Lane
Morgan, TN 30610

Dear Mr. Reeves:

We were very sorry to learn that you found a fly in your tuna fish.

Here at Sea-Tasty we are very careful about the hygiene of our plant. The tuna are scrubbed thoroughly as soon as we receive them. After they are processed, they are inspected visually at three different points. Before we pack them, we rinse and sterilize the cans to ensure that no foreign material is sealed in them.

Because of these stringent controls, we really don't see how you could have found a fly in the can. Nevertheless, we are enclosing coupons good for two cans of Sea-Tasty tuna.

We hope this letter restores your confidence in us.

Truly yours,

Valarie Lumaris

Valarie Lumaris
Customer Service Representative

VL/ck

Enclosures

i.

Hefty Industries, Inc.
19 Central Avenue
Dover, TX 76104

December 11, 19--

Mr. Peter Arlen, Manager
Handyman Hardware, Inc.
Millersville, AL 61304

Dear Mr. Arlen:

Thank you for bringing this matter to our attention.

In answer to your question--yes, we have had a few complaints about the handle snapping on our 9" pipe wrench.

Our engineers brought the wrench back to the lab and discovered a design flaw that accounts for the problem. We have redesigned the wrench and have not had any complaints since.

We are not selling the old model anymore because of the risk. Therefore we have no use for your two dozen. However, since you have been a good customer, we would be willing to exchange the old ones for the new design.

We trust this will be a satisfactory solution.

Sincerely,

Robert Panofsky, President
Hefty Industries, Inc.

CHAPTER NINETEEN

THE JOB-APPLICATION PROCEDURE

For most of you, the first non-academic test of your technical writing skills will come when you make up your job-application materials. These materials will provide employers with information about your academic and employment experience, personal characteristics, and reasons for applying for the position. But they provide a lot more, too. Employers have learned that one of the most important skills an employee can bring to a job is the ability to communicate effectively. Therefore, potential employers look carefully for evidence of writing skills. Job-application materials pose a double hurdle: employers want to know both what you can do and how well you can communicate.

Some students think that once they get a satisfactory job, they will never again have to worry about résumés and application letters. Statistics show that they are wrong. The typical professional changes jobs more than ten times in the course of his or her career. Although this chapter pays special attention to the student's first job hunt, the skills and materials discussed here also apply to established professionals who wish to change jobs.

FIVE WAYS TO GET A JOB

There are five traditional ways to get a professional-level position:

1. through a college or university placement office

2. through a professional placement bureau
3. through a published job ad
4. through an unsolicited letter to an organization
5. through connections

Almost all colleges and universities have placement offices, which bring hiring organizations and students together. Generally, students submit a résumé—a brief listing of credentials—to the placement office. The résumés are then made available to representatives of business and industry. After studying the résumés, these representatives use the placement office to arrange on-campus interviews with selected students. Those who do best in the campus interviews are then invited by the representatives to visit the organization for a tour and another interview. Sometimes a third interview is scheduled; sometimes an offer is made immediately or shortly thereafter. The advantage of this system is twofold: first, it is free; second, it is easy. The student merely has to deliver a résumé to the placement office and wait to be contacted.

A professional placement bureau offers essentially the same service as a college placement office, but it charges either the employer (or, rarely, the new employee) a fee—often, 10% of the first year's salary—when the client accepts a position. Placement bureaus cater primarily to more advanced professionals who are looking to change jobs.

Published job ads generally offer the best opportunity for both students and professionals. Organizations advertise in three basic kinds of publications: public-relations catalogs (such as *College Placement Annual*), technical journals, and newspapers. If you are looking for a job, you should regularly check the major technical journals in your field and the large metropolitan newspapers—especially the Sunday editions. In responding to an ad, you must include with the résumé a job-application letter—one that highlights the crucial information on the résumé.

You need not wait for a published ad. You can write unsolicited letters of application to organizations you would like to work for. The disadvantage of this technique is obvious: there might not be an opening. Yet many professionals favor this technique, because there are fewer competitors for those jobs that do exist, and sometimes organizations do not advertise all available positions. And sometimes an impressive unsolicited application can prompt an organization to create a position for you. Before you write an unsolicited application, make sure you learn as much as you can about the organization (see Chapter 3): current and anticipated major projects, hiring plans, and so forth. You should know as much as you can about any organization you are applying to, of course, but when you are submitting an unsolicited application you have no other source of information on which to base your résumé and letter.

The use of connections to get a job is effective if you have a relative or an acquaintance in a position to exert some influence or at least point out that a position might be opening up. Other good sources of contacts include employers from your past summer jobs and faculty members in your field.

The college placement office, published ads, unsolicited letters, and connections are most useful for students. Too many students rely solely on the placement office, thereby limiting the range of possibilities to those organizations that choose to visit the college. The best system is to use the placement office and respond to published ads. If an attractive organization is not advertising, send an unsolicited application letter after telephoning them to find out the name of the appropriate person and department.

THE RÉSUMÉ

Whether it is to be submitted to a college placement office or sent along with a job-application letter, the résumé communicates in two ways: by its appearance and by its content.

APPEARANCE OF THE RÉSUMÉ

Because in almost all cases potential employers see the résumé before they see the person who wrote it, the résumé has to make a good first impression. Employers believe—often correctly—that the appearance of the résumé reflects the professionalism of the writer. When employers see a sloppy résumé, they assume that the writer would do sloppy work. A neat résumé implies that the writer would do professional work. When employers look at a résumé, therefore, they see more than a single piece of paper; they see dozens of documents they will be reading if they hire the writer.

Some colleges and universities advise students to have their résumés professionally printed. A printed résumé is attractive, and that's good—provided, of course, that the information on it is consistent with its professional appearance. Most employers agree, however, that a neatly typed résumé photocopied on good-quality paper is just as effective.

People who type and photocopy their résumés are more likely to tailor different versions to the needs of the organizations to which they apply—a good strategy. People who go to the trouble and expense of a professional printing job are far less likely to make up different résumés; they tend to submit the printed one to all of their prospects. The résumé looks so good that they feel it is not worth tinkering with. This strategy is dangerous, for it encourages the writer to underestimate the importance of directing the content of the résumé to a specific audience.

However they are reproduced, résumés should appear neat and professional. They should have

1. *Generous margins.* Leave a one-inch margin on all four sides.
2. *Clear type.* Use a typewriter with clear, sharp, unbroken letters. Avoid strikeovers and obvious corrections.
3. *Symmetry.* Arrange the information so that the page has a balanced appearance.
4. *Adequate white space.* Avoid a cluttered, packed arrangement of information.

CONTENT OF THE RÉSUMÉ

Although different experts advocate different approaches to résumé writing, everyone agrees on two things.

First, the résumé must be completely free of errors. Grammar, punctuation, usage, and spelling errors undercut your professionalism by casting doubt on the accuracy of the information contained in the résumé. Ask for assistance after you have written the draft, and proofread the finished product at least twice.

Second, the résumé must provide clear and specific information, without generalizations or self-congratulation. Your résumé is a sales document, but you are both the salesperson and the product. You cannot say, "I am a terrific job candidate," as if the product were a toaster or an automobile. Instead, you have to provide the specific details that will lead the reader to the conclusion that you are a terrific job candidate. You must *show* the reader. Telling the reader is graceless and, worse, unconvincing.

A résumé should be long enough to include all of the pertinent information but not so long as to bore or irritate the reader. Generally, a student's résumé should be kept to one page. If, however, the student has special accomplishments to describe—such as journal articles or patents—a two-page résumé is appropriate. If your information comes to just over one page, either eliminate or condense some material to make it fit onto one page, or modify the physical layout of the résumé so that it fills a substantial part of the second page.

ELEMENTS OF THE RÉSUMÉ

Almost every résumé has five basic sections:

1. identifying information
2. education
3. employment
4. personal information
5. references

But your résumé should be geared to one particular person: you. Many people have some special skills or background that could be conveyed in additional sections. These sections, as discussed below, should be used where appropriate.

IDENTIFYING INFORMATION Your full name, address, and phone number should always be placed at the top of the page. Generally, you should present your name in full capitals, centered at the top. Use your complete address, with the zip code. The two-letter state abbreviations used by the Post Office (see Appendix B) are now preferred. Also use your complete phone number, with the area code. For the telephone exchange, numbers are preferred to letters.

Students who have two addresses and phone numbers should make sure that both are listed and identified clearly. Often, an employer will try to call a student during an academic holiday to arrange an interview.

EDUCATION The education section usually follows the identifying information on the résumé of a student or a recent graduate. People with substantial professional experience usually place the employment experience section before the education section.

The following information is included in the education section: the degree, the institution, its location, and the date of graduation. After the degree abbreviation (such as B.S., B.A., A.A., or M.S.), list the academic major (and, if you have one, the minor)—for example, "B.S. in Materials Engineering." Identify the institution by its full name: "The Pennsylvania State University," not "Penn State." Also include the city and state of the institution. If your degree has not yet been granted, write "Anticipated date of graduation" or "Degree expected in" before the month and year.

You should also list any other institutions you attended beyond the high school level—even those at which you did not earn a degree. Students are sometimes uneasy about listing community colleges or junior colleges; they shouldn't be. Employers are generally impressed to learn that a student began at a smaller or less advanced school and was able to transfer to a four-year college or university. The listing for other institutions attended should include the same information as the main listing. Arrange the entries in reverse chronological order: that is, list first the school you attended most recently.

In addition to this basic information, many students and recent graduates like to include more details about their educational experiences. They feel, quite correctly, that the basic information alone implies that they merely endured an institution and received a degree. Of the several ways for you to fill out the education section, perhaps the easiest is to list your grade-point average (provided that it is substantially above the median for the graduating class). Or list your average in the courses in your major, if that is more impressive.

You can also expand the education section by including a list of courses that would be of particular interest to the reader. Advanced courses in an area of your major concentration might be appropriate, especially if the potential employer has mentioned that area in the job advertisement. Also useful would be a list of communications courses—technical writing, public speaking, organizational communications, and the like. Employers are looking for people who can communicate what they know. A listing of business courses on an engineer's résumé—or the reverse—also shows special knowledge and skills. The only kind of course listing that is *not* particularly helpful is one that merely names the traditional required courses for your major. Make sure to list courses by title, not by number; employers won't know what Chemistry 250 is.

Another useful way to amplify the education section of the résumé is to describe a special accomplishment. If you did a special senior design or research project, for example, a two- or three-line description would be informative. Include in this description the title and objective of the project, any special or advanced techniques or equipment you used, and—if they are known—the major results. Such a description might be phrased as follows: "A Study of Composite Substitutes for Steel—a senior design project intended to formulate a composite material that can be used to replace steel in car axles." Even a traditional academic course in which you conducted a sophisticated project and wrote a sustained report can be described profitably. A project discussion makes you look less like a student—someone who takes courses—

and more like a professional—someone who designs and conducts projects.

Finally, you also can list in the education section any honors and awards you received. Scholarships, internships, and academic awards all offer evidence of an exceptional job candidate. If you have received a number of such honors, or some that were not exclusively academic, it might be more effective to list them separately (in a section called "Honors" or "Awards") rather than in a subsection of the education section. Often, some body of information could logically be placed in two or even three different locations, and you must decide where it will make the best impression.

The education section is the easiest part of the résumé to adapt for different positions. For example, a student majoring in electrical engineering who is applying for a position that calls for strong communications skills can draw up a list of communications courses in one version of his or her résumé. The same student can use a list of advanced electrical engineering courses in a résumé directed to another potential employer. As you compose the education section of your résumé, consider carefully what aspects of your background can be emphasized to address the needs of the particular job opening.

EMPLOYMENT The employment section, like the education section, conveys at least the basic information about each job you've held: the dates of employment, the organization's name and location, and your position or title. This information is self-explanatory.

However, a skeletal listing of nothing more than these basic facts would not be very informative or impressive. As in the education section, you should provide carefully selected details.

What readers want to know, after they have learned where and when you were employed, is what you actually did. Therefore, you should provide a two- to three-line description for each position. Focus the description on one or more of the following factors:

1. *Reports.* What kinds of documents did you write or assist in writing? List, especially, various governmental forms and any long reports.

2. *Clients.* What kinds of, and how many, clients did you transact business with as a representative of your organization?

3. *Skills.* What kinds of technical skills did you learn or practice in doing the job?

4. *Equipment.* What kinds of technical equipment did you operate or supervise? Mention, in particular, any computer skills you demonstrated, for they can be very useful in almost every kind of position.

5. *Money.* How much money were you responsible for? Even if you considered your bookkeeping position fairly simple, the fact that the organization grossed, say, $2 million a year shows that the position involved real responsibility.

6. *Personnel.* How many personnel were under your supervision? Students sometimes supervise small groups of other students or technicians. Supervising, naturally, shows maturity and responsibility.

Here is a sample listing:

```
June-September 19--: Millersville General Hospital, Millersville,
                     TX. Student Dietician. Gathered dietary histo-
                     ries and assisted in preparing menus for a 300-
                     bed hospital. Received "excellent" on all
                     items in evaluation by head dietician.
```

In just a few lines, you can show that you sought and accepted responsibility and that you acted professionally. Do not write, "I accepted responsibility"; rather, present facts that lead inevitably to that impression.

Naturally, not all jobs entail such professional skills and responsibilities. Many students find summer work as laborers, sales clerks, short-order cooks, and so forth. If you have not had a professional position, list the jobs you have had, even if they were completely unrelated to your career plans. If the job title is self-explanatory—such as waitress or service station attendant—don't elaborate. Every job is valuable. You learn that you are expected to be someplace at a specific time, wear appropriate clothes, and perform some specific duties. Also, every job helps pay college expenses. If you can say that you earned, say, 50% of your annual expenses through a job, employers will be impressed by your self-reliance. Most of them probably started out with nonprofessional positions. And don't forget that any job you have held can yield a valuable reference.

The various jobs should be listed and described in reverse chronological order on the résumé, to highlight those positions you have held most recently.

One further word: if you have held a number of nonprofessional positions as well as several professional positions, the nonprofessional ones can be grouped together in one listing:

```
Other Employment: Cashier (summer, 1983), salesperson (part-time,
                  1983), clerk (summer, 1982).
```

This technique prevents the nonprofessional positions from drawing the reader's attention away from the more important positions.

PERSONAL INFORMATION This section of the résumé has changed considerably in recent years. Only a decade ago, most résumés would list the writer's height, weight, date of birth, and marital status. Most résumés today include none of these items. One explanation for this change is that federal legislation now prohibits organizations from requiring this information from applicants. Perhaps a more important explanation is that most people have come to feel that such personal information is irrelevant to a person's ability to perform a job effectively.

The personal information section of the résumé *is* the appropriate place for a few items about your outside interests. Participation in community service organizations—such as Big Brothers—or volunteer work in a hospital is, of course, an extremely positive factor. Hobbies that are in some way related to your career interests—for example, amateur electronics for an engineer—are useful, too. You also should list any sports, especially those that might be socially useful in your professional career, such as tennis, racquetball, and golf. Any university-sanctioned activity—such as membership on a team, participation on the college newspaper, or election to a responsible position in an academic organization or a residence hall—should also be pointed out. In general, do not include activities that might create a negative impression in the reader: hunting, gambling, performing in a rock band. And always omit such activities as reading and meeting people—everybody reads and meets people.

REFERENCES You may list the names of three or four referees—people who have written letters of recommendation or who have agreed to speak on your behalf—on your résumé. Or you may simply say that you will furnish the names of the referees upon request. The length of your résumé sometimes dictates which style to use. If the résumé is already long, the abbreviated form might be preferable. If it does not fill out the page, the longer form might be the one to use. However, each style has advantages and disadvantages that you should consider carefully.

Furnishing the referees' names appears open and forthright. It shows that you have already secured your referees and have nothing to hide. If one or several of the referees are prominent people in their fields, the reader is likely to be impressed. And, perhaps most important, the reader can easily phone the referees or write them a letter. Listing the referees makes it easy for the prospective employer to proceed with the hiring process. The only disadvantage of this style is that it takes up space on the résumé that might be needed for other information.

Writing "References will be furnished upon request," on the other hand, takes up only one line on the résumé. In addition, it leaves you in a flexible position. You can still secure referees after you have sent out the résumé. Also, you can send selected letters of reference to prospective employers, based on your analysis of what they are looking for. Using different references for different positions is sometimes just as valuable as sending different résumés. However, some readers will interpret the nonlisting style as evasive or secretive or assume that you have not yet asked prospective referees. A greater disadvantage is that if the readers are impressed by the résumé and want to learn more about you, they cannot do so quickly and directly.

If you decide to include the listing, identify every referee by name, title, organization, mailing address, and phone number. For example:

```
Dr. Robert Ariel
Assistant Professor of Biology
Central University
Portland, OR 97202
(503) 666-6666
```

Remember that a careful choice of referees is as important as careful writing of the résumé. Solicit references only from persons who know your work best and for whom you have done the best work—for instance, a professor from whom you have received A's. It is unwise to ask prominent professors who do not know your work well, for the advantage of having a famous name on the résumé will be offset by the referee's brief and uninformative letter. Often, a young and less-well-known professor can write the most informative letter or provide the best recommendation. Try also to have at least one reference letter from an employer, even if the job was not professional.

After you have decided whom to ask, give the potential referee an opportunity to decline the request gracefully. Sometimes the person has not been as impressed with your work as you think. Also, if you simply ask, "Would you please write a reference letter for me?" the potential referee might accept and then write a lukewarm letter. It is better to follow up the first question with

"Would you be able to write an enthusiastic letter for me?" or "Do you feel you know me well enough to write a strong recommendation?" If the potential referee shows any signs of hesitation or reluctance, you can withdraw the invitation at that point. The scene is a little embarrassing, but it is better than receiving a half-hearted recommendation.

OTHER ELEMENTS The discussion so far has concentrated on the sections that appear on virtually everyone's résumé. Other sections are either discretionary or are appropriate only for certain writers.

A statement of objectives, in the form of a brief sentence—for example, "Objective: Entry-level accounting position with medium-size public accounting firm"—can be placed near the top of the résumé, generally right below the identifying information. A statement of objectives gives the appearance that the student has a clear sense of direction and goals, which is usually good. Sometimes, however, employers think it is unrealistic for students to state their goals with such self-assurance and certainty. Most students, in fact, have had relatively little experience doing the job for which they have been studying. As a result, their career goals often change radically soon after they leave school. An additional disadvantage of the statement of objectives is that it can sometimes eliminate you from consideration. An employer who knows that the vacant position is not the one you want might conclude that you would not be interested in the job and therefore reject your application.

If you are a veteran of the armed forces, include a military service section on the résumé. Define your military service as if it were any other job, citing the dates, locations, positions, ranks, and tasks. Often a serviceperson receives regular evaluations from a superior; these "marks" can work in your favor.

If you have a working knowledge of a foreign language, your résumé should include a *Language Skills* section. Language skills are particularly relevant if the potential employer has international interests and you could be useful in translation or foreign service.

Figures 19–1 and 19–2 provide examples of effective résumés. The job-application letters that accompany these résumés are shown in Figure 19–3 and Figure 19–4.

Notice that résumés often omit the *I* at the start of sentences. Rather than writing, "I prepared bids, . . ." many people will write, "Prepared bids. . . ." Whichever style you use, be consistent.

FIGURE 19-1

Résumé

KENNETH CHAING
(215) 525-6881

753 Westborn Drive
Ardmore, PA 19316

EDUCATION

B.S. in Civil Engineering
Eastern University, Lynwood, PA
Anticipated June 1984

Grade-Point Average: 3.35 (of 4.0)

Advanced Business Courses

Financial Accounting
Legal Options in Decision Making
Manpower Management
Budgeting
Advanced Accounting
Labor Law

EMPLOYMENT

May 1983-September 1983
Gilmore Construction, Redford, PA
Prepared bids and estimates for storm and sanitary piping.
Revised drafting for 700-unit housing complex.
Ordered piping materials amounting to $325,000.

June 1982-September 1982
Gilmore Construction, Redford, PA
Worked in drafting and reproductions department.
Supervised piping sales.

June 1981-September 1981
Pertwell Construction, Salford, PA
Laborer.

PERSONAL INFORMATION

Member, American Society of Civil Engineers
Eagle Scout

THE JOB-APPLICATION LETTER

The job-application letter is crucial because it is the first thing your reader sees. If the letter is ineffective, the reader probably will not bother to read the résumé.

If job candidates had infinite time and patience, they would make up a different résumé for each prospective employer, highlighting their particular appropriateness for that one job. But because most candidates don't have unlimited time and patience, they make up one or two different versions of their résumés. As a

REFERENCES

Mr. Allen Chrome, President
Gilmore Construction
Frazer Park
Redford, PA 18611
(306) 912-1773

Mr. Len Lefkowitz, P.E.
Chief Engineer, Pertwell
Construction
1911 Market Street
Redford, PA 18611
(306) 432-1814

Dr. Harold Murphy
Professor of Civil Engineering
Eastern University
Lynwood, PA 19314
(215) 669-4300

result, the typical résumé makes a candidate look only relatively close to the ideal the employer has in mind. Thus, candidates must use job-application letters to appeal directly and specifically to the needs and desires of particular employers.

APPEARANCE OF THE JOB-APPLICATION LETTER

Like the résumé it introduces, the letter must be error-free and professional-looking. A good job-application letter has all of the virtues of any business letter: adequate margins, clear and uniform type, no strikeovers or broken letters. And, of course, it must

FIGURE 19-2

Résumé

ANDREA SARNO

Home Address:
1314 Old Oaks St.
Wynnewood, WA 99123
(206) 612-1414

School Address:
311 Hamilton Hall
Western University
Warren, WA 99314
(206) 669-4136

Objective: Entry-level position in accounting.

Education: Bachelor of Science in Accounting
Anticipated June 1984
Western University, Warren, WA 99314

Advanced Business Courses

Investment Analysis
Advanced Accounting
Principles of Management Accounting
Management Information Systems
Federal Tax Law
Collective Bargaining

Experience: Harmon Kline, Inc. Washington, D.C. June-September 1983
Worked in the audit department. Supervised the auditing of seven multinational corporations. Wrote audit reports (average length, 18 pages) for each of the clients. Recommended internal improvements in audit department that saved over $40,000 annually.

Franklin Warner, Inc., Spokane. June-September 1982
Worked as a clerk in the accounts receivable department.

Honors and Awards: Brackenbury Scholarship, 1983
Outstanding Senior Woman, 1983-84
President, American Accounting Society Student Chapter, Western University, 1983
Member, Lacrosse Team

References: Furnished upon request.

conform to one of the basic letter formats (see Chapter 18, "Correspondence").

Such advice is easy to give but hard to follow. The problem is that every job-application letter must be typed individually, because its content is unique. (Even if you have access to a word processor, part of the letter must be typed by hand.) In creating the job-application letter, therefore, you have a double burden: the one page that makes the greatest impression has to be typed separately each time you apply for a position. Compared to job-application letters, résumés are simple: you have to type them perfectly only once.

CONTENT OF THE JOB-APPLICATION LETTER

Like the résumé, the job-application letter is a sales document. Its purpose is to convince the reader that you are an outstanding candidate and should be called in for an interview. Of course, you accomplish this through skillful use of hard evidence, not through empty self-praise.

The job-application letter is *not* an expanded version of the résumé. In the letter, you choose from the résumé two or three points that will be of greatest interest to the potential employer and develop them into paragraphs. If one of your previous part-time positions called for specific skills that the employer is looking for, that position might be the subject of a substantial paragraph in the letter, even though the résumé devotes only a few lines to it. The key to a good application letter is selectivity. If you try to cover every point on your résumé, the letter will be fragmented. The reader then will have a hard time forming a clear impression of you, and the purpose of the letter will be thwarted.

In most cases, a job-application letter should fill up the better part of a page. Like all business correspondence, it should be single-spaced, with double spaces between paragraphs. A full page gives you space to develop a substantial argument. For more-experienced candidates, the letter may be longer, but most students find that they can adequately summarize their credentials in one page. A long letter is not necessarily a good letter. If you write at length on a minor point, you end up being boring. Worse still, you appear to have poor judgment. Employers are always seeking candidates who can say much in a limited space, not the reverse.

ELEMENTS OF THE JOB-APPLICATION LETTER

Among the mechanical elements of the job-application letter, the inside address—the name, title, organization, and address of the recipient—is most important. If you know the correct form of this information from an ad, there is no problem. However, if you are uncertain about any of the information—the recipient's name, for example, might have an unusual spelling—you should verify it by phoning the organization. Many people are very sensitive about such matters, so you should not risk beginning the letter with a misspelling or incorrect title. When you do not know who should receive the letter, do not address it to a department of the company—unless the job ad specifically says to do so—because nobody in that department might feel responsible for dealing with it. Instead, phone the company to find out who manages the department. If you are unsure of the appropriate department or division to write to, address the letter to a high-level executive, such as the president of the organization. The letter will be directed to the right person. Also, because the application includes both a letter and a résumé, the enclosure notation (see Chapter 18) should be typed in the lower left-hand corner of the letter.

The four-paragraph letter that will be discussed here is, naturally, only a very basic model. Because every substantial job-application letter has an introductory and a concluding paragraph, the minimum number of paragraphs for the job-application letter has to be four. But there is no reason that an application letter cannot have five or six paragraphs. The four-paragraph letter, however, is a useful model for most situations. The four paragraphs can be categorized as follows:

1. introductory paragraph
2. education paragraph
3. employment paragraph
4. concluding paragraph

THE INTRODUCTORY PARAGRAPH The introductory paragraph establishes the tone for the rest of the letter: it is your opportunity to capture the reader's attention. Specifically, the introductory paragraph has four functions:

1. It identifies your source of information.
2. It identifies the position you are interested in.
3. It states that you wish to be considered for that position.
4. It forecasts the rest of the letter.

Your source of information is usually a published advertisement or an employee already working for the organization. If it is an ad, identify specifically the publication and its date of issue. If it is an employee, identify that person by name and title. (In an unsolicited application, all you can do is ask if a position is available.) In addition to naming the source of information, you should clearly identify the position for which you are applying. Often, the reader of the letter will be unfamiliar with all of the different ads that the organization has placed. Without the title of the position, he or she will not know which position you are interested in. Also, specifically state that you would like to be considered for the position. And finally, carefully choose a few phrases that forecast the body of the letter by summarizing its main points. These four aspects of the introductory paragraph need not appear in any particular order, nor need they each be covered in a single sentence. The following examples of introductory paragraphs demonstrate different ways to provide the necessary information.

I am writing in response to your notice in the May 13 New York Times. I would like to be considered for the position in systems programming. My studies at Eastern University in computer science, along with my programming experience at Airborne Instruments, would qualify me, I believe, for the position.

My academic training in hotel management and my experience with Sheraton International have given me a solid background in the hotel industry. I would like to be considered for any management trainee position that might be available.

Mr. Howard Alcott of your Research and Development Department has suggested that I write. He feels that my organic chemistry degree and my practical experience with Brown Laboratories might be of value to you. Do you have an entry-level position in organic chemistry for which I might be considered?

As these sample paragraphs indicate, the important information can be conveyed in any number of ways. The difficult part of the introductory paragraph—and of the whole letter as well—is to achieve the proper tone. You must appear quietly self-confident. Because the letter will be read by someone who is professionally superior to you, the tone must be modest, but it should not be self-effacing or negative. Never say, for example, "I do not have a very good background in civil engineering, but I'm willing to learn." The reader will take this kind of statement at face value and probably stop reading right there. You should show pride in your education and experience, while at the same time suggesting by your tone that you have much to learn about your field.

THE EDUCATION PARAGRAPH For most students the education paragraph should come before the employment paragraph, because the content of the former will be stronger. If, however, your work experience is more pertinent than your education, discuss your work first.

In devising your education paragraph, take your cue from the job ad. What aspect of your education most directly responds to the job requirements? If the ad stresses the need for versatility, you might structure your paragraph around the range and diversity of your courses. Also, you might discuss course work in a field related to your major, such as business or communications skills. Extracurricular activities are often very valuable; if you were an officer in a student organization in your field, you could discuss the activities and programs that you coordinated. Perhaps the most popular strategy for developing the education paragraph is to discuss skills and knowledge gained from advanced course work in the major field.

Whatever information you provide, the key to the education paragraph is to develop one unified idea, rather than to toss a series of unrelated facts onto the page. Notice how each of the following education paragraphs develops a unified idea:

At Eastern University, I have taken a wide range of courses in the sciences, but my most advanced work has been in chemistry. In one laboratory course, I developed a new aseptic technique that lowered the risk of infection by over 40%. This new technique was the subject of an article in the Eastern Science Digest. Representatives from three national breweries have visited our laboratory to discuss the technique with me.

To broaden my education at Southern, I took eight business courses in addition to my requirements for the Civil Engineering degree. Because your ad mentions that the position will require substantial client contact, I feel that my work in marketing, in particular, would be of special value. In an advanced marketing seminar, my project culminated in a 20-page sales brochure on the different kinds of building structures for sale to industrial customers in our section of the city.

The most rewarding part of my education at Western University took place outside the classroom. My entry in a fashion design competition sponsored by the university won second place. More important, through the competition I met the chief psychologist at Eastern Regional Hospital, who invited me to design clothing for handicapped persons. This project has given me an interest in an aspect of design that has up to now received little attention. I would hope to be able to pursue this interest once I start work.

Notice that each of these paragraphs begins with a topic sentence—a forecast of the rest of the paragraph—and uses considerable detail and elaboration to develop the main idea. Notice, too, a small point: if you haven't already specified your major and college or university in the introductory paragraph, be sure to do so in the discussion of your education.

THE EMPLOYMENT PARAGRAPH Like the education paragraph, the employment paragraph should begin with a topic sentence and then elaborate a single idea. That idea might be that you have had a broad background or that a single position has given you special skills that make you particularly suitable for the available position. The following examples show effective experience paragraphs.

For the past three summers and part-time during the academic year, I have worked for Redego, Inc., a firm that specializes in designing and planning industrial complexes. I began at Redego as an assistant in the drafting room. By the second summer, I was accompanying a civil engineer on field inspections. Most recently, I have been involved with an engineer in designing and drafting the main structural supports for a 15-acre, $30 million chemical facility.

Although I have worked every summer since I was fifteen, my most recent position, as a technical editor, has been the most rewarding. I was chosen by Digital Systems, Inc., from among 30 candidates because of my dual background in computer science and writing. My job was to coordinate the editing of computer manuals. Our copy editors, who are generally not trained in computer science, need someone to help verify the technical accuracy of their revisions. When I was unable to answer their questions, I was responsible for getting the correct answer from our systems analysts and making sure the computer novice could follow it. This position gave me a good understanding of the process by which operating manuals are created.

I have worked in merchandising for three years as a part-time and summer salesperson in men's fashions and accessories. I have had experience in inventory control and helped one company to switch from a manual to an on-line system. Most recently, I assisted in clearing $200,000 in out-of-date men's fashions: I coordinated a campaign to sell half of the merchandise at cost and was able to convince the manufacturer's representative to accept the other half for full credit. For this project, I received a certificate of appreciation from the company president.

Notice how these writers carefully define their duties to give their readers a clear idea of the nature and extent of their responsibilities.

Although you will discuss your education and experience in separate paragraphs, try to link these two halves of your background. If an academic course led to an interest that you were able to follow up in a job, make that point clear in the transition from one paragraph to the other. Similarly, if a job experience helped shape your academic career, tell the reader about it.

THE CONCLUDING PARAGRAPH The concluding paragraph of the job-application letter is like the ending of any sales letter: its function is to stimulate action. In this case, you want the reader to invite you for an interview. In the preceding paragraphs of the letter, you provided the information that should have convinced your reader to give you another look. In the last paragraph, you want to make it easy for him or her to do so. The concluding paragraph contains three main elements:

1. a reference to your résumé
2. a request for an interview
3. your phone number

If you have not yet referred to the enclosed résumé, do so at this point. Then, politely but confidently request an interview—

making sure to use the phrase "at your convenience." Don't make the request sound as though you're asking a personal favor. And be sure to include your phone number and the time of day you can be reached—even though the phone number is also on your résumé. Many employers will pick up a phone and call promising candidates personally, so you should make it as easy as possible for them to get in touch with you.

Following are examples of effective concluding paragraphs.

The enclosed résumé fills in the details of my education and experience. Could we meet at your convenience to discuss further the skills

FIGURE 19-3

Job-application Letter

753 Westborn Drive
Ardmore, PA 19316

November 23, 19--

Mr. Arnold Peck
Director of Personnel
Lientz Construction, Inc.
119 Westview Drive
Willoughby, OH 44094

Dear Mr. Peck:

With my civil engineering degree from Eastern University and my practical experience in the field, I feel I could be of value to Lientz Construction. I would like to be considered for the junior civil engineering position described in the November 22 New York Times.

Your notice calls for a candidate with "business sense." While pursuing my civil engineering studies at Eastern, I took as many business courses as I could. Of particular value were three such courses: manpower management, budgeting, and labor law. These advanced seminars allow the students to research particular topics of their own choice. I was intrigued by the sometimes conflicting goals of high productivity and profits and good labor-management relations and did my research on the recent history of such problems and their solutions in the construction industry.

In three summers' work with two different construction firms, I had a chance to see the practical side of this issue. I started as a laborer, and I think this experience will be of great value to me in every aspect of my career in construction. My experience preparing bids and estimates has given me a good background in customer relations. In addition, I was given the responsibility for revising the entire drafting of a 700-unit housing complex in southern New Jersey.

and experience I could bring to Pentamax? A message can be left for me anytime at (333) 444-5555.

More information about my education and experience is included on the enclosed résumé, but I would appreciate the opportunity to meet with you at your convenience to discuss my application. I can be reached after noon Tuesdays and Thursdays at (333) 444-5555.

Figures 19–3 and 19–4 show examples of effective job-application letters corresponding to the résumés presented in Figures 19–1 and 19–2.

Mr. Arnold Peck
page 2
November 23, 19--

The enclosed résumé fills in the details of my skills and experience. Would I be able to meet with you at your covenience to discuss my qualifications for this position? You can leave a message for me any weekday at (215) 525-6681.

Very truly yours,

Kenneth Chaing

Enclosure (1)

FIGURE 19–4

Job-application Letter

April 19, 19--

311 Hamilton Hall
Western University
Warren, WA 99314

Mr. Thomas Gilligan, Senior Partner
Sampson, Thrall and Co.
43 Hawthorne Place
Sawyersville, MD 21301

Dear Mr. Gilligan:

Ms. Jane Phillips of your Audit Division suggested that I write to you to inquire if you will be hiring any new graduates in accounting this June. If so, would you please consider my application?

I met Ms. Phillips when she came to Western University to speak to our student chapter of the American Accounting Society. As president of the chapter, I arranged seven guest lectures and two trips to local accounting firms. I have long felt that an academic education in a field such as accounting can be enhanced tremendously by contact with professionals in the field. Ms. Phillips's talk, "New Auditing Techniques," was well received by over 50 students and faculty members.

My accounting skills were put to their most severe test in my most recent position, with Harmon Kline in Washington, D.C. I coordinated a number of audits and wrote the reports for the clients. In addition, at my suggestion an improvement was made in the audit procedures that saved the company over $40,000 annually.

THE FOLLOW-UP LETTER

The final element in a job search is a follow-up letter, which is sent to a potential employer after an interview or plant tour. Your purpose in writing a follow-up letter is to thank the representative for taking the time to see you and to remind him or her of your particular qualifications for the job. You also can take this opportunity to restate your interest in the position.

The follow-up letter can do more good with less effort than any other step in the job-application procedure. The reason is sim-

Mr. Thomas Gilligan
page 2
April 19, 19--

I would be grateful for the opportunity to discuss my accounting experience and any other aspects of my background that the enclosed résumé does not cover. Could we meet at your convenience? Please write to me at the above address or call me any day after 2:30 p.m. at (206) 669-4136.

Yours truly,

Andrea Sarno

Enclosure

ple: so few candidates take the time to write the letter. Figure 19–5 provides an example of a follow-up letter.

WRITER'S CHECKLIST

The following checklist covers the résumé, the job-application letter, and the follow-up letter.

RÉSUMÉ

1. Does the résumé have a professional appearance, with generous margins, clear type, a symmetrical layout, and adequate white space? ____

FIGURE 19-5
Follow-up Letter

1901 Chestnut Street
Phoenix, AZ 63014

July 13, 19--

Mr. Daryl Weaver
Director of Operations
Cynergo, Inc.
Spokane, WA 92003

Dear Mr. Weaver:

I would like to thank you for taking the time yesterday to show me your facilities and to introduce me to your colleagues.

Your advances in piping design were particularly impressive. As a person with hands-on experience in piping design, I can fully appreciate the advantages your design will have.

The vitality of your projects and the obvious good fellowship among your employees further confirm my initial belief that Cynergo would be a fine place to work. I would look forward to joining your staff.

Sincerely yours,

Albert Rossman

Albert Rossman

2. Is the résumé completely free of errors? ____

3. Does the identifying information section contain your name, address, and phone number? ____

4. Does the education section include your degree, your institution and its location, and your anticipated date of graduation, as well as any other information that will help your reader appreciate your qualifications? ____

5. Does the employment section include, for each job, the dates of employment, the organization's name and location, and your position or title, as well as a description of your duties and accomplishments? ____

6. Does the personal information section include relevant hobbies or activities, including extracurricular interests? ____

7. Does the references section include the names, job title, organization, mailing address, and phone number of three or four referees? If you are not listing this information, does the strength of the rest of the résumé offset the omission? ____

8. Does the résumé include any other appropriate sections, such as military service, language skills, or honors? ____

JOB-APPLICATION LETTER

1. Does the letter have a professional appearance? ____

2. Does the introductory paragraph identify your source of information and the position you are applying for, state that you wish to be considered, and forecast the rest of the letter? ____

3. Does the education paragraph respond to your reader's needs with a unified idea introduced by a topic sentence? ____

4. Does the employment paragraph respond to your reader's needs with a unified idea introduced by a topic sentence? ____

5. Does the concluding paragraph include a reference to your résumé, a request for an interview, and your phone number? ____

6. Does the letter include an enclosure notation? ____

FOLLOW-UP LETTER

Does the letter thank the interviewer and remind him or her of your qualifications? ____

EXERCISES

1. In a newspaper or journal, find a job advertisement for a position in your field for which you might be qualified. Write a résumé and a job-application letter in response to the ad.

2. Revise the following résumés to make them more effective. Make up any reasonable details.

a.

Bob Jenkins
2319 Fifth Avenue
Waverly, CT 01603
Phone: 611-3356

Personal Data:		22 years old Height 5'11" Weight 176 lbs.
Education:		B.S. in Electrical Engineering University of Connecticut, June 19--
Experience:	6/---9/--	Falcon Electronics Examined panels for good wiring. Also, I revised several schematics.
	6/---9/--	MacDonalds Electrical Supply Co. Worked parts counter.
	6/---9/--	Happy Burger Made hamburgers, fries, shakes, fish sandwiches, and fried chicken.
	6/---9/--	Town of Waverly Outdoor maintenance man. In charge of cleaning up McHenry Park and Municipal Pool picnic grounds. Did repairs on some electrical equipment.
Background:		Born and raised in Waverly. Third baseman, Fisherman's Rest softball team. Hobbies: jogging, salvaging and repairing appliances, reading magazines, politics.
References:		Will be furnished upon request.

b.

Harold Perkins

1415 Ninth Street
Altoona, Pa. 16013
667-1415

319 W. Irvin Avenue
State College, Pa.
304-9316

Education:

Economics Major
Penn State, 19--
Cum: 3.03

Important courses: Econ. 412, 413, 501, 516.
Fin. 353, 614 (seminar), Mgt. 502-503.

Experience:	Radio Shack, State College Bookkeeper. Maintained the books.
	Radio Shack, State College Clerk
	Holy Redeemer Hospital Volunteer
Organizations:	Scuba Club Alpha Sigma Sigma--Pledge Master Treasurer, Penn State Economics Association

References:

George Williams Economics Dept. McKee Hall State College, Pa.	Arthur Lawson Radio Shack State College, Pa.	Dr. John Tepper Holy Redeemer Hospital Boalsburg, Pa.

3. Revise the following job-application letters to make them more effective. Make up any reasonable details.

a.

April 13, 19--

Wayne Grissert
Best Department Store
113 Hawthorn
Atlanta, Georgia

Dear Mr. Grissert:

As I was reading the Sunday Examiner, I came upon your ad for a buyer. I have always been interested in learning about the South, so would you consider my application?

I will receive my degree in fashion design in one month. I have taken many courses in fashion design, so I feel I have a strong background in the field.

Also, I have had extensive experience in retail work. For two summers I sold women's accessories at a local clothing store. In addition, I was a temporary department head for two weeks.

I have enclosed a résumé and would like to interview you at your convenience. I hope to see you in the near future. My phone number is 436-6103.

Sincerely,

Brenda Adamson

Brenda Adamson

b.

113 Holloway Drive
Nanuet, Oklahoma 61403

May 11, 19--

Miss Betty Richard
Manager of Technical Employment
Scott Paper Co.
Philadelphia, Pa. 19113

Dear Miss Betty,

I am writing this letter in response to your ad in the Philadelphia Inquirer, March 23, regarding a job opening in research and development.

I am expecting a B.S. degree in chemical engineering this June. I am particularly interested in design and have a strong background in it. I am capable of working independently and am able to make certain important decisions pertaining to certain problems encountered in chemical processes. I am including my résumé with this letter.

If you think I can be of any help to you at your company, please contact me at your convenience. Thank you.

Sincerely,

Glenn Corwin

4. Revise this follow-up letter to make it more effective. Make up any reasonable details.

914 Imperial Boulevard
Durham, NC 27708

November 13, 19--

Mr. Ronald O'Shea
Division Engineering
Safeway Electronics, Inc.
Holland, MI 49423

Dear Mr. O'Shea:

Thanks very much for showing me around your plant. I hope I was able to convince you that I'm the best person for the job.

Sincerely yours,

Robert Wilcox

APPENDIX A

HANDBOOK

This handbook concentrates on grammar, style, punctuation, and mechanics. Where appropriate, common errors are defined directly after the correct usage is discussed.

Many of the usage recommendations made here are only suggestions. If your organization or professional field has a style guide that makes different recommendations, you should of course follow it.

Also, note that this is a selective handbook. It cannot replace full-length treatments, such as the handbooks often used in composition courses.

STYLE

USE MODIFIERS EFFECTIVELY

Technical writing is full of modifiers—phrases and clauses that describe other elements in the sentence. To make your meaning clear, you must use modifiers effectively. You must clearly communicate to your readers whether a modifier provides necessary information about its referent (the word or phrase it refers to) or whether it simply provides additional information. Further, you must make sure that the referent itself is always clearly identified.

RESTRICTIVE AND NONRESTRICTIVE MODIFIERS **A restrictive modifier**, as the term implies, restricts the meaning of its referent: that is, it provides information necessary to identify the referent. In the following example, the restrictive modifiers are italicized.

The aircraft *used in the exhibitions* are slightly modified.

Please disregard the notice *you just received from us.*

In most cases, the restrictive modifier doesn't require a pronoun, such as *that* or *which*. If you choose to use a pronoun, however, use *that*: "The aircraft that are used in the exhibits are slightly modified." (If the pronoun refers to a person or persons, use *who*.) Notice that restrictive modifiers are not set off by commas.

A **nonrestrictive modifier** does not restrict the meaning of its referent: in other words, it provides information that is not necessary to identify the referent. In the following examples, the nonrestrictive modifiers are italicized.

The space shuttle, *described by NASA as a space bus*, made its first flight in 1980.

When you arrive, go to the Registration Area, *which is located on the second floor.*

Like the restrictive modifier, the nonrestrictive modifier usually does not require a pronoun. If you use one, however, choose *which* (except, of course, when referring to a person or persons). Note that nonrestrictive modifiers are separated from the rest of the sentence by commas.

MISPLACED MODIFIERS The placement of the modifier often determines the meaning of the sentence. Notice, for instance, how the placement of *only* changes the meaning in the following sentences.

Only Turner received a cost-of-living increase last year.
(*Meaning:* Nobody else received one.)

Turner received only a cost-of-living increase last year.
(*Meaning:* He didn't receive a merit increase.)

Turner received a cost-of-living increase only last year.
(*Meaning:* He received a cost-of-living increase as recently as last year.)

Turner received a cost-of-living increase last year only.
(*Meaning:* He received a cost-of-living increase in no other year.)

Misplaced modifiers—those that appear to modify the wrong referent—pose a common problem in technical writing. The solution is, in general, to make sure the modifier is placed as close as possible to its intended referent. Frequently, the misplaced modifier is a phrase or a clause:

MISPLACED

The subject of the meeting is the future of geothermal energy in the downtown Webster Hotel.

CORRECT

The subject of the meeting in the downtown Webster Hotel is the future of geothermal energy.

MISPLACED

Jumping around nervously in their cages, the researchers speculated on the health of the mice.

CORRECT

The researchers speculated on the health of the mice jumping around nervously in their cages.

A special kind of misplaced modifier is called a **squinting modifier**—one that is placed ambiguously between two potential referents, so that the reader cannot tell which one is being modified:

UNCLEAR

We decided immediately to purchase the new system.

CLEAR

We immediately decided to purchase the new system.

CLEAR

We decided to purchase the new system immediately.

UNCLEAR

The men who worked on the assembly line reluctantly picked up their last paychecks.

CLEAR

The men who worked reluctantly on the assembly line picked up their last paychecks.

CLEAR

The men who worked on the assembly line picked up their last paychecks reluctantly.

DANGLING MODIFIERS A **dangling modifier** is one that has no referent in the sentence:

Searching for the correct answer to the problem, the instructions seemed unclear.

In this sentence, the person doing the searching has not been identified. To correct the problem, rewrite the sentence to put the clarifying information either *within* the modifier or *next to* the modifier:

As I was searching for the correct answer to the problem, the instructions seemed unclear.

Searching for the correct answer to the problem, I thought the instructions seemed unclear.

A writer sometimes can correct a dangling modifier by switching from the indicative mood (a statement of fact) to the imperative mood (a request or command):

DANGLING

To initiate the procedure, the BEGIN button should be pushed.

CORRECT

To initiate the procedure, push the BEGIN button.

In the imperative, the referent—in this case, *you*—is understood.

KEEP PARALLEL ELEMENTS PARALLEL

A sentence is **parallel** if its coordinate elements are expressed in the same grammatical form: that is, all its clauses are either passive or active, all its verbs are either infinitives or participles, and so forth. By creating and sustaining a recognizable pattern for the reader, parallelism makes the sentence easier to follow.

Notice how faulty parallelism weakens the following sentences.

NONPARALLEL

Our present system is costing us profits and reduces our productivity. (*nonparallel verbs*)

PARALLEL

Our present system is costing us profits and reducing our productivity.

NONPARALLEL

The dignitaries watched the launch, and the crew was applauded. (*nonparallel voice*)

PARALLEL

The dignitaries watched the launch and applauded the crew.

NONPARALLEL

The typist should follow the printed directions; do not change the originator's work. (*nonparallel mood*)

PARALLEL

The typist should follow the printed directions and not change the originator's work.

A subtle form of faulty parallelism often occurs with the correlative constructions, such as *either . . . or*, *neither . . . nor*, and *not only. . . but also*:

NONPARALLEL

The new refrigerant not only decreases energy costs but also spoilage losses.

PARALLEL

The new refrigerant decreases not only energy costs but also spoilage losses.

In this example, "decreases" applies to both "energy costs" and "spoilage losses." Therefore, "decreases" should precede the first half of the correlative construction. Note that if the sentence contains two different verbs, the first half of the correlative construction should precede the verb:

The new refrigerant not only decreases energy costs but also prolongs product freshness.

When creating parallel constructions, make sure that parallel items in a series do not overlap, thus changing or confusing the meaning of the sentence:

CONFUSING

The speakers will include partners of law firms, businessmen, and civic leaders.

CLEAR

The speakers will include businessmen, civic leaders, and partners of law firms.

The problem with the original sentence is that "partners" appears to apply to "businessmen" and "civic leaders." The revision solves the problem by rearranging the items so that "partners" cannot apply to the other two groups in the series.

Parallelism should be maintained not only within individual sentences, but also among sentences in paragraphs. When you establish a pattern in a paragraph, follow it through. The pattern may be as simple as numbering the steps in a process:

PARALLEL

Correlating the two results is a three-part procedure. First, ... Second, ... And third, . . .

Keep the voice consistent—either active or passive:

PARALLEL

The sample is placed in the Petri dish. Two drops of culture are added. . . . After two days, the growth rate is examined by. . . . Finally, the sample is weighed.

AVOID AMBIGUOUS PRONOUN REFERENCE

Pronouns must refer clearly to the words or phrases they replace. **Ambiguous pronoun references** can lurk in even the most innocent-looking sentences:

UNCLEAR

Remove the cell cluster from the medium and analyze it. (*Analyze what, the cell cluster or the medium?*)

CLEAR

Analyze the cell cluster after removing it from the medium.

CLEAR

Analyze the medium after removing the cell cluster from it.

CLEAR

Remove the cell cluster from the medium. Then analyze the cell cluster.

CLEAR

Remove the cell cluster from the medium. Then analyze the medium.

Ambiguous references can also occur when a relative pronoun such as *which*, or a subordinating conjunction, such as *where*, is used to introduce a dependent clause:

UNCLEAR

She decided to evaluate the program, which would take five months. (*What would take five months, the program or the evaluation?*)

CLEAR

She decided to evaluate the program, a process that would take five months. (*By replacing "which" with "a process that," the writer clearly indicates that the evaluation will take five months.*)

CLEAR

She decided to evaluate the five-month program. (*By using the adjective "five-month," the writer clearly indicates that the program will take five months.*)

UNCLEAR

This procedure will increase the handling of toxic materials outside the plant, where adequate safety measures can be taken. (*Where can adequate safety measures be taken, inside the plant or outside?*)

CLEAR

This procedure will increase the handling of toxic materials outside the plant. Because adequate safety measures can be taken only in the plant, the procedure poses risks.

CLEAR

This procedure will increase the handling of toxic materials outside the plant. Because adequate safety measures can be taken only outside the plant, the procedure will decrease safety risks.

As the last example shows, sometimes the best way to clarify an unclear pronoun is to split the sentence in two, eliminate the problem, and add clarifying information. Clarity is always the primary characteristic of good technical writing. If more words will make your writing clearer, use them.

Ambiguity can also occur at the beginning of a sentence:

UNCLEAR

Allophanate linkages are among the most important structural components of polyurethane elastomers. They act as cross-linking sites. (*What act as cross-linking sites, allophanate linkages or polyurethane elastomers?*)

CLEAR

Allophanate linkages, which are among the most important structural components of polyurethane elastomers, act as cross-linking sites. (*The writer has changed the second sentence into a clear nonrestrictive modifier.*)

Your job is to use whichever means—restructuring the sentence or dividing it in two—will best assure that the reader will know exactly which word or phrase the pronoun is replacing.

COMPARE ITEMS CLEARLY When comparing or contrasting items, make sure your sentence clearly communicates the relationship. A simple comparison between two items often causes no problems: "The X3000 has more storage than the X2500." However, don't let your reader confuse a comparison and a simple statement of fact. For example, in the sentence "Trout eat more than minnows," does the writer mean that trout don't restrict their diet to minnows or that trout eat more than minnows eat? If a comparison is intended, a second verb should be used: "Trout eat more than minnows do." And if three items are introduced, make sure that the reader can tell which two are being compared:

AMBIGUOUS

Trout eat more algae than minnows.

CLEAR

Trout eat more algae than they do minnows.

CLEAR

Trout eat more algae than minnows do.

Beware of comparisons in which different aspects of the two items are compared:

ILLOGICAL

The resistance of the copper wiring is lower than the tin wiring.

LOGICAL

The resistance of the copper wiring is lower than that of the tin wiring.

In the illogical construction, the writer contrasts "resistance" with "tin wiring" rather than with the resistance of tin wiring. In the revision, the pronoun *that* is used to substitute for the repetition of "resistance."

USE ADJECTIVES CLEARLY In general, adjectives are placed before the nouns they modify: "the plastic washer." Technical writing, however, often requires clusters of adjectives. To prevent confusion, use commas to separate coordinate adjectives, and use hyphens to link compound adjectives.

Adjectives that describe different aspects of the same noun are known as coordinate adjectives:

a plastic, locking washer

In this case, the comma replaces the word *and*.

Note that sometimes an adjective is considered part of the noun it describes: "electric drill." When one adjective is added to "electric drill," no comma is required: "a reversible electric drill." The addition of two or more adjectives, however, creates the traditional coordinate construction: "a two-speed, reversible electric drill."

The phrase "two-speed" is an example of a compound adjective—one made up of two or more words. Use hyphens to link the elements in compound adjectives that precede nouns:

a variable-angle accessory

increased cost-of-living raises

The hyphens in the second example prevent the reader from momentarily misinterpreting "increased" as an adjective modifying "cost" and "living" as a participle modifying "raises."

A long string of compound adjectives can be confusing even if hyphens are used appropriately. To ensure clarity in such a case, put the adjectives into a clause or phrase following the noun:

UNCLEAR

an operator-initiated, default-prevention technique

CLEAR

a technique initiated by the operator for preventing default

In turning a string of adjectives into a phrase or clause, make sure the adjectives cannot be misread as verbs. Use the pronouns *that* and *which* to prevent confusion:

CONFUSING

The good experience provides is often hard to measure.

CLEAR

The good that experience provides is often hard to measure.

MAINTAIN NUMBER AGREEMENT

Number disagreement commonly takes one of two forms in technical writing: (1) the verb disagrees in number with the subject when a prepositional phrase intervenes; (2) the pronoun disagrees in number with its antecedent or referent when the latter is a collective noun.

SUBJECT-VERB DISAGREEMENT A prepositional phrase does not affect the number of the subject and the verb. The following examples show that the object of the preposition can be plural in a singular sentence, or singular in a plural sentence. (The subjects and verbs are italicized.)

INCORRECT

The *result* of the tests *are* promising.

CORRECT

The *result* of the tests *is* promising.

INCORRECT

The *results* of the test *is* promising.

CORRECT

The *results* of the test *are* promising.

Don't be misled by the fact that the object of the preposition and the verb don't sound natural together, as in *tests is* or *test are*. Grammatical agreement of subject and verb is the primary consideration.

PRONOUN-ANTECEDENT DISAGREEMENT The pronoun-antecedent disagreement problem crops up most often when the antecedent, or referent, is a collective noun—one that can be interpreted as either singular or plural, depending on its usage:

INCORRECT

The company is proud to announce a new stock option plan for their employees.

CORRECT

The company is proud to announce a new stock option plan for its employees.

In this example, "the company" acts as a single unit; therefore, the singular verb, followed by a singular pronoun, is appropriate. When the individual members of a collective noun are stressed, however, plural pronouns and verbs are appropriate: "The inspection team have prepared their reports."

PUNCTUATION

THE PERIOD Periods are used in the following instances.

1. At the end of sentences that do not ask questions or express strong emotion:

The lateral stress still needs to be calculated.

2. After most abbreviations:

M.D.
U.S.A.
etc.

(For a further discussion of abbreviations, see p. 518.)

3. With decimal fractions:

4.056
$6.75
75.6%

THE EXCLAMATION POINT

The exclamation point is used at the end of a sentence that expresses strong emotion, such as surprise or doubt:

The nuclear plant, which was originally expected to cost $1.6 billion, eventually cost more than $4 billion!

Because technical writing requires objectivity and a calm, understated tone, technical writers rarely use exclamation points.

THE QUESTION MARK

The question mark is used at the end of a sentence that asks a direct question:

What did the commission say about effluents?

Do not use a question mark at the end of a sentence that asks an indirect question:

He wanted to know whether the procedure had been approved for use.

When a question mark is used within quotation marks, the quoted material needs no other end punctuation:

"What did the commission say about effluents?" she asked.

THE COMMA The comma is the most frequently used punctuation mark, as well as the one about whose usage many writers most often disagree. Following are the basic uses of the comma.

1. To separate the clauses of a compound sentence (one composed of two or more independent clauses) linked by a coordinating conjunction (*and, or, nor, but, so, for, yet*):

Both methods are acceptable, but we have found that the Simpson procedure gives better results.

In many compound sentences, the comma is needed to prevent the reader from mistaking the subject of the second clause for an object of the verb in the first clause:

The RESET command affects the field access, and the SEARCH command affects the filing arrangement.

Without the comma, the reader is likely to interpret the coordinating conjunction "and" as a simple conjunction linking "field access" and "SEARCH command."

2. To separate items in a series composed of three or more elements:

The manager of spare parts is responsible for ordering, stocking, and disbursing all spare parts for the entire plant.

The comma following the second-to-last item technically is not required, because of the presence of the conjunction. It is a good idea, however, to use the comma anyway, to clarify the separation and prevent possible misreading.

3. To separate introductory words, phrases, and clauses from the main clause of the sentence:

However, we will have to calculate the effect of the wind.

To facilitate trade, the government holds a yearly international conference.

Whether the workers like it or not, the managers have decided not to try the flextime plan.

In each of these three examples, the comma helps the reader follow the sentence. Notice in the following example how the comma actually prevents misreading:

Just as we finished eating, the rats discovered the treadmill.

The comma is optional if the introductory text is brief and cannot be misread.

CORRECT

First, let's take care of the introductions.

CORRECT

First let's take care of the introductions.

4. To separate the main clause from a dependent clause:

The advertising campaign was canceled, although most of the executive council saw nothing wrong with it.

Most accountants wear ties, whereas few engineers do.

5. To separate nonrestrictive modifiers (parenthetical clarifications) from the rest of the sentence:

Jones, the temporary chairman, called the meeting to order.

6. To separate interjections and transitional elements from the rest of the sentence:

Yes, I admit your findings are correct.

Their plans, however, have great potential.

7. To separate coordinate adjectives:

The finished product was a sleek, comfortable cruiser.

The heavy, awkward trains are still being used.

The comma here takes the place of the conjunction *and*. If the adjectives are not coordinate—that is, if one of the adjectives modifies the combination of the adjective and the noun—do not use a comma:

They decided to go to the first general meeting.

8. To signal that a word or phrase has been omitted from an elliptical expression:

Smithers is in charge of the accounting; Harlen, the data management; Demarest, the publicity.

In this example, the commas after "Harlen" and "Demarest" show that the phrase "is in charge of" has been omitted.

9. To separate a proper noun from the rest of the sentence in direct address:

John, have you seen the purchase order from United?
What I'd like to know, Betty, is why we didn't see this problem coming.

10. To introduce most quotations:

He asked, "What time were they expected?"

11. To separate towns, states, and countries:

Bethlehem, Pennsylvania, is the home of Lehigh University.
He attended Lehigh University in Bethlehem, Pennsylvania, and the University of California at Berkeley.

Note the use of the comma after "Pennsylvania."

12. To set off the year in dates:

August 1, 1983, is the anticipated completion date.

Note the use of the comma after "1983." If the month separates the date from the year, the commas are not used:

The anticipated completion date is 1 August 1983.

13. To clarify numbers:

12,013,104

(European practice is to reverse the use of commas and periods in writing numbers: periods are used to signify thousands, and commas to signify decimals.)

14. To separate names from professional or academic titles:

Harold Clayton, Ph.D.
Marion Fewick, CLU
Joyce Carnone, P.E.

COMMON ERRORS

1. No comma between the clauses of a compound sentence:

INCORRECT

The mixture was prepared from the two premixes and the remaining ingredients were then combined.

CORRECT

The mixture was prepared from the two premixes, and the remaining ingredients were then combined.

2. No comma (or just one comma) to set off a nonrestrictive modifier:

INCORRECT

The phone line, which was installed two weeks ago had to be disconnected.

CORRECT

The phone line, which was installed two weeks ago, had to be disconnected.

3. No comma separating introductory words, phrases, or clauses from the main clause, when misreading can occur:

INCORRECT

As President Canfield has been a great success.

CORRECT

As President, Canfield has been a great success.

4. No comma (or just one comma) to set off an interjection or a transitional element:

INCORRECT

Our new statistician, however used to work for Konaire, Inc.

CORRECT

Our new statistician, however, used to work for Konaire, Inc.

5. Comma splice (a comma used to "splice together" independent clauses not linked by a coordinating conjunction):

INCORRECT

All the motors were cleaned and dried after the water had entered, had they not been, additional damage would have occurred.

CORRECT

All the motors were cleaned and dried after the water had entered; had they not been, additional damage would have occurred.

CORRECT

All the motors were cleaned and dried after the water had entered. Had they not been, additional damage would have occurred.

6. Superfluous commas:

INCORRECT

Another of the many possibilities, is to use a "first in, first out" sequence. (*In this sentence, the comma separates the subject, "Another," from the verb, "is."*)

CORRECT

Another of the many possibilities is to use a "first in, first out" sequence.

INCORRECT

The schedules that have to be updated every months are, 14, 16, 21, 22, 27, and 31. (*In this sentence, the comma separates the verb from its complement.*)

CORRECT

The schedules that have to be updated every month are 14, 16, 21, 22, 27, and 31.

INCORRECT

The company has grown so big, that an informal evaluation procedure is no longer effective. (*In this sentence, the comma separates the predicate adjective "big" from the clause that modifies it.*)

CORRECT

The company has grown so big that an informal evaluation procedure is no longer effective.

INCORRECT

Recent studies, and reports by other firms confirm our experience. (*In this sentence, the comma separates the two elements in the compound subject.*)

CORRECT

Recent studies and reports by other firms confirm our experience.

INCORRECT

New and old employees who use the processed order form, do not completely understand the basis of the system. (*In this sentence, a comma separates the subject and its restrictive modifier from the verb.*)

CORRECT

New and old employees who use the processed order form do not completely understand the basis of the system.

THE SEMICOLON

Semicolons are used in the following instances.

1. To separate independent clauses not linked by a coordinating conjunction:

The second edition of the handbook is more up-to-date; however, it is more expensive.

2. To separate items in a series that already contains commas:

The members elected three officers: Jack Resnick, president; Carol Wayshum, vice-president; Ahmed Jamoogian, recording secretary.

In this example, the semicolon acts as a "supercomma," keeping the names and titles clear.

COMMON ERROR Use of a semicolon when a colon is called for:

INCORRECT

We still need one ingredient; luck.

CORRECT

We still need one ingredient: luck.

THE COLON

Colons are used in the following instances.

1. To introduce a word, phrase, or clause that amplifies or explains a general statement:

The project team lacked one crucial member: a project leader.

Here is the client's request: we are to provide the preliminary proposal by November 13.

We found three substances in excessive quantities: potassium, cyanide, and asbestos.

The week had been productive: fourteen projects had been completed and another dozen had been initiated.

Note that the text preceding a colon should be able to stand on its own as a main clause:

INCORRECT

We found: potassium, cyanide, and asbestos.

CORRECT

We found potassium, cyanide, and asbestos.

2. To introduce items in a vertical list, if the introductory text would be incomplete without the list:

We found the following:

potassium
cyanide
asbestos

3. To introduce long or formal quotations:

The president began: "In the last year . . ."

COMMON ERROR Use of a colon to separate a verb from its complement:

INCORRECT

The tools we need are: a plane, a level, and a T-square.

CORRECT

The tools we need are a plane, a level, and a T-square.

CORRECT

We need three tools: a plane, a level, and a T-square.

THE DASH Dashes are used in the following instances.

1. To set off a sudden change in thought or tone:

The committee found—can you believe this?—that the company bore *full* responsibility for the accident.

That's what she said—if I remember correctly.

2. To emphasize a parenthetical element:

The managers' reports—all ten of them—recommend production cutbacks for the coming year.

Arlene Kregman—the first woman elected to the board of directors—is the next scheduled speaker.

3. To set off an introductory series from its explanation:

Wetsuits, weight belts, tanks—everything will have to be shipped in.

When a series *follows* the general statement, a colon replaces the dash:

Everything will have to be shipped in: wetsuits, weight belts, and tanks.

Note that typewriters do not have a key for the dash. In typewritten text, a dash is represented by two uninterrupted hyphens. No space precedes or follows the dash.

COMMON ERROR Use of a dash as a "lazy" substitute for other punctuation marks:

INCORRECT

The regulations—which were issued yesterday—had been anticipated for months.

CORRECT

The regulations, which were issued yesterday, had been anticipated for months.

INCORRECT

Many candidates applied—however, only one was chosen.

CORRECT

Many candidates applied; however, only one was chosen.

PARENTHESES Parentheses are used in the following instances.

1. To set off incidental information:

Please call me (x3104) when you get the information.

Galileo (1564–1642) is often considered the father of modern astronomy.

H. W. Fowler's *Modern English Usage* (New York: Oxford University Press, 2nd ed., 1965) is still the final arbiter.

2. To enclose numbers and letters that label items listed in a sentence:

To transfer a call within the office, (1) place the party on HOLD, (2) press TRANSFER, (3) press the extension number, and (4) hang up.

COMMON ERROR Use of parentheses instead of brackets to enclose the writer's interruption of a quotation (see the discussion of brackets):

INCORRECT

He said, "The new manager (Farnham) is due in next week."

CORRECT

He said, "The new manager [Farnham] is due in next week."

THE HYPHEN Hyphens are used in the following instances.

1. In general, to form compound adjectives that precede nouns:

general-purpose register
meat-eating dinosaur
chain-driven saw

Note that hyphens are not used after words that end in *-ly*:

newly acquired terminal

Also note that hyphens are not used when the compound adjective follows the noun:

The Woodchuck saw is chain driven.

Many organizations have their own preferences about hyphenating compound adjectives. Check to see if your organization has a preference.

2. To form some compound nouns:

vice-president
editor-in-chief

3. To form fractions and compound numbers:

one-half
fifty-six

4. To attach some prefixes and suffixes:

post-1945
president-elect

5. To divide a word at the end of a line:

We will meet in the pavil-
ion in one hour.

Whenever possible, avoid such breaks; they annoy some readers. When you do use them, check the dictionary to make sure you have divided the word *between* syllables.

THE APOSTROPHE

Apostrophes are used in the following instances.

1. To indicate the possessive case:

the manager's goals
the foremen's lounge
the employees' credit union
Charles's T-square

To indicate joint possession, add the apostrophe and the *s* to only the last noun or proper noun:

Watson and Crick's discovery

To indicate separate possession, add an apostrophe and an *s* to each of the nouns or pronouns:

Newton's and Galileo's ideas

Make sure you do not add an apostrophe or an *s* to possessive pronouns: *his, hers, its, ours, yours, theirs.*

2. To form contractions:

I've
can't
shouldn't
it's

The apostrophe usually indicates an omitted letter or letters. For example, *can't* is *can(no)t*, *it's* is *it(i)s*.

Some organizations discourage the use of contractions; others have no preference. Find out the policy your organization follows.

3. To indicate special plurals:

three 9's
two different JCL's
the why's and how's of the problem

As in the case of contractions, it is a good idea to learn the stylistic preferences of your organization. Usage varies considerably.

COMMON ERROR Use of the contraction *it's* in place of the possessive pronoun *its*.

INCORRECT

The company does not feel that the problem is it's responsibility.

CORRECT

The company does not feel that the problem is its responsibility.

QUOTATION MARKS

Quotation marks are used in the following instances.

1. To indicate titles of short works, such as articles, essays, or chapters:

Smith's essay "Solar Heating Alternatives"

2. To call attention to a word or phrase that is being used in an unusual way or in an unusual context:

A proposal is "wired" if the sponsoring agency has already decided who will be granted the contract.

Don't use quotation marks as a means of excusing poor word choice:

The new director has been a real "pain."

3. To indicate direct quotation: that is, the words a person has said or written:

"In the future," he said, "check with me before authorizing any large purchases."

As Breyer wrote, "Morale *is* productivity."

Do not use quotation marks to indicate indirect quotation:

INCORRECT

He said that "third quarter profits would be up."

CORRECT

He said that third quarter profits would be up.

CORRECT

He said, "Third quarter profits will be up."

RELATED PUNCTUATION Note that if the sentence contains a "tag"—a phrase identifying the speaker or writer—a comma is used to separate it from the quotation:

John replied, "I'll try to fly out there tomorrow."

"I'll try to fly out there tomorrow," John replied.

Informal and brief quotations require no punctuation before the quotation marks:

She said "Why?"

In the United States (but not in most other English-speaking nations), commas and periods at the end of quotations are placed within the quotation marks:

The project engineer reported, "A new factor has been added."

"A new factor has been added," the project engineer reported.

Question marks, dashes, and exclamation points, on the other hand, are placed inside the quotation marks when they apply only to the quotation and outside the quotation marks when they apply to the whole sentence:

He asked, "Did the shipment come in yet?"

Did he say, "This is the limit"?

Note that only one punctuation mark is used at the end of a set of quotation marks:

INCORRECT

Did she say, "What time is it?"?

CORRECT

Did she say, "What time is it?"

BLOCK QUOTATIONS When quotations reach a certain length—generally, more than four lines— writers tend to switch to a block format. In typewritten manuscript, a block quotation is usually

1. indented ten spaces from the left-hand margin
2. single-spaced
3. typed without quotation marks
4. introduced by a colon

(Different organizations observe their own variations on these basic rules.)

McFarland writes:

> The extent to which organisms adapt to their environment is still being charted. Many animals, we have recently learned, respond to a dry winter with an automatic birth control chemical that limits the number of young to be born that spring. This prevents mass starvation among the species in that locale.

Heather concurs. She writes, "Biological adaptation will be a major research area during the next decade."

MECHANICS

ELLIPSES Ellipses (three spaced periods) indicate the omission of some material from a quotation. A fourth period with no space before it precedes ellipses when the sentence in the source has ended and you are omitting material that follows or when the omission follows a portion of the source's sentence that is in itself a grammatically complete sentence:

"Send the updated report . . . as soon as you can."

Larkin refers to the project as "an attempt . . . to clarify the issue of compulsory arbitration. . . . We do not foresee an end to the legal wrangling . . . but perhaps the report can serve as a definition of the areas of contention."

In the second example, the writer has omitted words after "attempt" and after "wrangling." In addition, she has used a sentence period plus three spaced periods after "arbitration," which ends the original writer's sentence; and she has omitted the following sentence.

BRACKETS Brackets are used in the following instances.

1. To indicate words added to a quotation:

The minutes of the meeting note that "He [Pearson] spoke out against the proposal."

2. To indicate parentheses within parentheses:

(For further information, see Charles Houghton's *Civil Engineering Today* [New York: Arch Press, 1982].)

ITALICS Italics (or underlining) are used in the following instances.

1. For words used as words:

In this report, the word *operator* will refer to any individual who is actually in charge of the equipment, regardless of that individual's certification.

2. To indicate titles of long works (books, manuals, etc.), periodicals and newspapers, films, plays, and long musical works:

See Houghton's *Civil Engineering Today*.

We subscribe to the *New York Times*.

3. To indicate the names of ships, trains, and airplanes:

The shipment is expected to arrive next week on the *Penguin*.

4. To set off foreign expressions that have not become fully assimilated into English:

The speaker was guilty of *ad hominem* arguments.

5. To emphasize words or phrases:

Do not press the ERASE key.

If your typewriter or word processor does not have italic type, indicate italics by underlining.

`Darwin's Origin of Species is still read today.`

NUMBERS The use of numbers varies considerably. Therefore, you should find out what guidelines your organization or research area follows in choosing between words and numerals. Many organizations use the following guidelines.

1. Use numerals for technical quantities, especially if a unit of measurement is included:

3 feet
12 grams
43,219 square miles
36 hectares

2. Use numerals for nontechnical quantities of 10 or more:

300 persons
12 whales
35% increase

3. Use words for nontechnical quantities of less than 10:

three persons
six whales

4. Use both words and numerals

a. For back-to-back numbers:

six 3-inch screws

fourteen 12-foot ladders
3,012 five-piece starter units

In general, use the numeral for the technical unit. If the nontechnical quantity would be cumbersome in words, use the numeral.

b. For round numbers over 999,999:

14 million light-years
$64 billion

c. For numbers in legal contracts or in documents intended for international readers:

thirty-seven thousand dollars ($37,000)
five (5) relays

d. For addresses:

3801 Fifteenth Street

SPECIAL CASES

1. If a number begins a sentence, use words, not numerals:

Thirty-seven acres was the agreed-upon size of the lot.

Many writers would revise the sentence to avoid this problem:

The agreed-upon size of the lot was 37 acres.

2. Don't use both numerals and words in the same sentence to refer to the same unit:

On Tuesday the attendance was 13; on Wednesday, 8.

3. Write out fractions, except if they are linked to technical units:

two-thirds of the members
3½ hp.

4. Write out approximations:

approximately ten thousand people
about two million trees

5. Use numerals for titles of figures and tables and for page numbers:

Figure 1
Table 13
page 261

6. Use numerals for decimals:

3.14
1,013.065

Add a zero before decimals of less than one:

0.146
0.006

7. Avoid expressing months as numbers, as in "3/7/84": in the United States, this means March 7, 1984; in many other countries, it means July 3, 1984. Use one of the following forms:

March 7, 1984
7 March 1984

8. Use numerals for times if A.M. or P.M. are used:

6:10 A.M.
six o'clock

ABBREVIATIONS

Abbreviations provide a useful way to save time and space, but you must use them carefully; you can never be sure that your readers will understand them. Many companies and professional organizations have lists of approved abbreviations.

Analyze your audience in determining whether and how to abbreviate. If your readers include nontechnical people unfamiliar with your field, either write out the technical terms or attach a list of abbreviations. If you are new in an organization or are writing for publication for the first time in a certain field, find out what abbreviations are commonly used. If for any reason you are unsure about whether or how to abbreviate, write out the word.

The following are general guidelines about abbreviations.

1. You may make up your own abbreviations. For the first reference to the term, write it out and include, parenthetically, the ab-

breviation. In subsequent references, use the abbreviation. For long works, you might want to write out the term at the start of major units, such as chapters.

The heart of the new system is the self-loading cartridge (slc).

This technique is also useful, of course, in referring to existing abbreviations that your readers might not know:

The cathode-ray tube (CRT) is your control center.

2. Most abbreviations do not take plurals:

1 lb
3 lb

3. Most abbreviations in scientific writing are not followed by periods:

lb
cos
dc

If the abbreviation can be confused with another word, however, use a period:

in.
Fig.

4. Spell out the unit if the number preceding it is spelled out or if no number precedes it:

How many square meters is the site?

CAPITALIZATION

For the most part, the conventions of capitalization in general writing apply in technical writing.

1. Capitalize proper nouns, titles, trade names, places, languages, religions, and organizations:

William Rusham
Director of Personnel
Quick Fix Erasers

Bethesda, Maryland
Methodism
Italian
Society for Technical Communication

In some organizations, job titles are not capitalized unless they refer to specific persons:

Alfred Loggins, Director of Personnel, is interested in being considered for vice-president of marketing.

2. Capitalize headings and labels:

A Proposal to Implement the Wilkins Conversion System
Section One
The Problem
Figure 6
Mitosis
Table 3
Rate of Inflation, 1960–1980

APPENDIX B METRIC CONVERSION TABLES

The United States is one of the few nations in which the metric system (formally called Le Système International d'Unités, and abbreviated SI) is not used in everyday life. However, the metric system *is* used in most technical and scientific contexts throughout the world, including the United States.

By using the following table, which is accurate to the third decimal place, you can convert U.S. units to metric units, and vice versa.

	IF YOU KNOW	YOU CAN FIND	IF YOU MULTIPLY BY
LENGTH	**Inches**	**Millimeters**	**25.400**
	Feet	**Centimeters**	**30.480**
	Yards	**Meters**	**0.914**
	Miles	**Kilometers**	**1.609**
	Millimeters	**Inches**	**0.039**
	Centimeters	**Inches**	**0.394**
	Meters	**Yards**	**1.094**
	Kilometers	**Miles**	**0.621**
AREA	**Square inches**	**Square centimeters**	**6.451**
	Square feet	**Square meters**	**0.093**
	Square yards	**Square meters**	**0.836**
	Square miles	**Square kilometers**	**2.590**

	Acres	Square hectometers (hectares)	0.405
	Square centimeters	Square inches	0.155
	Square meters	Square yards	1.196
	Square kilometers	Square miles	0.386
	Square hectometers (hectares)	Acres	2.471
MASS	Ounces	Grams	28.349
	Pounds	Kilograms	0.454
	Short tons	Megagrams (metric tons)	0.907
	Grams	Ounces	0.035
	Kilograms	Pounds	2.205
	Megagrams (metric tons)	Short tons	1.103
LIQUID VOLUME	Ounces	Milliliters	29.573
	Pints	Liters	0.473
	Quarts	Liters	0.946
	Gallons	Liters	3.785
	Milliliters	Ounces	0.034
	Liters	Pints	2.114
	Liters	Quarts	1.057
	Liters	Gallons	0.264
TEMPERATURE	Degrees Fahrenheit	Degrees Celsius	5/9, after subtracting 32
	Degrees Celsius	Degrees Fahrenheit	9/5, then add 32

Names and Symbols of Metric Prefixes

PREFIX	SYMBOL	MULTIPLE
deka-	da	10
hecto-	h	10^2
kilo-	k	10^3
mega-	M	10^6
giga-	G	10^9
deci-	d	10^{-1}
centi-	c	10^{-2}
milli-	m	10^{-3}
micro-	μ	10^{-6}
nano-	n	10^{-9}
pico-	p	10^{-12}
femto-	f	10^{-15}
atto-	a	10^{-18}

APPENDIX C

POST OFFICE ABBREVIATIONS FOR THE STATES

The following abbreviations for the states (as well as the District of Columbia and Puerto Rico) have been approved by the United States Postal Service. These abbreviations should be used in all correspondence. Note that each abbreviation consists of two capital letters, without periods or spaces. Also note that some of the abbreviations—such as those of Maine, Minnesota, and Maryland—might not be what you would expect.

Alabama	AL	Illinois	IL
Alaska	AK	Indiana	IN
Arizona	AZ	Iowa	IA
Arkansas	AR	Kansas	KS
California	CA	Kentucky	KY
Colorado	CO	Louisiana	LA
Connecticut	CT	Maine	ME
Delaware	DE	Maryland	MD
District of Columbia	DC	Massachusetts	MA
Florida	FL	Michigan	MI
Georgia	GA	Minnesota	MN
Hawaii	HI	Mississippi	MS
Idaho	ID	Missouri	MO

Montana	MT	Puerto Rico	PR
Nebraska	NE	Rhode Island	RI
Nevada	NV	South Carolina	SC
New Hampshire	NH	South Dakota	SD
New Jersey	NJ	Tennessee	TN
New Mexico	NM	Texas	TX
New York	NY	Utah	UT
North Carolina	NC	Vermont	VT
North Dakota	ND	Virginia	VA
Ohio	OH	Washington	WA
Oklahoma	OK	West Virginia	WV
Oregon	OR	Wisconsin	WI
Pennsylvania	PA	Wyoming	WY

APPENDIX D

TECHNICAL WRITING

USAGE AND GENERAL WRITING

STYLE MANUALS

GRAPHIC AIDS

PROPOSALS

ORAL PRESENTATIONS

JOURNAL ARTICLES

SELECTED BIBLIOGRAPHY

TECHNICAL WRITING

Currently, more than one hundred technical writing texts and guides are on the market. Following is a representative selection.

Blicq, R. S. 1981. *Technically—write! Communication for the technical man*. 2nd ed. Englewood Cliffs, N.J.: Prentice-Hall.

Brusaw, C. T., G. J. Alred, and W. E. Oliu. 1982. *Handbook of technical writing*. 2nd ed. New York: St. Martin's.

Fear, D. E. 1978. *Technical writing*. 2nd ed. New York: Random House.

Houp, K. W., and T. E. Pearsall. 1980. *Reporting technical information*. 4th ed. Beverly Hills, Calif.: Glencoe.

Mathes, J. C., and D. W. Stevenson. 1976. *Designing technical reports*. Indianapolis: Bobbs-Merrill.

Mills, G. H., and J. A. Walter. 1978. *Technical writing*. 4th ed. New York: Holt, Rinehart and Winston.

Pickett, N. A., and A. A. Laster. 1980. *Technical English*. 3rd ed. New York: Harper & Row.

Sherman, T. A., and S. S. Johnson. 1983. *Modern technical writing*. 4th ed. Englewood Cliffs, N.J.: Prentice-Hall.

Wiseman, H. M. 1974. *Basic technical writing*. 3rd ed. Columbus, Ohio: Charles E. Merrill.

Also see the following journals:

IEEE Transactions on Professional Communication
Journal of Business Communication
Journal of Technical Writing and Communication
Technical Communication
The Technical Writing Teacher

USAGE AND GENERAL WRITING

Barzun, J. 1975. *Simple and direct: A rhetoric for writers*. New York: Harper & Row.

Bernstein, T. M. 1965. *The careful writer*. New York: Atheneum.

Corbett, E. P. J. 1971. *Classical rhetoric for the modern student*. 2nd ed. New York: Oxford University Press.

Flesch, R. 1980. *The ABC of style—A guide to plain English*. New York: Harper & Row.

Fowler, H. W. 1965. *A dictionary of modern English usage*. 2nd ed., rev. by Sir E. Gowers. New York: Oxford University Press.

Hayakawa, S. I. 1978. *Language in thought and action*. 4th ed. New York: Harcourt Brace Jovanovich.

Strunk, W., and E. B. White. 1978. *The elements of style*. 3rd ed. New York: Macmillan.

Trimble, J. R. 1975. *Writing with style: Conversations on the art of writing*. Englewood Cliffs, N.J.: Prentice-Hall.

STYLE MANUALS

American Chemical Society. 1978. *Handbook for authors*. Washington, D.C.: American Chemical Society.

American National Standards, Inc. 1979. *American national standard for the preparation of scientific papers for written or oral preparation*. ANSI Z39.16—1972. New York: American National Standards Institute.

CBE Style Manual Committee. 1978. *Council of biology editors style manual: A guide for authors, editors, and publishers in the biological sciences.* 4th ed. Washington, D.C.: Council of Biology Editors.

The Chicago manual of style. 1982. 13th ed., rev. Chicago: Univ. of Chicago Press.

ORNL style guide. 1974. Oak Ridge, Tenn.: Oak Ridge National Laboratory.

Pollack, G. 1977. *Handbook for ASM editors.* Washington, D.C.: American Society for Microbiology.

Skillin, M., R. Gay, et al. 1974. *Words into type.* 3rd ed. Englewood Cliffs, N.J.: Prentice-Hall.

U.S. Government Printing Office style manual. 1967. Rev. ed. Washington, D.C.: Government Printing Office.

Also, many private corporations, such as John Deere, DuPont, Ford Motor Company, General Electric, and Westinghouse, have their own style manuals.

GRAPHIC AIDS

Beakley, G. C., Jr., and D. D. Autore. 1973. *Graphics for design and visualization.* New York: Macmillan.

Hoelscher, R. P., C. H. Springer, and J. S. Dobrovolny. 1968. *Graphics for engineers: Visualizations, communication and design.* New York: Wiley.

MacGregor, A. J. 1979. *Graphics simplified: How to plan and prepare effective charts, graphs, illustrations, and other visual aids.* Toronto: Univ. of Toronto Press.

Morris, G. E. 1975. *Technical illustrating.* Englewood Cliffs, N.J.: Prentice-Hall.

Thomas, T. A. 1978. *Technical illustration.* 3rd ed. New York: McGraw-Hill.

Turnbull, A. T., and R. N. Baird. 1980. *The graphics of communication.* 4th ed. New York: Holt, Rinehart and Winston.

Also see the following journals:

Graphic Arts Monthly

Graphics: USA

PROPOSALS

Lefferts, R. 1980. *Getting a grant in the nineteen eighties: How to write successful grant proposals.* 2nd ed. Englewood Cliffs, N.J.: Prentice-Hall.

Society for Technical Communication. 1973. *Proposals and their preparation.* Vol. 1. Washington, D.C.: Society for Technical Communication.

ORAL PRESENTATIONS

Anastasi, T. E. Jr. 1972. *Communicating for results.* Menlo Park, Calif.: Cummings.

Howell, W. S., and E. G. Barmann. 1971. *Presentational speaking for business and the professions.* New York: Harper & Row.

Weiss, H., and J. B. McGrath, Jr. 1963. *Technically speaking: Oral communication for engineers, scientists, and technical personnel.* New York: McGraw-Hill.

JOURNAL ARTICLES

Day, R. A. 1979. *How to write and publish a scientific paper.* Philadelphia: ISI.

Graham, B. P. 1980. *Magazine article writing: Substance and style.* New York: Holt, Rinehart and Winston.

Michaelson, H. B. 1982. *How to write and publish engineering papers and reports.* Philadelphia: ISI.

Mitchell, J. H. 1968. *Writing for professional and technical journals.* New York: Wiley.

Zinsser, W. 1980. *On writing well.* 2nd ed. New York: Harper & Row.

Also see the following journals:

The Writer
Writer's Digest

Extract from Robert Jastrow UNTIL THE SUN DIES reprinted by permission of W. W. Norton & Company, Inc. Copyright © 1977 by Robert Jastrow.

Pie chart (working wives), p. 70 of "Women's Work," by Rae André. Across the Board 18, 7 (July, Aug. 81). Copyright The Conference Board.

Material from Energy Information Administration, *Monthly Energy Review*, June 1983. Reprinted with permission.

Material from PEDIATRICS, Volume 71, Number 4, Page A5, April 1983. Copyright American Academy of Pediatrics 1983.

Excerpt from "Hot Springs on the Ocean Floor" by John M. Edmond and Karen Von Damm. Copyright © 1983 by Scientific American Inc. All rights reserved.

Excerpt from "Industrial Microbiology" by Arnold L. Demain and Nadine A. Solomon. Copyright © 1981 by Scientific American Inc. All rights reserved.

David M. Blaies and Jeffrey F. Williams, "A Simplified Method for Staining Mast Cells with Astra Blue," *Stain Technology*, 56. Copyright 1981, the Williams & Wilkins Co. Reproduced by permission.

Excerpt based on Martin L. Harris, INTRODUCTION TO DATA PROCESSING, 2nd edition. Copyright © 1979. Reprinted by permission of John Wiley & Sons, Inc.

"Robot Trends at General Motors," p. 73, by R. C. Beecher and Robert Dewar, *American Machinist*, August 1979. © 1979 McGraw-Hill Inc.

From SOLAR HEATING: A Construction Manual by Total Environment Action, Inc. Copyright 1980, 1981 by the authors. Reprinted with the permission of the publisher, Chilton Book Co., Radnor, PA.

Installation instructions for tub sliding door © Novi American, Inc. 1980.

Ratio graph reproduced from *Survey of Current Business*, March 1982, V. 62, N. 3, p.5.

Extract from *Commerce Business Daily*, June 28, 1983, p. 3.

INDEX